98-HHE-5

CYANIDE FORMATION AND FATE IN COMPLEX EFFLUENTS AND ITS RELATION TO WATER QUALITY CRITERIA

by:

Michael Kavanaugh, Ph.D., P.E., DEE
Rula A. Deeb, Ph.D.
Daniel Markowitz, Ph.D.
Malcolm Pirnie, Inc.

David A. Dzombak, Ph.D., P.E., DEE
Anping Zheng, Ph.D.
Carnegie Mellon University

Thomas L. Theis, Ph.D.
University of Chicago at Illinois (formerly of Clarkson University)

Thomas C. Young, Ph.D.
Clarkson University

Richard G. Luthy, Ph.D., P.E., DEE
Stanford University (formerly of Carnegie Mellon University)

2003

Water Environment Research Foundation

The Water Environment Research Foundation, a not-for-profit organization, funds and manages water quality research for its subscribers through a diverse public-private partnership between municipal utilities, corporations, academia, industry, and the federal government. WERF subscribers include municipal and regional water and wastewater utilities, industrial corporations, environmental engineering firms, and others that share a commitment to cost-effective water quality solutions. WERF is dedicated to advancing science and technology addressing water quality issues as they impact water resources, the atmosphere, the lands, and quality of life.

For more information, contact:
Water Environment Research Foundation
635 Slaters Lane, Suite 300
Alexandria, VA 22314-1177
Tel: (703) 684-2470
Fax: (703) 299-0742
www.werf.org
werf@werf.org

This report was co-published by the following organization. For nonsubscriber sales information, contact:

IWA Publishing
Alliance House, 12 Caxton Street
London SW1H 0QS, United Kingdom
Tel: +44 (0) 20 7654 5500
Fax: +44 (0) 20 7654 5555
www.iwapublishing.com
publications@iwap.co.uk

Library of Congress Catalog Card Number: 2003114706

IWAP ISBN: 1-84339-632-7

Malcolm Pirnie, Inc.

The research on which this report is based was funded, in part, by the United States Environmental Protection Agency through Cooperative Agreement No. 825237-01 with the Water Environment Research Foundation (WERF). Unless an EPA logo appears on the cover, this report is a publication of WERF, not EPA. Funds awarded under the Cooperative Agreement cited above were not used for editorial services, reproduction, printing, or distribution.

ACKNOWLEDGMENTS

This report would not have been possible without the dedication of the following individuals and organizations: Prakasam Tata, Bernard Sawyer, William Lazouskas and George Richardson (Metropolitan Water Reclamation District of Greater Chicago, Cicero, IL); Michael Delaney, Nancy McSweeney, Larissa Zilitinkevitch, and Polina Epelman (Massachusetts Water Resources Authority, Charlestown, MA); William Ellgas and Jack Lim (East Bay Municipal Utility District, Oakland, CA); Jack Wahlstrom (Gulf Coast Waste Disposal Authority, Pasadena, TX); Dipak Patel and Reza Iranpour (City of Los Angeles); Rosanna Lacarra (City of Carlsbad - formerly with City of Sunnyvale); Regina Galicki and Stanley Tan (City of Sunnyvale); Mark Citriglia, Spain James, and Erwin Odeal (Northeast Ohio Regional Sewer District); David Brown (Allegheny County Sanitary Authority, Pittsburgh, PA); Sharon Drop (ALCOA, Alcoa Center, PA); John Sebroski and Rebecca Swartling (Bayer Corporation, New Martinsville, WV); and John Flaherty (Centre Analytical Laboratories, State College, PA). The project team thanks Kara Forman (formerly with Malcolm Pirnie, Oakland, CA), Jeff Moeller (WERF), Rajat Ghosh, and David Nakles (The RETEC Group, Monroeville, PA).

The project team acknowledges co-funders of this project including ALCOA, Inc. (Pittsburgh, PA), Consumers Energy (Jackson, MI), East Bay Municipal Utility District (Oakland, CA), Electric Power Research Institute (Palo Alto, CA), Gas Technology Institute (Chicago, IL), Michigan Consolidated Gas Company (Detroit, MI), Niagara Mohawk Power Corporation (Syracuse, NY), Wisconsin Electric Power Company (Milwaukee, WI), Northeast Ohio Regional Sewer District (Cuyahoga Heights, OH), and the U.S. Environmental Protection Agency (U.S. EPA) through Cooperative Agreement No. CR-825237-01 with the Water Environment Research Foundation.

Report Preparation

Principal Investigator:
Michael C. Kavanaugh, Ph.D., P.E., DEE
Malcolm Pirnie, Inc.

Project Team:
Rula A. Deeb, Ph.D.
Daniel Markowitz, Ph.D.
Malcolm Pirnie, Inc.

David A. Dzombak, Ph.D., P.E., DEE
Anping Zheng, Ph.D.
Carnegie Mellon University

Thomas L. Theis, Ph.D.
University of Illinois, Chicago (formerly with Clarkson University)

Thomas C. Young, Ph.D.
Clarkson University

Richard G. Luthy, Ph.D., P.E., DEE
Stanford University (formerly with Carnegie Mellon University)

Project Subcommittee

Philip B. Dorn, Ph.D., Chair
Shell Global Solutions US

James A. Buckley, Ph.D.
Metro-King County (Wash.) Environmental Laboratory

W. Dickinson Burrows, Ph.D., P.E., DEE
U.S. Army Center for Health Promotion and Preventative Medicine

Rick D. Cardwell, Ph.D.
Parametrix, Inc.

Bob Thurnau
U.S. EPA

Kevin Brix
EcoTox

Joseph W. Gorsuch
Eastman Kodak Company

Jim Pletl, Ph.D.
Hampton Roads (Va.) Sanitation District

Water Environment Research Foundation Staff

Director of Research: Daniel M. Woltering, Ph.D.
Senior Project Manager: Margaret Stewart

ABSTRACT AND BENEFITS

Abstract:

Cyanide occurs in many industrial and municipal wastewaters and is often an expected constituent of typical treatment plant wastewater streams. However, a growing number of wastewater treatment plants (WWTPs) across the United States have detected cyanide in chlorinated effluents at levels exceeding influent concentrations. Because water quality criteria and related discharge limits are typically low—in some cases as low as 5 µg/L—some of these WWTPs periodically exceed effluent cyanide standards. Potential causes include cyanide formation during wastewater chlorination processes, the presence of interferences that cause false negatives, and false positives caused by artifacts of sample handling or analytical techniques. The possible causes of the apparent cyanide formation phenomenon were investigated in this study.

The aqueous chemistry of cyanide is relatively complex. Cyanide can exist in its free forms (HCN and CN^-), in metal–cyanide coordination complexes, in organocyanide compounds, and in metal–cyanide solids. The toxicity, treatability, and environmental fate of cyanide strongly depend on cyanide speciation. In addition to concerns related to cyanide formation in wastewater, the management of cyanide-bearing wastewaters has been complicated by the lack of cyanide species-specific analytical techniques. Cyanide testing methods used for regulatory purposes are operationally derived and do not provide information on specific cyanide species.

This three-year research project was initiated by the Water Environment Research Foundation in 1998 to address research gaps associated with analytical techniques for monitoring and regulating cyanide in municipal wastewater and receiving waters, cyanide fate and formation in municipal wastewater treatment plants, fate of cyanide species discharged to receiving waters, and the relationship of water quality criteria and discharge standards for cyanide. First, the researchers evaluated the reliability of seven analytical methods for measurement of cyanide species at low (ppb level) concentrations for a range of municipal and industrial wastewaters. Second, the research team used laboratory, modeling, and field studies to evaluate the sources, transport, and fate of cyanide species in WWTPs and receiving waters. This included studies at WWTPs and in the laboratory to identify precursor compounds to cyanide formation in the chlorination process. Third, information on the toxicity of cyanide species to aquatic organisms was compiled and synthesized. Fourth, the researchers examined current federal and state cyanide water quality criteria and their relationship to discharge standards. Finally, the researchers developed various compliance strategies for WWTPs that experience difficulties in consistently meeting cyanide discharge permit limits.

Benefits:

- Evaluates seven methods for the analysis of low (ppb) levels of cyanide that for the most part yield accurate and reproducible results across a range of contaminated water matrices.
- Identifies and provides a model simulation of the effects of three processes that govern the fate of cyanide species discharged to receiving waters.
- Documents the occurrence of cyanide formation upon chlorination in six municipal wastewater treatment plants.

- Recommends that wastewater influent undergo frequent measurements of thiocyanate—an important precursor for cyanide formation—for WWTPs with a history of cyanide formation.
- Recommends chlorine dosing modification to destroy free cyanide formed in the chlorination process, and other strategies for complying with stringent effluent discharge limits for cyanide.
- Demonstrates the potential usefulness of developing a "multitiered" set of water quality standards for cyanide in which specific chemical species or classes of cyanide, as measured by appropriate analytical methods, are identified and used.

Keywords: Cyanide, formation, wastewater, treatment, fate, water quality criteria, analytical methods.

TABLE OF CONTENTS

LIST OF TABLES

LIST OF FIGURES

Water Environment Research Foundation

LIST OF ACRONYMS

AFT	dissolved air flotation tank
ALCOSAN	Allegheny County Sanitary Authority
ASTM	American Society for Testing and Materials
ATSDR	Agency for Toxic Substances and Disease Registry
BOD	biochemical oxygen demand
CATC	cyanide amenable to chlorination
CCC	chlorine contact channel
CMFR	completely mixed flow reactor
DBP	disinfection by-product
DITP	Deer Island Treatment Plant
DIW	deionized water
EBMUD	East Bay Municipal Utility District
ECA	effluent concentration allowances
ERA	Environmental Resource Associates
FGR	fixed growth reactor
GCA	Gulf Coast Waste Disposal Authority
IC	ion chromatography
LAGWRP	Los Angeles–Glendale Water Reclamation Plant
LPM	Lagrangian parcel method
MDL	method detection limit
MLE	maximum likelihood estimation
MWRA	Massachusetts Water Resources Authority
MWRDGC	Metropolitan Water Reclamation District of Greater Chicago
NPDES	National Pollutant Discharge Elimination System
NTU	nephelometric turbidity units
PFRD	plug flow reactor with dispersion
Ppb	parts per billion
QA/QC	quality assurance and quality control
RETEC	The RETEC Group, Inc.
RWQCB	San Francisco Bay/Delta Regional Water Quality Control Board
RSD	relative standard deviation
RTD	hydraulic residence time distribution
SIP	state implementation plan
TOC	total organic carbon
TOTCN	total cyanide
TRC	total residual chlorine
TSS	total suspended solids
UV	ultraviolet
U.S. EPA	U.S. Environmental Protection Agency
VBRS	Vince Bayou Receiving Station
WAD	weak acid dissociable
WQC	water quality criteria
WTF	Washburn Tunnel Wastewater Treatment Facility

WPCP	City of Sunnyvale Water Pollution Control Plant
WRP	Hanover Park Water Reclamation Plant
WWTP	wastewater treatment plant

EXECUTIVE SUMMARY

Numerous wastewater treatment plants (WWTPs) across the United States have detected cyanide in chlorinated effluents at levels exceeding those in influent waters. In some cases these levels exceed National Pollutant Discharge Elimination System (NPDES) discharge limits. Cyanide can be formed during wastewater chlorination. In addition, false positive results are possible, caused by artifacts of sample handling and analytical techniques. Interferences can also cause false negatives, underpredicting the amount of cyanide discharged by the WWTP. The management of cyanide wastewaters can therefore be complicated by method shortcomings. There is also a lack of species-specific analytical techniques. Cyanide methods used for regulatory purposes are operationally derived and do not provide information on specific cyanide species causing toxicity in receiving waters. Because of these and other issues there has been a strong need for a comprehensive evaluation of water quality problems related to cyanide in addition to an evaluation of alternative water quality management strategies. In response the Water Environment Research Foundation initiated a three-year research project in 1998 titled "Cyanide Formation and Fate in Complex Effluents and Its Relation to Water Quality Criteria."

The project team's research approach involved first the identification and evaluation of analytical methods for measuring cyanide species at detection limits appropriate for water quality management, and the testing of these methods at WWTPs participating in the study to determine the fate of cyanide during treatment. Second, the approach involved an assessment of the apparent sources and/or causes of cyanide formation and disappearance in WWTPs based on plant monitoring programs, compilation of data already collected, and laboratory tests of cyanide formation. Third, the approach involved a compilation of water quality criteria currently used for water quality management as well as a summary of existing aquatic toxicity data for cyanide species in receiving waters. Finally, the approach involved an assessment of the relative magnitude of processes influencing the fate of cyanide in receiving waters with the support of industrial participants. Specifically, the research was divided into four key tasks:

Task 1: Evaluation and testing of analytical methods for cyanide species in municipal and industrial waste waters
Task 2: Sources, transport and fate of cyanide species in WWTPs and receiving waters
Task 3: Relationship of cyanide speciation and cyanide toxicity
Task 4: Evaluation of water quality criteria for cyanide

To begin with, for Task 1, seven methods for analyzing cyanide in reagent water and different contaminated water matrices were compared and evaluated, including methods for weak acid dissociable (WAD) cyanide (Method 4500-CN-I), free cyanide by microdiffusion (ASTM D4282-95), available cyanide (U.S. EPA Method OIA-1677), automated WAD cyanide by thin film distillation (ASTM D4374-93), metal cyanides by ion chromatography (Dionex Application Note No.55), automated total cyanide by thin film distillation (ASTM D4374-93), and automated total cyanide by UV digestion (U.S. EPA Method 335.3). The first four methods are designed to analyze free and WAD cyanide. Five different contaminated water samples were used in this study. Results suggested that a number of species-specific cyanide analytical techniques could be employed reliably for low-concentration analysis of their target cyanide analytes or analyte groups with a range of contaminated water types. Exceptions include the available cyanide, the ion chromatography, and the thin film distillation methods which when applied to raw municipal wastewater because of significant interference problems and/or low recoveries. Also, there was recovery of some diffusible cyanide in microdiffusion tests with nickel-cyanide-spiked samples caused by the dissociation of this weak metal–cyanide complex during the test.

As part of Task 1, laboratory studies were performed to explore the potential contribution of organocyanide compounds in wastewater to elevated cyanide levels in the chlorinated effluents of WWTPs. Four model organocyanide compounds (acetonitrile, amygdalin, cyanocobalamin, and 2-acetoxy-3-butenenitrile) were selected and tested with varying chlorine dosages for release of CN by total and diffusible cyanide procedures. The coenzyme form of vitamin B_{12}, which does not contain cyanide, was also tested. Whereas acetonitrile and amygdalin were found to not yield cyanide with or without chlorination, cyanide was released with increasing chlorine dosages in the presence of cyanocobalamin. 2-acetoxy-3-butenenitrile had generally decreased release of cyanide with increased chlorine dosages. Both cyanocobalamin and coenzyme vitamin B_{12} gave evidence of the formation of the cobalt cyanide complex after chlorination although the molecular structure of the latter contains no cyanide.

Task 2 involved an extensive evaluation of cyanide sources and the implementation of a comprehensive monitoring program at several full-scale WWTPs to assess the fate of cyanide species during treatment. In addition, Task 2 objectives included the performance of laboratory studies to investigate the effect of chlorination on the transformation and production of cyanide species. Finally, Task 2 involved a study to predict the fate of cyanide species in surface waters.

An intensive WWTP monitoring program was implemented at six full-scale treatment facilities of participating municipal utilities during the second and third years of the project (from March 2000 to July 2001, varied for each WWTP). The objective was to investigate the fate and transport of cyanide in wastewater treatment processes with a focus on chlorination and dechlorination processes. A review of historical monitoring data for cyanide species in these WWTPs was also conducted. The aim of the WWTP monitoring program was to obtain a better understanding of the cyanide formation phenomenon as it occurs in the field. A database was developed to investigate cyanide formation in wastewater secondary treatment. Data from all participating WWTPs showed evidence of cyanide formation in this one-year monitoring effort although cyanide formation patterns varied significantly from one plant to another and from one season to another. Generally, the chlorination of thiocyanate seemed to be the most important mechanism for the formation of cyanide in wastewater treatment processes especially during chlorination/dechlorination. This observation is consistent with the findings of the laboratory study which was performed to investigate the mechanisms of cyanide formation in wastewater treatment processes.

Task 2 also involved several laboratory studies. First, studies were performed to investigate the effects of thiocyanate on free cyanide formation during chlorination and UV disinfection of WWTP secondary effluent. Thiocyanate (SCN^-) is a cyanide-related compound encountered in most WWTP influents. SCN^- may be decomposed to free cyanide (CN^-) in some circumstances. Effects of thiocyanate on the formation of cyanide during chlorination and UV disinfection were studied through the laboratory study with synthetic solutions and WWTP secondary effluents. Results indicated that total cyanide was detected in thiocyanate solutions after chlorination in which the chlorine dose was not sufficient to destroy SCN^- completely, thus ensuring no residual chlorine to destroy any free cyanide produced. It was also found that thiocyanate can be broken down to yield cyanide by UV irradiation. In addition, thiocyanate was observed to cause a significant positive error in the conventional total cyanide analysis technique involving acidic distillation and standard colorimetric measurement when nitrate was also present in the sample.

Second, experiments were performed to evaluate the potential formation of cyanide as a result of chloramination from hypochlorite reaction with residual ammonia, which may occur in partially nitrified or unnitrified wastewater. The potential generation of cyanide species in WWTPs wastewater upon chlorination in the presence of residual ammonia, resulting in chloramine formation, was investigated in experiments with synthetic solutions and WWTP secondary effluents. This study demonstrated that low concentrations (about 5–25 mg/L as CN) of cyanogen chloride (CNCl), a highly toxic cyanide species not measured in total or free cyanide analyses, could be detected as a result of chloramination reactions in WWTP secondary effluents. The potential for chloramination of nitrogen-bearing organic compounds to yield cyanogen chloride and/or free cyanide was demonstrated in experiments with synthetic solutions spiked with selected precursor organics: L-serine, benzene, catechin, and humic acid. The amino acid L-serine yielded the largest concentrations of CNCl upon chloramination. Additionally, detectable cyanide, around 10 mg/L, was observed in solutions of L-serine and in WWTP secondary effluents that were chloraminated followed by dechlorination to prevent destruction of any free cyanide produced. Thus, chlorination of WWTP secondary effluent containing residual ammonia can lead to chloramination of organic compounds and the resulting production of cyanogen chloride and free cyanide.

Third, experiments were performed to evaluate the potential formation of cyanide from the nitrosation process (i.e., reaction with nitrite). The presence of nitrite ion as a common constituent in domestic wastewater effluents may play an important role in the formation of cyanide through reaction with certain kinds of organic compounds, (e.g., 1,2,4-trihydroxybenzene). CN^- was identified as a product of nitrosation of 1,2,4-trihydroxybenzene at 25°C. Substantial CN^- formation was also observed at low pH in experiments with WWTP effluents when reactive organics and nitrite were both spiked. Formation of cyanide through nitrosation was strongly pH dependent, being most significant at low pH and negligible at neutral to high pH. This result points to HNO_2 as being more reactive than the dissociated NO_2^- ion. The reaction of these nitrite species with organics also occurs, and at greater rates, in conventional analyses for total cyanide that involve distillation. Sufficient sample pretreatment with sulfamic acid at the time of sampling, not at the time of analysis, is highly recommended.

Task 2 also identified and provided a successful numerical simulation of three processes that simultaneously appear to determine the fate of all cyanide species in industrially-impacted receiving water. These processes include photolysis of hexacyanoferrate [i.e., $Fe(CN)_6^{-4}$ and $Fe(CN)_6^{-3}$], volatilization of HCN and biotic utilization of HCN. The investigation required four interdependent activities, including literature review and data gap determination, site selection and periodic monitoring for relevant physical and chemical environmental variables in system, data gap elimination by experimental observation of fate-determining processes, and numerical model development and calibration.

The principal data gaps identified during the literature search included a lack of knowledge pertaining to the volatilization and biotic utilization of HCN that would be sufficient to permit adequate representation of these processes in a numerical modeling framework. To remedy these deficiencies, a modest program involving field (*in situ*) and laboratory (*in vitro*) observations and experimentation was undertaken. The program included the use of a flux chamber to directly observe the flux of HCN from the water surface to air at two surface water field sites affected by hexacyanoferrate. The results demonstrated a range of HCN volatilization fluxes of 2-15 μg HCN· $m^{-2} \cdot hr^{-1}$. The specific value of the flux depended on both the water phase concentration (i.e., concentration gradient for volatilization) and the temperature.

Monitoring of cyanide concentrations in water samples collected from the field site demonstrated a significant decline in total cyanide concentrations—greater than would result from measured volatilization losses—during continuous loading by cyanide-contaminated inflowing water. The specific monitoring site that showed this behavior was a pilot-scale constructed wetland that was installed to test the feasibility of using wetlands for the removal of cyanide and other contaminants of concern. The magnitude of cyanide losses was viewed as presumptive evidence for biotic utilization of HCN in the system inasmuch as other potential fate-determining processes (e.g., volatilization, precipitation or sorption, and sediment accumulation) did not appear to function at a magnitude that could account for the majority of cyanide losses. Accordingly, laboratory studies were undertaken to assess aerobic bioutilization of HCN by microflora obtained from the constructed wetland. Preliminary experiments using KCN-spiked slurries of wetland sediments and water to assess nutritional needs revealed that the use of amendments such as municipal wastewater and standard heterotrophic microbiological growth media (nutrient broth) did not enhance the bioutilization rate of HCN. These results demonstrated several features that were consistent with microbial utilization of CN when it was the only nitrogen source available in the medium. It was noted that the observed *in vitro* HCN utilization was strong presumptive evidence for bioutilization of HCN in the constructed wetland and remaining receiving water system. Nevertheless, it was also noted that the rates *in vitro* could not be directly used in the simulation model because the laboratory assessment was undertaken using microbiota in suspended culture whereas the field situation would be a combination of suspended and fixed film organisms. Accordingly, calibration of the model to estimate the biodecay rate was still necessary.

A model was then developed to predict the fate of cyanide in surface water. Model development began with a Eulerian framework applied to the state variables (hexacyanoferrate and HCN) and the fate-determining processes (photolysis, volatilization, and biotic utilization). A conventional temporal and spatial discretization was applied to formulate a finite difference scheme for solution of the governing equations. However, early results suggested that a Lagrangian framework would be better suited to the advection-dominated conditions that prevailed at the field site. Indeed, the calibrated Lagrangian model reproduced concentration magnitudes and spatial trends in the coupled cyanide species.

The results from the modeling work led to the overall conclusion that the cyanide fate process knowledge is such that site-specific numerical modeling of cyanide species fate can be accomplished when aided by collection of routine monitoring data and simple availability bioassays. More specific conclusions may also be drawn. For example, the receiving water system did not accumulate HCN from hexacyanoferrate photolysis. Instead, the combination of biotic utilization and volatilization prevented HCN from accumulating. Indeed, the modeling results suggested that exposure to HCN never began to approach the initial analytical total of cyanide species.

Stated differently, the characteristics of a longitudinal concentration profile for HCN downstream from a point source would involve development of a maximum exposure concentration, a "critical point" for biotic exposure that is reminiscent of the behavior of oxygen deficit as modeled by the classical Streeter-Phelps approach. This means numerical experiments could be undertaken with the model to identify conditions that would be necessary for HCN accumulation to become an issue at a specific site, including providing an estimate of the maximum likely exposure to HCN. With a small amount of speculation, the success of the modeling approach indicates that this conclusion may be carried forward an additional step. This is to suggest that the potential exists to develop and apply a waste load allocation (WLA) strategy for cyanide (i.e., WLA/TMDL) much as is the current practice for managing the discharge of more conventional pollutants to receiving waters (e.g., oxygen demand and phosphorus). Such an application would appear possible because the underlying model described here is essentially generic in that the cyanide fate-determining processes therein not only are the most significant ones for the studied system, but the processes themselves are expected to be functional and likely of greatest importance in the vast majority of all surface waters systems where cyanide contamination is an issue, at least for systems where HCN/CN^- and hexacyanoferrates predominate the cyanide speciation.

Task 3 involved the compilation of studies in which the toxicity of cyanide species was reported. This included a brief evaluation of fresh and marine cyanide water quality criteria and chemistry, and a discussion of cyanide speciation in experimental design. Formal recommendations for future cyanide toxicity testing protocols, including specific test procedures and tested species, were made by the project team.

Finally, Task 4 focused on compiling utility toxicity data in addition to existing water quality criteria data to identify correlations between cyanide discharge concentrations and toxicity tests.

The results of this study suggest that there are several alternative strategies that can be pursued by municipal and industrial dischargers. This is especially true if it can be shown that the standard based on total cyanide measurements is too conservative and does not provide a benefit commensurate with the increase in costs associated with monitoring and/or treatment to meet the total cyanide standard. The goal of any discharger should be to protect human health and the environment, but at a cost commensurate with the benefits obtained. Standards should be established that provide sufficient protection to human health and the environment without requiring expenditures that provide only limited additional benefits and limited additional risk reduction. It must be stressed, however, that each wastewater discharge case involving cyanide species may have site-specific factors that must be accounted for to determine whether an alternative strategy as discussed below should be pursued.

Alternative compliance strategies for WWTPs include the following:

- Reduction of cyanide precursors to sanitary sewers from industrial or commercial dischargers;
- Obtaining a site-specific discharge permit;
- Requesting a change in the point of monitoring where compliance would be evaluated;

- Modification of the activities within a wastewater plant causing production of cyanide, namely the disinfection process; and
- Requesting a change in the permit requirements with specific attention to alternative analytical methods measuring the concentration of the toxic form of cyanide.

The results of this study are significant to water industry. The information on analytical methods, cyanide toxicity, and fate and transport of cyanide in WWTPs and in receiving waters can be used to evaluate water quality criteria for cyanide discharges. It is anticipated that this comprehensive evaluation will provide a more scientifically rigorous framework for the management of cyanide in WWTPs and industrial wastewaters.

Please note: Data files (Appendix B) for this research may be available through WERF and/or the third-party publishers listed on page ii and the back cover of this report. For more information, call the organization from which you purchased this report or contact WERF at 703-684-2470 or visit our website at www.werf.org.

CHAPTER 1.0

PROJECT BACKGROUND

1.1 Background

Influent at municipal wastewater treatment plants (WWTPs) typically contains cyanide at low (<5 μg/L) to nondetectable concentrations. Cyanide in municipal WWTP influent is derived from industrial sources, such as electroplating, steel production, or photographic finishing facilities.

Although influent cyanide concentrations usually are low, many municipal WWTPs have reported periodic detection of elevated concentrations of cyanide in final effluent, and at concentrations that exceed influent concentrations. This has been observed at treatment plants across the United States (Environ, 1996; Delaney et al., 1997; Forman and Kavanaugh, 1999). Effluent cyanide concentrations have been observed at 5-20 parts per billion (ppb) of total cyanide or greater. Effluent cyanide concentrations of this magnitude typically represent a permit level exceedance, as water quality criteria for cyanide and related discharge limits are very low, in many cases as low as 5 μg/L. Studies of the apparent cyanide formation (e.g., Environ, 1996; Delaney et al., 1997) have pointed to the chlorination process as the location in WWTPs where the cyanide concentration increase occurs.

Potential causes include cyanide formation during wastewater chlorination processes, the presence of interferences that cause false negatives, and false positives caused by artifacts of sample handling or analytical techniques. Investigation of the causes of the apparent cyanide formation was a primary focus of this project.

Concern about control of cyanide concentrations in WWTP effluents to meet low regulatory discharge limits has been coupled with concern about accurate, reproducible, and appropriate cyanide measurements. Routine measurement of cyanide species at ppb levels is challenging, especially in complex matrices like wastewater. In addition, the commonly used "total cyanide" analytical method for monitoring cyanide in wastewater is not appropriate in many situations. This method captures relatively low-toxicity metal-cyanide species in addition to the toxic free cyanide species (HCN and CN^-). However, water quality criteria, and often discharge standards, for cyanide frequently are focused on free cyanide species.

Analytical methods are available for distinguishing among free cyanide, weak metal-cyanide complexes, and strong metal-cyanide complexes, but such methods have not yet gained widespread use, in large part because of concern about reliability for measuring low concentrations in complex matrices. Hence, in this project the performance of several leading, species-specific cyanide analysis techniques was evaluated with respect to low concentrations in wastewater matrices.

Another issue involved with very low discharge limits for cyanide is the appropriateness of the limits. There is inconsistency across the state and federal water quality regulatory authorities in the United States with respect to surface water quality criteria and the effluent discharge limits that are linked to them. This inconsistency has led to questions about the scientific basis for low discharge limits for cyanide. The inconsistency occurs in several ways, including the forms of cyanide measured, the aquatic organisms protected with the criteria, the kinds of tests used to assess aquatic toxicity, and the nature and units of the quantitative criteria specified. Related to the issue of appropriateness of discharge limits is the extent of knowledge about the fate of cyanide species upon discharge to receiving waters. Little has been reported in the scientific literature on cyanide fate and transport in receiving waters. Assessment of ecological risk of cyanide in aquatic systems is often based on assumptions about processes affecting cyanide in water (ATSDR, 1997).

To help improve the scientific basis for decisions about appropriate water quality criteria and discharge standards for cyanide, this project examined processes governing cyanide fate in receiving waters through laboratory and field studies. In addition, summaries of the toxicological data underlying current federal water quality criteria for cyanide, and of state and federal water quality criteria for cyanide were prepared. This work was undertaken to provide insight into the strengths and weaknesses of the scientific foundation of current water quality criteria and discharge standards for cyanide in the United States.

To investigate the apparent formation of cyanide in municipal WWTPs, to improve understanding of the fate of cyanide species upon discharge to receiving waters, and to assess the foundation for current water quality criteria and discharge standards for cyanide, the Water Environment Research Foundation initiated a three-year research project in 1998. A diverse research team was formed comprising consultants (Malcolm Pirnie, Inc.), university researchers (Carnegie Mellon University, Clarkson University), three industrial organizations, and 13 utilities managing WWTPs with reported cyanide-related problems.

This research was divided into four key tasks as follows:

- *Task 1* – Evaluation and Testing of Analytical Methods for Cyanide Species in Municipal and Industrial Waste Waters
- *Task 2* – Sources Transport and Fate of Cyanide Species in WWTPs and Receiving Waters
- *Task 3* – Toxicity of Cyanide Species to Aquatic Organisms
- *Task 4* – Evaluation of Water Quality Criteria for Cyanide

1.2 Overview of Report

The results of this research effort are presented in this report by task number. Chapters 2.0 and Chapter 3.0 present the results of Task 1 investigations. Chapters 4.0–8.0 present the results of Task 2 investigations. Chapter 9.0 presents the results of Task 4 investigations. Finally, Chapter 10.0 presents the major conclusions of the study as well as recommendations for future research. In addition, this final chapter discusses the challenges faced by WWTPs in meeting cyanide discharge requirements, and offers strategic approaches to help WWTPs attain permit compliance.

CHAPTER 2.0

EVALUATION OF ANALYTICAL METHODS FOR CYANIDE SPECIES IN MUNICIPAL AND INDUSTRIAL WASTEWATERS*

2.1 Introduction

Cyanide is common in many industrial wastewaters, and can sometimes be found in municipal wastewaters. The determination of cyanide species is an important test in assessing these wastewaters. However, identification and quantification of various cyanide species is not commonly practiced. Rather, water quality professionals usually employ bulk measurements such as "total cyanide." Complicating the regulation of cyanide is the lack of analytical methods capable of unambiguously distinguishing among cyanide forms at detection limits appropriate for water quality management. Approved methods for the analysis of cyanide typically rely on either total cyanide, which does not attempt to differentiate cyanide fractions, or on operational definitions of various cyanide fractions such as "weak acid dissociable" (WAD), "amenable to chlorination," and "available."

With respect to analysis, cyanide implies the sum of all cyanide groups (-C ≡N) in compounds that can be determined analytically as the cyanide ion (CN^-), using colorimetric, amperometric, or electrochemical measurement. Cyanide has been recognized by the U.S. Environmental Protection Agency (U.S. EPA) as a method-defined analyte; that is, cyanide measured is defined by the method used (U.S. EPA, 1999a,b). Confusion arises when assigning a reliable degree of risk to the presence of cyanide because the methods and results are inconsistent and difficult to interpret.

Cyanide compounds are generally classified as simple and complex cyanides. Simple cyanides are those compounds that are readily converted into hydrogen cyanide (HCN) under acidic conditions or into CN^- under basic conditions, such as NaCN, KCN, NH_4CN, etc. The cyanide ion, with a great tendency to act as a ligand, can be complexed with metals to form metal

---

* Reprinted in part with permission from *Environ. Sci. Technol.* **2003**, *37*, 107-115.

cyanide complexes [e.g., $Cu(CN)_3^-$, $Ni(CN)_4^{2-}$, $Zn(CN)_4^{2-}$]. These metal-complexed cyanides have widely varying stabilities and may dissociate depending on several factors. Therefore, complex cyanides can be further classified as WAD cyanides and strongly complexed cyanides. The WAD cyanides include zinc, nickel, copper, and cadmium cyanide complexes and the strongly complexed cyanides consist of cobalt and iron cyanide complexes.

Cyanide species vary in their toxicity, reactivity, and environmental fate and transport properties (Boening and Chew, 1999; Ghosh et al., 1999; Shifrin et al., 1996). Free cyanide, in particular HCN, is the primary toxic form. Because WAD cyanide includes free cyanide, it is often used as the measure of readily available cyanide for regulatory and assessment purposes, in addition to the total cyanide requirement. However, the most common U.S. EPA-approved method for measuring free and weakly complexed cyanide—cyanide amenable to chlorination [CATC; Method 4500-CN-G (APHA/AWWA/WEF, 1998); D2036-98-B (ASTM, 1998b)]—has many well-documented shortcomings (Ingersol et al., 1983; Altmayer, 1997).

This study assesses the accuracy, precision, and reliability of cyanide species analytical methods with a broad spectrum of contaminated water matrices such as WWTP influents, WWTP effluents, and industrial contaminated groundwater. Seven test methods were evaluated by conducting intralaboratory studies at five laboratories. The primary focus was to test and compare various cyanide analytical methods at low concentration ranges (<100 ppb) in municipal and industrial water matrices with the objective of identifying alternative methods for use in routine wastewater monitoring. The secondary focus was to compare and evaluate four out of the seven methods for the analysis of free and weakly complexed cyanide (i.e., WAD cyanide) in reagent water and different contaminated water matrices.

2.2 Overview of Cyanide Analytical Methods

The most commonly used U.S. EPA-approved analytical methods for cyanide are the total cyanide method [Method 4500-CN-C (APHA/AWWA/WEF, 1998); Method D2036-98-A (ASTM, 1998b); Method 335.4 (U.S. EPA, 1983)], and the CATC method [Method 4500-CN-G (APHA/AWWA/WEF, 1998); Method D2036-98-B (ASTM, 1998b)]. Methods evaluated in this study were WAD cyanide [Method 4500-CN-I (APHA/AWWA/WEF, 1998)], free cyanide by microdiffusion [Method D4282-95 (ASTM, 1998a)], available cyanide [Method OIA-1677 (U.S. EPA, 1999c)], automated WAD cyanide by thin film distillation [Method 4500-CN-I (APHA/AWWA/WEF, 1998)], metal cyanides by ion chromatography (IC; Dionex Corp., 1988), automated total cyanide by thin film distillation [Method D4374-93 (ASTM 1998d)], and automated total cyanide by low-power ultraviolet (UV) digestion [Method 335.3 (U.S. EPA, 1983)].

Industries and municipalities have intensively used these methods for cyanide analysis in wastewater, drinking water, and other aqueous samples. Moreover, the available cyanide method (Method OIA-1677) and the total cyanide by low-power UV digestion method are approved methods under National Pollutant Discharge Elimination System (NPDES). The first four methods are designed for the analysis of free and weakly complexed cyanide, defined herein as WAD cyanide. A summary of the analytical methods and laboratory assignments is presented in Table 2-1; more details on the methods are provided in Zheng et al. (2003a).

2.3 Methods

2.3.1 Waters Tested and Sampling Procedures

The method evaluation study was conducted in reagent (deionized) water and five different types of contaminated waters (Table 2-2). Total cyanide concentrations varied from nondetectable to 182 ppb in preserved unfiltered samples. An unfiltered 32-L sample was collected for each contaminated water. Each sample was then pretreated, if necessary, to remove sulfide (with powdered lead carbonate) and oxidants (with sodium thiosulfate) following procedures specified in *Standard Methods* [Method 4500-CN-B (APHA/AWWA/WEF, 1998)]. Split fractions (5 L) were distributed to all participating laboratories. Separate aliquots (2 L) of each contaminated water sample were also collected for a range of characterization analyses including major ions, metals, alkalinity, organic content [biochemical oxygen demand (BOD), total organic carbon (TOC)], and total cyanide. Table 2-3 summarizes the characteristics of these contaminated waters. During the study, split samples of the contaminated waters were analyzed with each of the analytical methods evaluated. Additional details about the sample collection protocol and sample characterization analyses are available in Appendix A.

2.3.2 Analysis of Unspiked and Spiked Samples

Samples were initially analyzed at each laboratory for total cyanide by the conventional manual distillation technique [Method 4500-CN-C (APHA/AWWA/WEF, 1998)]. Samples were also analyzed at Carnegie Mellon as received for WAD cyanide and free cyanide by microdiffusion. Samples were then spiked at different concentration levels for analysis of cyanide species in each contaminated water with the methods selected for study. Reagents used

Table 2-1. Cyanide Analytical Methods Evaluated in the Study.*

Method	Method code	CN species measured	Reference	Laboratory
Available cyanide, U.S. EPA Method OIA-1677	Avail-CN	WAD-CN	U.S. EPA, 1999c	Bayer Corp.
Free cyanide by microdiffusion, ASTM D 4282-95	FCN-MD	Free CN	ASTM, 1998a	Carnegie Mellon
Metal cyanides by ion chromatography	MCN-IC	Metal cyanides	Dionex, 1988	Alcoa [a]
Total cyanide by low power UV digestion	TCN-LPUD	Total CN	U.S. EPA, 1983	MWRA [b]
Total cyanide by thin film distillation, ASTM D4374-93	TCN-TFD	Total CN	ASTM, 1998d	MWRDGC [c]
WAD cyanide, SM 4500-CN-I	WAD-CN	WAD-CN	APHA/AWWA/WEF, 1998	Carnegie Mellon
WAD cyanide by thin film distillation, SM 4500-CN-I, automated	WAD-TFD	WAD-CN	APHA/AWWA/WEF, 1998	MWRDGC [c]

[a] Centre Analytical Laboratories, Inc. (State College, PA) performed the analysis for Alcoa.
[b] Massachusetts Water Resources Authority.
[c] Metropolitan Water Reclamation District of Greater Chicago.
* Reprinted with permission from *Environ. Sci. Technol.* 2003, *37*, 107-115. Copyright 2003 American Chemical Society.

Table 2-2. Contaminated Water Matrices Evaluated in the Study.*

No.	Site/plant name	Site/ plant type	Sample type	Sampling date
CW1	Manufactured gas plant (MGP) site groundwater, NY	Former MGP site	Contaminated groundwater	3/15/99
CW2	Hanover Park Water Reclamation Plant, Metropolitan Water Reclamation District, Hanover Park, IL	WWTP	Unchlorinated WWTP secondary effluent	4/19/99
CW3	Hanover Park Water Reclamation Plant, Metropolitan Water Reclamation District, Hanover Park, IL	WWTP	Chlorinated WWTP secondary effluent	5/17/99
CW4	Deer Island Treatment Plant of the Massachusetts Water Resources Authority (MWRA), Boston, MA	WWTP	WWTP primary clarifier effluent	6/14/99
CW5	Aluminum smelting plant, TN	Waste disposal site	Contaminated groundwater	7/14/99

for spiking were sodium cyanide (NaCN) (Aldrich Catalog 43159-1), potassium tetra-cyanonickelate [$K_2Ni(CN)_4$] (Aldrich Catalog 41515-4), and potassium ferrocyanide [$K_4Fe(CN)_6$] (Aldrich Catalog 22768-4). The spiking studies were conducted to evaluate matrix-dependent method detection limits (MDLs), precision, and bias (Zheng et al., 2003a). All participants were able to obtain reagents from the same production lots of the above cyanide chemicals from Aldrich Chemical Company (Milwaukee, WI).

2.3.3 Method Detection Limits

Samples spiked with sodium cyanide, nickel cyanide, and iron cyanide complexes determined MDLs for all water matrices. Seven replicate analyses were conducted at cyanide concentrations between one and five times the estimated MDL, as specified in the U.S. EPA protocol (U.S. EPA, 1994a). The MDL was calculated according to the following equation:

$$\text{MDL} = 3.143 \times SD \tag{2-1}$$

where:

SD = standard deviation of the seven replicate measurements, ppb.

Note that the MDL is a theoretical lower limit. The practical quantitation limit is typically taken as 5-10 times the MDL.

2.3.4 Bias, Precision, and Percent Recovery

Bias, precision, and percent recovery were also determined and used to assess the performance of the various analytical methods by spiking with three different cyanide species in different water matrices, with analyses of seven replicates at three concentration levels (Zheng, et al., 2003a). Doses of sodium cyanide, nickel cyanide, and ferrocyanide were added in separate spiking experiments to yield concentrations at the following levels: 5-20 ppb, 30-50 ppb, and 100–150 ppb. The difference between a true value (obtained by calculation for spiking) and the related mean of seven measurements is an estimate of the average systematic error (i.e., bias of

Table 2-3. Characterization Data for Contaminated Waters.[a]*

Analyte	Contaminated water					Units
	CW1	CW2	CW3	CW4	CW5	
Alkalinity	547	202	161	166	NA [b]	mg/L as $CaCO_3$
BOD_5	24.8	< 2	4	102	NA	mg/L
$CBOD_5$	NA	< 2	3	NA	NA	mg/L
Chloride	14	124	114	688	6.6	mg/L
Chlorine - residual	0.01	0	0.7	<0.030	NA	mg/L
Chlorine / oxidants - test strip	NA	Neg [c]	Neg	NA	NA	
Cyanide - total, manual dist	0.182	0.002	0.004	<0.010	0.0295	mg/L
Cyanide - WAD	NA	0.0002	0.0033	NA	NA	mg/L
Hardness	NA	NA	280	NA	NA	mg/L
Nitrogen - ammonia (NH_3-N)	NA	0.10	0.00	25.40	<0.293	mg/L
Nitrogen - nitrate (NO_3-N)	<0.1	7.21	10.98	<0.005	0.26	mg/L
Nitrogen - nitrite (NO_2-N)	<0.1	0.00	0.10	<0.005	<0.1	mg/L
Nitrogen - total Kjeldahl (TKN)	NA	0.70	1.54	NA	NA	mg/L
ORP	215	360	406	-190	NA	mV
Oxygen, dissolved	NA	7.1	7.3	NA	NA	mg/L
pH	7.41	7.0	7.3	6.8	NA	pH units
Phenol - automated dist	NA	NA	9	NA	NA	µg/L
Phenol - manual dist	NA	< 2	< 2	NA	NA	µg/L
Phosphorous - total	NA	1.90	3.00	NA	NA	mg/L
Solids - suspended	NA	< 2	< 2	NA	NA	mg/L
Solids - total	NA	650	536	NA	NA	mg/L
Sulfate (SO_4^{2-})	6.04	66.24	66	83.3	18.9	mg/L
Sulfide (S^{2-})	<0.02	0.07	0.03	4.09	<0.02	mg/L
Sulfide - test strip	NA	Neg	Neg	NA	NA	
Thiocyanate	<0.005	0.10	NA	0.0468	0.005	mg/L
Total arsenic (As)	NA	0.00	0.00	NA	NA	mg/L
Total aluminum (Al)	0.0137	NA	NA	0.353	NA	mg/L
Total barium (Ba)	NA	0.034	0.026	NA	NA	mg/L
Total cadmium (Cd)	<0.0003	0.000	0.001	<0.002	NA	mg/L
Total calcium (Ca)	159.0	NA	NA	29.0	78.7	mg/L
Total chromium (Cr)	NA	0.000	0.005	NA	NA	mg/L
Total cobalt (Co)	1.60	NA	NA	<4.00	NA	µg/L
Total copper (Cu)	<0.001	0.012	0.036	0.0456	NA	mg/L
Total iron (Fe)	12.8	0.09	0.13	1.32	NA	mg/L
Total lead (Pb)	<0.0003	0.000	0.006	<0.015	NA	mg/L
Total magnesium (Mg)	14.71	NA	NA	38.3	7.29	mg/L
Total manganese (Mn)	1.71	0.03	0.011	0.132	NA	mg/L
Total mercury (Hg)	<0.2	NA	0.06	0.0657	NA	ug/L
Total molybdenum (Mo)	NA	0.00	0.01	NA	NA	mg/L
Total nickel (Ni)	0.00428	0.000	0.007	<0.003	NA	mg/L
Total potassium (K)	12.77	NA	NA	22.7	3.78	mg/L
Total selenium (Se)	NA	0.00	0.00	NA	NA	mg/L
Total silver (Ag)	<0.0003	0.002	0.001	0.00315	NA	mg/L
Total sodium (Na)	867	NA	NA	383	45.6	mg/L
Total zinc (Zn)	0.00146	0.065	0.069	0.0637	NA	mg/L
Total organic carbon	9.45	9	11	89.3	4.15	mg/L

[a] Analyses performed on preserved, unfiltered samples.
[b] NA, Not analyzed. [c] Neg, Negative.
* Adapted with permission from *Environ. Sci. Technol.* **2003**, *37*, 107-115. Copyright, 2003 American Chemical Society.

the test method). Negative bias results when the measured result is less than the true value, whereas positive bias results when the measured result is greater than the true value. Calculation of bias was based on procedures specified in *Standard Methods* (APHA/AWWA/WEF, 1998) and performed as follows:

$$Bias = \overline{X} - C \qquad (2\text{-}2)$$

where:

$\overline{X}$ = the mean of seven replicates for a matrix, cyanide species, and spiking concentration, ppb;

C = the true concentration, as determined from the sample spiking preparation, ppb.

Precision is a measure of the reproducibility and random error associated with a test method. The approach used to assess the precision of a test method was the relative standard deviation (RSD) of the seven measurements obtained for a particular spiking compound and concentration in a test water. A higher value of RSD indicates more random error and lower precision in the method.

$$RSD = SD / \overline{X} \times 100\% \qquad (2\text{-}3)$$

where:

SD = the standard deviation of the seven replicate measurements, ppb;
$\overline{X}$ = the mean of seven replicates for a matrix, cyanide species, and spiking concentration, ppb.

The percent recovery, *P*, of the spike, is another measure of accuracy and is calculated as follows:

$$P = M / T \times 100\% \qquad (2\text{-}4)$$

where:
M = the measured amount of the matrix spike found through analysis in the spiked sample, mg;
T = the actual amount of the matrix spike by weights and measures during the preparation of the spiking solution, mg.

2.3.5 Quality Assurance and Quality Control

A number of quality assurance and quality control (QA/QC) procedures were implemented over the course of this study.

- Laboratory control samples were analyzed with each batch of seven replicate samples in the whole process of study. Percent recovery data for all laboratory control samples were reported and were within the acceptable limits established by U.S. EPA (i.e., ±25% for cyanide) (U.S. EPA, 1997a).
- A detailed sampling protocol was developed and used consistently in the collection of the contaminated water samples.
- Consistent procedures for storing samples and analyzing them within specified holding times were followed.

- Standards prepared and certified by an independent organization were analyzed by each participating laboratory to check the accuracy of each species-specific cyanide analytical method.

In addition, sample uniformity was checked upon arrival at the participating laboratories. Each laboratory analyzed the sample for total cyanide immediately upon arrival using the conventional distillation method [Method 4500-CN-C (APHA/AWWA/WEF, 1998)]. These data, presented in Table 2-4, exhibit consistency of total cyanide concentrations obtained by the different laboratories, except for a somewhat higher total cyanide measurement for primary clarifier effluent by the MWRDGC laboratory.

2.4 Results

2.4.1 Method Detection Limits

Table 2-5 lists the results of MDLs determined from the analysis of spiked samples. It was shown that almost all the analytical methods were capable of achieving low MDLs (low ppb levels), for their target cyanide analytes or analyte groups in all water matrices tested. The typical MDL range for all waters was 1–2 ppb.

Table 2-4. Total Cyanide Analyses for Various Contaminated Waters.

Laboratory	Replicates	Cyanide concentrations in contaminated waters (ppm)				
		CW1	CW2	CW3	CW4	CW5
Alcoa [a]	1	0.19	0.010	<0.010	<0.010	0.029
	2	0.20	0.010	<0.010	<0.010	0.030
	3	0.21	0.010	<0.010	<0.010	0.030
	Mean	0.20	0.010	<0.010	<0.010	0.030
	SD	0.01	0	NA [d]	NA	0.001
Bayer	1	0.180	<0.008	<0.008	<0.008	0.029
	2	0.178	<0.008	<0.008	<0.008	0.027
	3	0.185	<0.008	<0.008	<0.008	0.030
	Mean	0.181	<0.008	<0.008	<0.008	0.029
	SD	0.004	NA	NA	NA	0.002
Carnegie Mellon	1	0.199	0.007	0.002	<0.002	0.031
	2	0.202	0.006	0.002	<0.002	0.028
	3	0.199	NA	0.002	<0.002	0.027
	Mean	0.200	0.007	0.002	<0.002	0.029
	SD	0.002	0.001	0.000	NA	0.002
MWRA [b]	1	NA	<0.01	<0.01	<0.01	0.028
	2	NA	<0.01	<0.01	<0.01	0.028
	3	NA	<0.01	<0.01	<0.01	0.029
	Mean	NA	<0.01	<0.01	<0.01	0.028
	SD	NA	<0.01	<0.01	<0.01	0.001
MWRDGC [c]	1	0.169	0.001	0.003	0.031	0.028
	2	NA	NA	NA	NA	NA
	3	NA	NA	NA	NA	NA
	Mean	NA	NA	NA	NA	NA
	SD	NA	NA	NA	NA	NA

[a] Centre Analytical Laboratories, Inc. (State College, PA) performed the analysis for Alcoa.
[b] Massachusetts Water Resources Authority.
[c] Metropolitan Water Reclamation District of Greater Chicago.
[d] NA = Not available.

Table 2-5. Method Detection Limits for Various Techniques and Water Matrices.*

Laboratory technique	Water matrix	Method detection limits (ppb)		
		NaCN	$K_2Ni(CN)_4$	$K_4Fe(CN)_6$
Alcoa MCN-IC	DIW [a]	NA[b]	1.82	0.81
	CW1	NA[b]	10.4	0.94
	CW2	NA[b]	4.02	0.49
	CW3	NA[b]	8.27	1.22
	CW4	NA[b]	10.7	0.24
	CW5	NA[b]	NA[c]	NA[c]
Bayer Avail-CN	DIW	0.43	0.38	NA[e]
	CW1	NA[c]	NA[c]	NA[e]
	CW2	5.14	0.99	NA[e]
	CW3	1.53	3.72	NA[e]
	CW4	NA[d]	NA[d]	NA[e]
	CW5	1.10	0.63	NA[e]
Carnegie Mellon WAD-CN	DIW	0.72	0.64	NA[e]
	CW1	NA[c]	NA[c]	NA[e]
	CW2	0.87	1.05	NA[e]
	CW3	0.53	0.64	NA[e]
	CW4	0.88	1.17	NA[e]
	CW5	1.68	1.76	NA[e]
Carnegie Mellon FCN-MD	DIW	0.44	0.85	NA[e]
	CW1	1.16	0.75	NA[e]
	CW2	1.15	0.00[g]	NA[e]
	CW3	0.00	0.42	NA[e]
	CW4	0.53	2.01	NA[e]
	CW5	0.87	1.50	NA[e]
MWRA TCN-LPUD	DIW	0.42	0.99	0.86
	CW1	NA[f]	NA[f]	NA[f]
	CW2	0.73	0.37	0.22
	CW3	1.34	0.88	0.69
	CW4	0.43	0.36	0.36
	CW5	NA[c]	NA[c]	NA[c]
MWRDGC TCN-TFD	DIW	2.47	1.68	1.19
	CW1	NA[c]	NA[c]	NA[c]
	CW2	2.17	1.53	6.50
	CW3	1.19	2.47	0.00
	CW4	NA[c]	NA[c]	NA[c]
	CW5	NA[c]	NA[c]	NA[c]
MWRDGC WAD-TFD	DIW	0.30	0.46	NA[e]
	CW1	NA[c]	NA[c]	NA[e]
	CW2	0.36	0.49	NA[e]
	CW3	0.81	0.47	NA[e]
	CW4	0.43	0.54	NA[e]
	CW5	0.49	1.00	NA[e]

[a] DIW = Reagent (deionized) water meeting the Type II specifications as defined in ASTM D 1193-91 (ASTM, 1998g).
[b] NA = Not available because NaCN is the main component in the eluent for ion chromatography technique.
[c] NA = Not available because the measured level of analyte in the sample was greater than five times the detection limit.
[d] NA = Not available because interferences prevented samples from being analyzed under the specifications of this project.
[e] NA = Not available because this method is not intended to detect iron cyanide.
[f] NA = Not available because the study on the MGP site groundwater was not conducted because of manufacturer delay in shipment of a new autoanalyzer.
[g] The MDL of 0.00 resulted from identical results in the seven replicates, yielding a standard deviation SD = 0.00, thus a MDL = 0.00 (see Equation 2-1).
* Adapted with permission from Zheng et al. , *Environ. Sci. Technol.* **2003**, *37*, 107-115. Copyright 2003 American Chemical Society.

2.4.2 Method Bias and Precision

Tables 2-6 through 2-8 provide a summary of bias and precision results for the various methods. The complete data set is in Appendix B and is available by request.

Method Avail-CN experienced substantial difficulty when it was applied for the analysis of WWTP primary clarifier effluent. The total cyanide results, obtained from the conventional test method [Method 4500-CN-C (APHA/AWWA/WEF, 1998)], showed <8.0 ppb cyanide. The available cyanide test showed a result that was greater than the calibration limit (400 ppb range), which indicated a significant positive interference. To evaluate the possibility of sulfide interference, the sample was treated with lead carbonate (as directed in the Method OIA-1677) and filtered immediately with a mean test result of 9.4 ppb. This shows a significant reduction in available cyanide after lead carbonate treatment indicating the possibility of sulfide interference. It is important to note that the standard test of sulfide interference using lead acetate paper, as employed in this study, may not be sufficiently sensitive for samples containing low (ppb) levels of sulfide.

Method FCN-MD achieved satisfactory results for its target cyanide species, HCN. However, there was partial recovery of diffusible cyanide in the tests for samples spiked with high levels of nickel cyanide species, resulting in a relatively large negative bias. This phenomenon reflects the apparent dissociation of this weak metal cyanide complex during the testing process. Nonetheless, RSD values were uniformly low, which indicated acceptable precision even under conditions of incomplete recovery.

Method MCN-IC generally exhibited the largest negative bias of all the test methods when applied to nickel cyanide-spiked samples. However, RSD values were relatively low, similar to Method FCN-MD.

Method TCN-LPUD, the other method that uses UV digestion besides the TCN-TFD method, performed well in all water matrices tested. It should be noted that the UV digestion for TCN-TFD is performed under alkaline conditions whereas the digestion for TCN-LPUD is conducted under slightly acidic conditions. This indicates that UV digestion can be effective in some method configurations.

Method TCN-TFD yielded relatively high negative bias in all water matrices, especially when applied to contaminated water with a high background cyanide concentration. This apparent underestimation of cyanide is consistent with U.S. EPA's concern about limitations of UV digestion for cyanide analysis of some samples.

Method WAD-CN achieved low bias in all water matrices indicating no apparent systematic errors in the method.

Method WAD-TFD performed well for all samples except the MGP site groundwater sample. This method performed extremely well for low levels of WAD cyanide in WWTP effluents (i.e., in samples spiked with 5 ppb of CN^-) because it was designed for an upper limit of 25 ppb of CN^- in undiluted samples. Avoidance of sample dilution can potentially eliminate some sources of error to obtain extremely low levels of WAD cyanide.

Overall, all methods demonstrated satisfactory precision with correspondingly low RSD values for their target cyanide analytes or analyte groups in all contaminated water matrices.

2.4.3 Cyanide Species Recovery Study

The data for cyanide species recovery in different water matrix types are presented in Tables 2-9 through 2-11.

Method Avail-CN achieved satisfactory recovery for two cyanide spiking species (not intended to measure Fe–CN complex) in almost all water matrices and all cyanide spiking levels, except for the analysis of WWTP primary clarifier effluent, as mentioned previously. The method achieved more than 130% recovery for some samples spiked with the lowest concentration level (i.e., 5 ppb, of cyanide species).

Method FCN-MD achieved a mean recovery of only 76% for nickel cyanide spiked at the 100 ppb level. This method is not designed to recover nickel cyanide, a weak metal cyanide complex, or ferrocyanide, a strong metal cyanide complex.

Method MCN-IC, one of the three methods tested that could detect the strongly complexed metal cyanide compound, ferrocyanide $Fe(CN)_6^{4-}$, was able to achieve satisfactory results in all water matrices except the primary clarifier effluent where low recoveries of 5.9% and 16% were observed for ferrocyanide spiked at 5 ppb and 100 ppb, respectively. For nickel cyanide spiked at the 100 ppb level, this method exhibited two negative recoveries and only recovered 56–73% of nickel cyanide in the other three water matrices.

Method TCN-LPUD performed well in ferrocyanide spiked wastewaters, with only one exception of an 8.5% recovery in the WWTP chlorinated secondary effluent.

Method TCN-TFD only recovered about 45% of 100 ppb of simple cyanide species (NaCN) spiked in MGP site groundwater and 53% in WWTP chlorinated secondary effluent. For the other waters tested, the method recovered about 85% of free cyanide, which was low compared to other methods. For NaCN spiked at the 5 ppb level, two negative values were observed. In nickel cyanide spiking tests, the method achieved variable recoveries in all five contaminated water matrices, ranging from 60-103%, with a mean recovery of 79.9% at the 100 ppb spiking level and a standard deviation of 18%. The method TCN-TFD did not achieve acceptable results for nickel cyanide spiked at the 5 ppb level in almost all wastewater samples tested. This method also underestimated ferrocyanide in all but two water matrices.

Method WAD-CN achieved satisfactory recovery for two cyanide species (not intended to measure Fe-CN complex) spiked in all water matrices at all concentration levels.

Method WAD-TFD achieved satisfactory recovery for two cyanide species (not intended to measure Fe-CN complex) spiked in almost all water matrices at all concentration levels, except for a negative value observed when spiking NaCN at the 5 ppb level.

Table 2-6. Bias and Precision of Cyanide Analytical Methods When Spiking 5 ppb of Cyanide Species in Different Water Matrices.*

Method		NaCN spiking		$K_2Ni(CN)_4$ spiking		$K_4Fe(CN)_6$ spiking	
		Bias (ppb)	RSD (%)	Bias (ppb)	RSD (%)	Bias (ppb)	RSD (%)
Avail-CN	DIW [a]	0.41	6.65	0.14	6.61	NA	
	CW1	1.51	8.88	0.40	4.97		
	CW2	1.92	19.59	0.55	4.43		
	CW3	0.51	4.18	1.64	9.27		
	CW4	NA [b]	NA	NA	NA		
	CW5	1.95	4.00	1.49	2.41		
FCN-MD	DIW	0.12	2.73	0.22	5.09	NA	
	CW1	1.12	5.74	-0.52	4.59		
	CW2	0.31	6.50	-0.47	0.00		
	CW3	0.43	0.00	-0.24	1.35		
	CW4	-1.72	3.19	-2.96	12.33		
	CW5	-0.15	5.60	0.10	9.21		
MCN-IC	DIW	NA		NA		NA	NA
	CW1					2.00	5.28
	CW2					-0.22	3.07
	CW3					0.75	6.27
	CW4					-4.70	8.42
	CW5					-0.93	2.99
TCN-LPUD	DIW	-0.48	4.20	-1.46	1.69	-1.11	6.17
	CW1	NA	NA	NA	NA	NA	NA
	CW2	0.11	4.49	-0.08	3.64	-0.76	1.68
	CW3	-4.12	3.33	-3.59	11.27	-4.58	5.33
	CW4	-0.20	1.92	-0.57	3.57	0.18	3.62
	CW5	0.78	2.17	-2.35	2.44	-0.53	2.89
TCN-TFD	DIW	-1.57	23.03	1.43	8.24	1.14	6.19
	CW1	-7.01	7.18	-41.59	6.10	-17.01	7.17
	CW2	-0.86	5.25	-3.29	4.58	0.43	14.35
	CW3	-6.86	6.19	-5.43	10.44	-4.00	0.00
	CW4	-1.69	2.22	2.88	4.55	-2.12	3.16
	CW5	-0.13	5.39	-0.42	3.01	1.72	4.90
WAD-CN	DIW	0.18	4.44	0.02	3.92	NA	
	CW1	0.91	9.14	2.59	14.16		
	CW2	-0.09	5.04	-1.81	5.16		
	CW3	0.12	2.13	-0.06	1.98		
	CW4	-1.02	5.16	-1.74	6.37		
	CW5	0.74	5.53	0.79	5.67		
WAD-TFD	DIW	0.03	1.99	0.69	2.64	NA	
	CW1	-5.47	0.70	-5.26	5.93		
	CW2	0.10	2.11	0.31	2.71		
	CW3	0.30	3.10	-0.34	1.93		
	CW4	0.23	2.58	1.00	2.74		
	CW5	0.19	2.32	0.29	4.58		

[a] Water matrix types: DIW = reagent (deionized) water; refer to Table 2-2 for CW1–CW5.

[b] NA = Not available; spiking not performed because of incompatibility with method or method limitation.

* Adapted with permission from Zheng et al., *Environ. Sci. Technol.* **2003**, *37*, 107-115. Copyright 2003 American Chemical Society.

Table 2-7. Bias and Precision of Cyanide Analytical Methods When Spiking 30 ppb of Cyanide Species in Different Water Matrices.[a]

Method		NaCN spiking		$K_2Ni(CN)_4$ spiking		$K_4Fe(CN)_6$ spiking	
		Bias (ppb)	RSD (%)	Bias (ppb)	RSD (%)	Bias (ppb)	RSD (%)
Avail-CN	DIW[b]	1.69	0.73	-1.10	0.55	NA	
	CW1	1.06	6.90	-2.09	1.82		
	CW2	5.28	2.59	1.89	2.34		
	CW3	2.08	3.45	3.62	2.06		
	CW4	NA[c]	NA	NA	NA		
	CW5	-1.80	1.37	-4.00	1.19		
FCN-MD	DIW	-1.07	1.62	-6.90	2.97	NA	
	CW1	1.37	2.70	-3.79	1.19		
	CW2	-0.13	2.39	-4.86	4.20		
	CW3	-2.83	1.63	-6.16	2.93		
	CW4	-1.48	1.37	-6.48	2.91		
	CW5	1.97	2.07	-9.94	7.73		
MCN-IC	DIW	NA		18.81	4.07	0.03	5.35
	CW1			-103.6	9.73	1.12	5.68
	CW2			-44.43	4.82	-2.21	8.21
	CW3			-31.13	7.69	-1.29	2.11
	CW4			-103.5	24.72	-12.62	5.52
	CW5			-27.26	12.08	-2.83	1.04
TCN-LPUD	DIW	-0.29	0.67	-0.14	2.51	-1.35	1.57
	CW1	NA	NA	NA	NA	NA	NA
	CW2	0.34	2.00	-2.28	0.83	-1.25	0.84
	CW3	0.65	1.38	-0.37	1.48	-0.04	4.27
	CW4	-0.96	2.10	-6.40	1.08	0.64	1.84
	CW5	-0.04	1.07	-6.65	1.82	-3.12	1.82
TCN-TFD	DIW	4.29	3.65	35.00	5.26	2.86	3.26
	CW1	-36.04	6.43	-32.47	9.94	-41.90	5.24
	CW2	-10.26	1.71	-14.68	4.57	-5.68	3.33
	CW3	-22.12	2.52	-23.98	10.00	-17.84	3.48
	CW4	-2.77	1.19	-0.48	2.32	-3.77	1.87
	CW5	-0.78	2.12	1.22	4.51	7.22	2.24
WAD-CN	DIW	0.50	1.67	1.60	1.53	NA	
	CW1	4.52	1.94	0.50	7.44		
	CW2	0.31	1.68	-1.44	1.26		
	CW3	-0.10	3.56	1.47	0.82		
	CW4	-1.13	2.40	-2.81	1.53		
	CW5	-0.08	5.83	-0.15	1.91		
WAD-TFD	DIW	1.44	1.15	2.03	1.09	NA	
	CW1	-5.19	1.34	-2.00	6.17		
	CW2	0.10	2.18	6.34	5.14		
	CW3	4.22	1.34	4.74	1.85		
	CW4	-6.23	2.00	3.50	2.05		
	CW5	1.80	1.13	2.55	1.99		

[a] ppb = µg/L
[b] Water matrix types: DIW = reagent (deionized) water; refer to Table 2-2 for CW1-CW5.
[c] NA = Not available; spiking not performed because of incompatibility with method or method limitation.

Table 2-8. Bias and Precision of Cyanide Analytical Methods When Spiking 100 ppb of Cyanide Species in Different Water Matrices.*

Method		NaCN Spiking		$K_2Ni(CN)_4$ Spiking		$K_4Fe(CN)_6$ Spiking [a]	
		Bias (ppb)	RSD (%)	Bias (ppb)	RSD (%)	Bias (ppb)	RSD (%)
Avail-CN	DIW [b]	-0.90	0.26	-4.63	0.81	NA	
	CW1	7.84	2.42	-4.01	2.59		
	CW2	11.06	0.48	4.62	1.07		
	CW3	6.15	1.23	4.43	1.80		
	CW4	NA [c]	NA	NA	NA		
	CW5	-7.14	1.65	-3.39	0.77		
FCN-MD	DIW	1.60	2.71	-42.34	2.68	NA	
	CW1	4.46	3.09	-25.72	7.05		
	CW2	0.54	0.91	-32.91	2.45		
	CW3	-7.62	0.61	-33.62	1.61		
	CW4	-5.48	1.50	-18.91	1.05		
	CW5	2.75	1.25	-19.90	1.21		
MCN-IC	DIW	NA		18.81	4.07	-1.07	2.71
	CW1			-103.58	9.73	1.12	5.68
	CW2			-44.43	4.82	-2.15	2.90
	CW3			-31.13	7.69	-4.47	3.85
	CW4			-103.48	24.72	-25.17	7.80
	CW5			-27.26	12.08	-5.57	0.94
TCN-LPUD	DIW	1.55	0.78	1.74	0.76	-5.01	0.80
	CW1	NA	NA	NA	NA	NA	NA
	CW2	0.36	2.26	2.96	0.37	-7.81	0.69
	CW3	-2.63	2.30	-6.81	1.29	0.36	4.19
	CW4	-3.37	0.67	-23.60	0.99	-1.61	1.41
	CW5	0.12	0.95	-10.61	1.36	-10.40	1.82
TCN-TFD	DIW	6.43	2.23	NA	NA	6.71	3.41
	CW1	-55.13	7.30	-33.70	8.30	-70.70	4.74
	CW2	-21.34	4.01	-30.20	5.12	-12.62	2.22
	CW3	-47.06	4.93	-40.49	6.53	-32.49	1.46
	CW4	-16.12	2.36	-8.12	2.66	-12.83	1.24
	CW5	-11.01	4.19	3.42	2.62	2.85	1.39
WAD-CN	DIW	3.94	0.95	6.87	1.16	NA	
	CW1	7.30	2.42	7.06	2.21		
	CW2	3.53	0.84	0.01	0.16		
	CW3	5.05	0.63	-7.73	1.73		
	CW4	-0.56	1.27	-1.57	0.53		
	CW5	2.73	0.72	0.01	0.31		
WAD-TFD	DIW	3.96	0.79	4.00	1.58	NA	
	CW1	-9.14	1.64	17.79	2.76		
	CW2	0.94	0.74	10.84	2.08		
	CW3	4.86	1.39	15.33	2.87		
	CW4	-16.14	1.61	-2.20	1.94		
	CW5	5.48	2.92	9.46	5.28		

[a] The concentration of Fe-CN spiked for MCN-IC method is 30 ppb.
[b] Water matrix types: DIW = Reagent (deionized) water; refer to Table 2-2 for CW1-CW5.
[c] NA = Not available; spiking not performed because of incompatibility with method or method limitation.
* Adapted with permission from Zheng et al., *Environ. Sci. Technol.* **2003**, *37*, 107-115. Copyright 2003 American Chemical Society.

Table 2-9. Percent Recovery of Cyanide Analytical Methods When Spiking NaCN in Different Water Matrices.*

Spiking concentration	Water types [a]	Avail-CN	FCN-MD	TCN-LPUD	TCN-TFD	WAD-CN	WAD-TFD
5 ppb [b] as CN	DIW	108.1	102.4	90.49	68.57	103.7	100.6
	CW1	130.3	121.9	NA	-17.39	117.9	-8.24
	CW2	138.6	106.2	102.1	83.76	98.18	102.1
	CW3	110.4	108.4	17.56	-36.34	102.4	106.3
	CW4	NA [c]	65.75	95.91	66.02	79.78	104.6
	CW5	139.0	97.13	115.6	97.42	114.2	103.7
30 ppb as CN	DIW	105.62	96.43	98.99	114.29	101.68	104.81
	CW1	103.52	104.5	NA	-18.79	114.79	82.69
	CW2	117.64	99.59	101.14	65.8	101.06	100.34
	CW3	106.92	90.89	102.19	26.27	99.67	114.08
	CW4	NA	95.07	96.82	90.79	96.26	79.24
	CW5	93.98	106.38	99.88	97.42	99.72	105.97
100 ppb as CN	DIW	99.10	101.6	101.6	106.4	103.9	104.0
	CW1	107.9	104.4	NA	45.26	128.4	90.98
	CW2	111.1	100.1	100.4	86.93	104.2	100.9
	CW3	106.2	92.64	97.37	52.94	111.3	104.9
	CW4	NA	94.54	96.63	83.88	101.3	83.86
	CW5	92.86	105.1	100.1	88.99	109.5	105.5

[a] Water matrix types: DIW = reagent (deionized) water; refer to Table 2-2 for CW1-CW5.

[b] ppb = μg/L. [c] NA = Not available; spiking not performed because of incompatibility with method or method limitation.

* Adapted with permission from Zheng et al., *Environ. Sci. Technol.* **2003**, *37*, 107-115. Copyright 2003 American Chemical Society.

Table 2-10. Percent Recovery of Cyanide Analytical Methods When Spiking $K_2Ni(CN)_4$ in Different Water Matrices.*

Spiking concentration	Water Types [a]	Avail-CN	FCN-MD	MCN-IC	TCN-LPUD	TCN-TFD	WAD-CN	WAD-TFD
5 ppb[b] as CN	DIW	102.7	104.3	NA [c]	70.71	128.6	100.4	113.7
	CW1	108.0	90.50	NA	NA	-708.8	147.0	-3.95
	CW2	111.0	93.77	NA	98.41	35.19	76.06	106.4
	CW3	132.9	96.76	NA	28.29	-7.77	99.15	93.45
	CW4	NA	51.65	NA	88.63	157.5	71.60	120.0
	CW5	129.9	102.0	NA	53.00	91.71	115.2	105.7
	DIW	96.33	77.38	118.81	99.71	216.67	105.31	106.76
30 ppb as CN	CW1	93.05	88.53	-3.58	NA	-6.88	101.51	93.31
	CW2	106.34	89.28	55.57	92.4	51.04	96.83	121.15
	CW3	112.06	86.24	68.87	98.74	20.08	103.28	115.79
	CW4	NA	82.35	-3.48	78.67	98.41	92.34	111.67
	CW5	86.65	68.23	72.74	77.83	104.09	99.51	108.49
100 ppb as CN	DIW	95.37	58.32	118.8	101.7	NA	106.8	104.0
	CW1	95.99	76.67	-3.58	NA	66.69	106.4	117.9
	CW2	104.6	78.24	55.57	103.4	78.08	100.0	110.8
	CW3	104.4	77.46	68.87	93.2	59.51	94.82	115.4
	CW4	NA	84.55	-3.48	76.41	91.88	98.72	97.80
	CW5	96.61	80.92	72.74	89.40	103.4	100.0	109.5

[a] Water matrix types: DIW = reagent (deionized) water; Refer to Table 2-2 for CW1-CW5.

[b] ppb = μg/L.

[c] NA = Not available

* Adapted with permission from Zheng et al., *Environ. Sci. Technol.* **2003**, *37*, 107-115. Copyright 2003 American Chemical Society.

Table 2-11. Percent Recovery of Cyanide Analytical Methods When Spiking $K_4Fe(CN)_6$ in Different Water Matrices.*

Spiking concentration	Water types [a]	MCN-IC	TCN-LPUD	TCN-TFD
5 ppb [b]	DIW	100.3	77.71	122.9
	CW1	56.70	NA [c]	-217.4
	CW2	95.52	84.74	109.5
	CW3	115.0	8.48	20.80
	CW4	5.92	103.6	57.45
	CW5	81.29	89.31	134.6
30 ppb				
	DIW	100.27	95.57	109.52
	CW1	69.26	NA	-38.31
	CW2	85.32	95.84	81.04
	CW3	91.38	99.87	40.56
	CW4	15.87	102.13	87.45
	CW5	81.15	89.61	124.09
100 ppb				
	DIW	95.15	95.00	106.7
	CW1	69.26	NA	29.29
	CW2	92.84	92.68	87.38
	CW3	85.11	100.4	67.51
	CW4	16.09	98.39	87.17
	CW5	81.45	89.61	102.9

[a] Water matrix types: DIW = reagent (deionized) water; Refer to Table 2-2 for CW1-CW5.
[b] ppb = µg/L.
[c] NA = Not available.
* Adapted with permission from Zheng et al., *Environ. Sci. Technol.* **2003**, *37*, 107-115. Copyright 2003 American Chemical Society.

2.4.4 Evaluation of Sample Preservation Issues

A preliminary sample preservation evaluation in conjunction with the Task 1 methods testing was conducted at Carnegie Mellon University. The effect of in-line filtration as a means of sample preservation was assessed using filtered and unfiltered samples from the MGP site during the March 1999 sampling event. The samples, collected by a representative of the Gas Research Institute (GRI) and Niagara Mohawk Power Corporation, contained about 200 ppb total cyanide and a small amount of organic material (TOC = 9.5 mg/L). Analyses for total cyanide, WAD cyanide, and free cyanide by microdiffusion were performed on filtered and unfiltered samples of the contaminated water. The filtration appeared to have no consistent effect on the concentration of total cyanide, WAD cyanide, and free cyanide.

2.4.5 Analyses of Performance Evaluation Samples

Performance evaluation samples were prepared by an independent organization, Environmental Resource Associates (ERA) of Arvada, Colorado, and analyzed by all the methods for further comparative evaluation of the methods. The sample had a total cyanide concentration of approximately 70 ppb consisting of potassium cyanide and potassium ferrocyanide cyanide. Potassium cyanide yielded a targeted WAD cyanide concentration of 23.0 ppb. Each participating laboratory analyzed triplicates of one blind sample using its species-based methods and also performed the conventional total cyanide analysis for comparison of total cyanide concentrations. The results for performance evaluation samples are listed in Table 2-12.

The results from the performance evaluation study (Table 2-12) showed that all laboratories were able to achieve satisfactory results within the acceptable limits identified by ERA in accordance with U.S. EPA guidelines.

Table 2-12. Analysis of Performance Evaluation Samples.[a] *

Laboratory	Total cyanide[b] ppb as CN	Target cyanide species[c], ppb as CN		
		CN^-	$Fe(CN)_6^{4-}$	$CN^- + Fe(CN)_6^{4-}$
Certified Values	70.0	23.0	47.0	70.0
Acceptance Limits[d]	51.1 – 88.9	16.8 – 29.2	34.3 – 59.7	51.1 – 88.9
Alcoa	73 ± 2	NA[e]	MCN-IC = 46 ± 1	NA
Bayer	67.4 ± 0.5	Avail-CN = 24.3 ± 2.5	NA	NA
Carnegie Mellon	68.9 ± 0.6	WAD-CN = 23.5 ± 0.7 FCN-MD = 23.3 ± 0.6	NA	NA
MWRA	57.7 ± 3.7	NA	NA	TCN-LPUD = 66.6 ± 2.3
MWRDGC	66[f]	WAD-TFD = 22.9 ± 0.2	NA	TCN-TFD = 80.3 ± 2.3

[a] Values are reported as the mean of three replicates unless otherwise stated. Concentrations are as CN^-, ppb =μg/L.
[b] Total cyanide analysis was performed by conventional distillation method such as *Standard Methods* 4500-CN-C (APHA/AWWA/WEF, 1995).
[c] Analyses of target cyanide species were performed by the designated species-based methods of each laboratory.
[d] Based on the criteria established by U.S.EPA: ±25% for cyanide (U.S. EPA, 1997a).
[e] NA = Not applicable.
[f] Only one measurement.
* Reprinted with permission from Zheng et al., *Environ. Sci. Technol.* **2003**, *37*, 107-115. Copyright 2003 American Chemical Society.

2.4.6 Overall Evaluation of Cyanide Analytical Methods

The seven methods evaluated in this study represent a subset of methods widely available for analysis of groups of cyanide species or individual cyanide species. A list of these widely available methods is provided in Table 2-13, which includes notes about the current approval status for each method by the U.S. EPA and by voluntary consensus standards bodies, in particular the professional society consortium that develops and maintains Standard Methods for the Examination of Water and Wastewater (APHA/AWWA/WEF, 1998), and the American Society of Testing and Materials (ASTM), which develops and maintains standard testing methods.

As indicated by the list of cyanide analytical methods in Table 2-13, the only methods currently approved by U.S. EPA are total cyanide by distillation, available cyanide (which measures free cyanide and weakly complexed cyanide species plus mercury cyanide species), and cyanide amenable to chlorination (which also measures free cyanide and weakly complexed cyanide species). Of the other methods, several have been approved by the *Standard Methods* group and/or ASTM, (i.e., WAD cyanide by distillation, free cyanide by microdiffusion, total cyanide by thin film distillation, and WAD cyanide by thin film distillation). The remaining methods—metal cyanides by ion chromatography and total cyanide by low power UV digestion—have not been approved by U.S. EPA, *Standard Methods,* or ASTM. Currently, there are industry group efforts in progress to obtain U.S. EPA approval for the microdiffusion method and for the ion chromatography method. Alcoa is currently pursuing an inter-laboratory validation study for low-level and high-level metal cyanide complex determination via the ion chromatography method with the intent of submitting the resulting validation data to the U.S. EPA Office of Water and to ASTM Committee D-19 on Water for method approval. In a separate effort, a consortium of electric utility organizations and Alcoa is pursuing another inter-laboratory validation study to obtain validation data for a modified version of the ASTM-approved microdiffusion method for free cyanide with the intent of submitting the resulting data to the U.S. EPA Office of Solid Waste for approval. RETEC is providing assistance with both efforts.

The method performance data obtained in this study indicate that all of the seven methods examined in this study have the potential to be applied effectively in contaminated waters with varying degrees of limitations for matrices of different complexity. All of the methods demonstrated very low detection limits (MDLs), and met the U.S. EPA detection limit requirements for cyanide (40 CFR 141.23, July 1, 2001; U.S. EPA, 2001), which was 0.005–0.05 ppm depending on the method. It was also shown that each method yielded acceptable results in all water matrices in terms of precision and bias, except for the available cyanide technique and the IC techniques when applied to raw WWTP wastewater.

Several of the methods have more potential for routine application than others. Some of the methods are more robust with respect to application to highly contaminated waters, and there are significant differences among the methods in cost and ease of implementation. These differences are summarized in a general way in Table 2-14. As the table indicates, there are trade-offs involved with the selection of a particular method. The newer more automated techniques (available cyanide, and the automated analysis techniques involving UV digestion and thin film distillation) provide for rapid accurate analyses for most types of waters but require the acquisition of expensive instrumentation. The older techniques that involve manual distillation are slower to implement, but are low cost and very robust—except for the CATC method. The free cyanide by microdiffusion method is a relatively new method that is simple, inexpensive and robust, though slower to implement. For routine use with contaminated waters, the methods having the greatest potential for widespread, robust application are total cyanide by distillation (the most widely used method today), WAD cyanide by distillation and free cyanide by microdiffusion. Of course, depending on the specifics of a particular situation, any of the other methods may be easily justified. Consideration of the trade-offs between performance, cost, ease of implementation, and speed of analysis necessarily must be done on a case-by-case basis. The results of this study will help in such assessments.

Water Environment Research Foundation

Table 2-13. Approval Status for Cyanide Analytical Methods.

Method	CN species measured	Method description	Approval status			
			U.S. EPA	ASTM [g]	SM [h]	Others [i]
Total cyanide by distillation	Total CN	Strong acidification, followed by distillation, to liberate HCN into NaOH scrubbing solution. CN is determined by colorimetric and potentiometric procedures	335.4 [a]	2036-98A	4500-CN-C, E, F	
WAD cyanide by distillation	WAD-CN	Mild acidification, followed by distillation, to liberate HCN into NaOH scrubbing solution. CN is determined by colorimetric and potentiometric procedures		2036-98C	4500-CN-I, E, F	
Free cyanide by microdiffusion	Free CN	HCN volatilizes from a sample buffered at pH 6 and diffuses through enclosed headspace into NaOH solution. CN is determined colorimetrically	ILV in progress [b]	4282-95		
Available cyanide	WAD-CN + Hg-CN	Samples are treated with ligand exchange reagents to liberate free cyanide from weakly complexed cyanide and mercury cyanide species, followed by amperometric measurement of the liberated free cyanide	OIA-1677 [c]	Draft		
Metal cyanides by ion chromatography	Metal cyanides	The anionic cyanometallates are separated on an IC column using gradient elution and are then detected by UV/Vis absorbance at approximately 215 nm	ILV in progress [d]	Draft- ILV in progress [d]		Dionex Corp. – No.55
Cyanide amenable to chlorination (CATC)	WAD-CN	Part of a sample is chlorinated to decompose cyanides. Both chlorinated and unchlorinated sample fractions are subjected to distillation. The difference is expressed as CNTC	Note [e]	2036-98B	4500-CN-G, E, F	
Total cyanide by thin film distillation	Total CN	Acidification, followed by ultraviolet light radiation, to dissociate strong complexes, and rapid exposure to heat treatment via continuous thin-film distillation		4374-00		
Total cyanide by low-power UV digestion	Total CN	Cyanide is released as HCN from cyanide complexes by means of UV digestion and distillation and is determined by a colorimetric procedure	Note [f]			Skalar Analytical
WAD cyanide by thin film distillation	WAD-CN	Acidification to dissociate weak metal-cyano complexes, and rapid exposure to heat treatment via continuous thin-film distillation		4374-00		

[a] U.S. EPA (1983)
[b] Interlaboratory validation study in progress by Alcoa and electric utility consortium; data to be submitted to U.S. EPA Office of Solid Waste.
[c] *Federal Register,* December 30, 1999 (Vol. 64, No. 250). This method was approved under the Office of Water's Alternate Test Procedure (ATP) program (U.S. EPA 1999a).
[d] Interlaboratory validation study in progress by Alcoa; data to be submitted to U.S. EPA Office of Water and ASTM.
[e] Approved formerly as U.S. EPA Method 335.1, withdrawn in 1994. U.S. EPA currently approves use of the ASTM and SM methods (*Technical Notes on Drinking Water Methods*, U.S. EPA, 1994c, EPA-600-R-94-173).
[f] Approved formerly as U.S. EPA Method 335.3, withdrawn in 1994 because of concern about underestimation of cyanide concentration in some samples by UV digestion.
[g] *Annual Book of ASTM Standards*, Vols. 14.01 and 14.02, American Society for Testing and Materials (ASTM), Philadelphia, PA, 2000. "Draft" means that a draft method has been proposed to ASTM and is under consideration by an ASTM technical committee, but has not received all approvals required to become an ASTM method.
[h] APHA/AWWA/WEF (1998).
[i] "Others" refers to other organizations such as a private-sector company that has developed an analytical method.

Table 2-14. Comparison of Cyanide Species Analytical Techniques.

Method	Time for analysis	Method complexity	Instrument cost	Performance with settled WWTP raw wastewater
Total cyanide by distillation	2 hours	low	less expensive	good spike recovery
WAD cyanide by distillation	2 hours	low	less expensive	good spike recovery
Free CN by microdiffusion	4 hours	low	inexpensive	good spike recovery
Cyanide amenable to chlorination	3 hours	low	less expensive	Note [1]
Available cyanide	minutes	low	expensive	interference and poor spike recovery
Metal cyanides by ion chromatography	minutes	high	expensive	poor spike recovery for some compounds
Total cyanide by low-power UV digestion	minutes	low	expensive	good spike recovery
Total cyanide by thin film distillation	minutes	low	expensive	moderate recovery
WAD cyanide by thin film distillation	minutes	low	expensive	good spike recovery

[a]The CATC method was not evaluated in this study, but much previous experience indicates that the method has problems with recovery, in particular, negative concentration determinations.

2.5 Conclusions

This study demonstrated that all seven of the cyanide analytical methods (including four methods for analysis of WAD cyanide) could achieve low MDLs, reasonably good accuracy and precision for their target cyanide analytes or analyte groups in all but the most contaminated water matrices. A brief summary of the major findings is presented below and more details can be found in Zheng et al. (2003a).

- The method Avail-CN encountered a significant interference problem for the analysis of WWTP primary clarifier effluent. The interference came from sulfide (S^{2-}) at ppb levels, an ion commonly found in municipal and industrial wastewaters. Upon collection of the sample researchers performed a pretreatment testing of sulfide that used lead acetate test paper, but failed to detect this low level of sulfide.
- The method FCN-MD recovered about 76% of diffusible cyanide in the testing with nickel cyanide-spiked samples, apparently reflecting dissociation of this weak metal cyanide complex during the testing process.
- The method TCN-TFD achieved low recoveries in several wastewater matrices with three cyanide species spiking at three concentration levels.
- The method TCN-LPUD, which utilizes a UV digestion technique, performed well in all water matrices. However, less-than-complete recoveries of spiked cyanide species were observed for another method utilizing UV digestion-TCN-TFD— for some contaminated water matrices. This indicates that UV irradiation for digestion in cyanide analysis can be effective in some method configurations, but requires careful evaluation of method performance.
- Among the four WAD cyanide methods evaluated, the conventional WAD-CN method, along with its automated version, the WAD-TFD method, was the best overall performer,

yielding acceptable cyanide recoveries in all water matrices, and was the best and the second best accuracy for the performance evaluation samples.

Overall, it was shown that the cyanide-species analytical techniques can be employed reliably for low-concentration analysis with a range of contaminated water types, with some exceptions noted above. In addition, the data obtained permit assessment of the potential for routine application of the particular methods with different types of waters. The WAD-CN method can be performed with instrumentation that is already available in many laboratories and that is inexpensive to acquire. The automated WAD-TFD method cannot only achieve low MDLs and good accuracy and precision, but can also reduce significantly the time for sample analysis. Therefore, the method WAD-CN and WAD-TFD would be the most appropriate alternative to the problematic CATC method to measure WAD cyanide in contaminated water samples requiring WAD cyanide determination.

CHAPTER 3.0

THE EFFECT OF CHLORINATION ON ORGANOCYANIDE COMPOUNDS

3.1 Introduction

This chapter summarizes results from experiments designed to explore the possibility of organocyanide compounds in wastewater contributing to elevated cyanide levels in the chlorinated effluents of publicly owned treatment works. Experimental details are provided in Yi et al. (2002). Four model compounds, acetonitrile, amygdalin, cyanocobalamin (vitamin B_{12}), and 2-acetoxy-3-butenenitrile were selected and tested according to accepted standard procedures (shown in Table 3-1), and also with varying chlorine dosages to check for the release of cyanide as determined by total and diffusible cyanide procedures. The coenzyme form of vitamin B_{12}, which does not contain cyanide, was also tested.

3.2 Results and Discussion

Results from standard testing procedures on clean water samples containing approximately 2 ppm of each compound are given in Table 3-2, where it is shown that acetonitrile and amygdalin exhibit no cyanide amenable to chlorination, 2-acetoxy-3-butenenitrile gives positive cyanide amenable to chlorination, and cyanocobalamin gives a negative amenable cyanide value.

Table 3-1. Detection Limits for Cyanide Analytical Methods Used in the Experiments.[*]

Cyanide analytical method	Detection limit (µg/L)
Total cyanide by distillation (U.S. EPA Method 335.2)	20 (500 mL sample)
Cyanide amenable to chlorination (U.S. EPA Method 335.1)	N/A [a]
Diffusible cyanide (ASTM Method D 4282)	2
Cobalt cyanide (Hexacyanocobalt) (Dionex Applic. Note 55)	0.2

[*] Reprinted with permission from Yi et al., *Water Environment Research*, **2002**, *74*(1), 51-56. Copyright 2002 Water Environment Federation.
[a] Not applicable.

Table 3-2. Percentage Recovery of Cyanide from Organocyanides by Total Distillation and Cyanide Amenable to Chlorination Procedures (Triplicates).*

Organocyanide compound	CN recovery without chlorination	CN recovery after excess chlorination
Acetonitrile	< 1.57	< 1.57
D-Amygdalin	< 1.38	< 1.38
2-Acetoxy-3-butenenitrile	73.40 ± 5.91	< 1.46
Cyanocobalamin	18.98 ± 7.67	41.06 ± 6.11

*Reprinted with permission from Yi et al., *Water Environment Research*, **2002**, *74*(1), 51-56. Copyright 2002 Water Environment Federation.

Results of cyanide released as a function of chlorine dosage are presented in Figures 3-1 through 3-3. 2-Acetoxy-3-butenenitrile had generally decreased release of cyanide with increased chlorine dosages, although a small increase in total cyanide was noted at the smallest chlorine dosage (Figure 3-1). Figure 3-2 shows the release of total cyanide and diffusible cyanide for cyanocobalamin as a function of chlorine dosage indicating an increasing trend for both forms at high chlorine levels. Both cyanocobalamin and coenzyme vitamin B_{12} gave evidence of the formation of the cobalt cyanide complex after chlorination, the latter even though the molecular structure contains no cyanide, as shown in Figure 3-3.

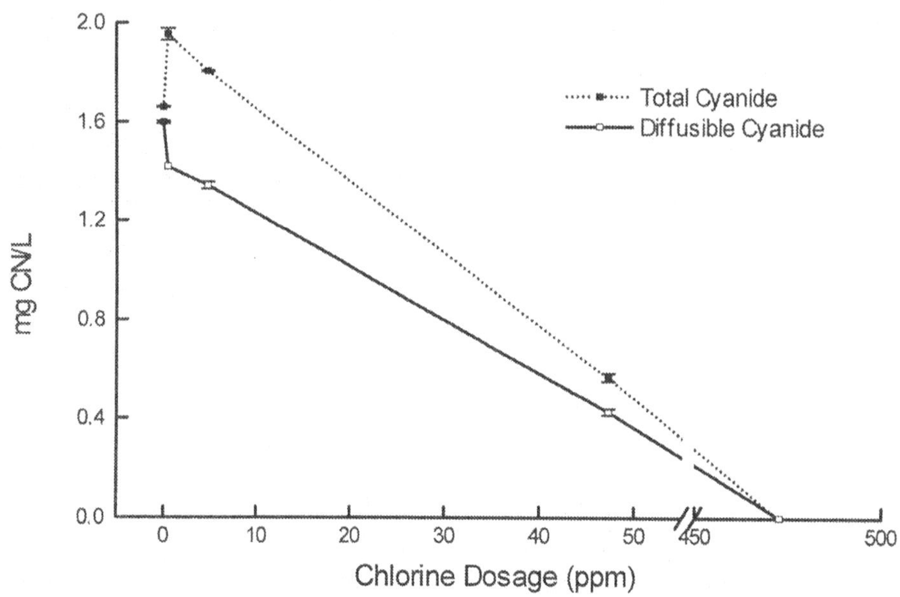

Figure 3-1. Total and Diffusible Cyanide for Chlorination of 2-Acetoxy-3-Butenenitrile. Initial Concentration = 2.14 mg CN/L. Bars represent one standard deviation about the mean.
*Reprinted with permission from Yi et al., *Water Environment Research*, **2002**, *74*(1), 51-56. Copyright 2002 Water Environment Federation.

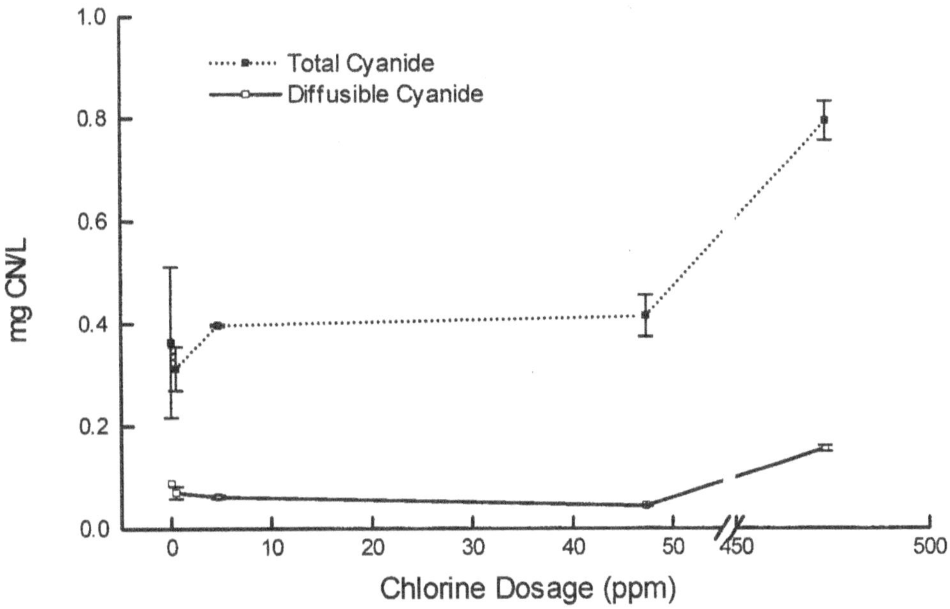

Figure 3-2. Total and Diffusible Cyanide for Chlorination of Cyanocobalamin. Initial Concentration = 1.92 mg CN/L. Bars represent one standard deviation about the mean.

*Reprinted with permission from Yi et al., *Water Environment Research*, **2002**, *74*(1), 51-56. Copyright 2002 Water Environment Federation.

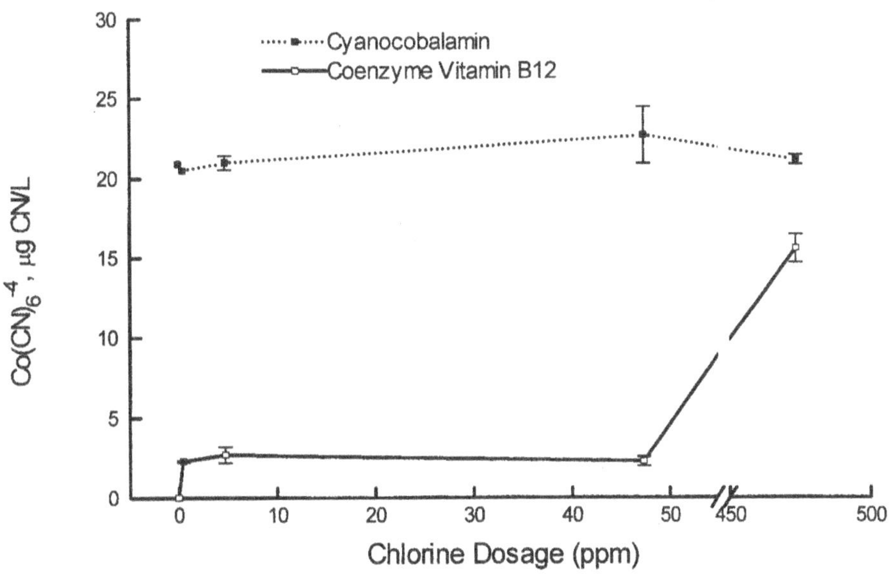

Figure 3-3. Release of Cobalt Cyanide for Chlorination of Cyanocobalamin and Coenzyme Vitamin B_{12}. Initial Concentrations = 100 mg/L. Bars represent one standard deviation about the mean.

*Reprinted with permission from Yi et al., *Water Environment Research*, **2002**, *74*(1), 51-56. Copyright 2002 Water Environment Federation.

The results for coenzyme vitamin B_{12} are especially interesting. This form contains no cyanide, yet small quantities are generated in the form of $Co(CN)_6^{-3}$ as a result of chlorination. Likely precursors are C=N moieties found within the corrin ring surrounding Co and the attached adenosine structure as shown in Figure 3-2, although reaction sequences for generating cyanide are not clear. The appearance of the liberated cyanide in the form of the cobalt complex, in distinction to other forms, can be attributed to the high stability of this complex. Although the amount of cyanide produced from the chlorination of coenzyme B_{12} is small in comparison to the initial concentration of the coenzyme, wastewater may contain other organic compounds with C=N heterocyclic structures that can also serve as cyanide precursors during chlorination.

The results of this research should be viewed with the context of current water quality standards for cyanide and the state of approved testing methods. Discharge permits at WWTPs generally limit allowable cyanide, as detected by U.S. EPA Methods 335.1 or 335.2, to 1–20 mg CN/L. With such low concentration limits, even small changes in cyanide speciation can be important. To the extent that cyanide-labile organocyanide compounds are present in wastewater, interpretation of test results, particularly for U.S. EPA Method 335.1 (Cyanide Amenable to Chlorination), can be confounded by the effect of postchlorination. The results indicate that the presence of cyanide-bearing organic compounds that are not captured in the total cyanide method can liberate upon chlorination free or metal-complexed cyanide compounds that are measurable.

CHAPTER 4.0

FATE OF CYANIDE IN WWTPS AND WWTP MONITORING PROGRAMS

4.1 Introduction

Raw wastewater at WWTPs sometimes contains cyanide at low concentrations, primarily from industrial sources. Because cyanide causes acute toxicity to various life forms (including bacteria in activated sludge plants), and because cyanide is a U.S. EPA priority pollutant limited by regulation to low concentrations in discharges, pretreatment for the removal of cyanide is usually required of industrial sources by WWTPs. Thus, influent concentrations of total cyanide at WWTPs are typically very low.

Despite the low influent concentrations, cyanide has been periodically detected at elevated concentrations in WWTP final effluent, often at concentrations that exceed the influent cyanide concentrations. This has been observed at treatment plants across the United States (Environ, Inc., 1996; Delaney et al., 1997; Sawyer et al., 1998; Forman and Kavanaugh, 1999). The phenomenon is of concern because the water quality criteria and related discharge limits of cyanide are typically very low; in many cases as low as 5 μg/L. There are several potential causes of cyanide formation in WWTPs and this is the focus of this study.

The study of cyanide formation in WWTPs is challenging because WWTP wastewater is composed of a complicated mixture of inorganic and organic compounds. Some of these compounds, such as thiocyanate (SCN^-), nitrite (NO_2^-), nitrate (NO_3^-), and sulfide (S^{2-}), etc., either are or have potential to become cyanide precursors or interfere with various cyanide analytical methods as discussed in Chapter 2.0 of this report. Some compounds are so closely interrelated in the formation mechanism of cyanide–such as cyanide, sulfide, and thiocyanate (Zheng et al., 2003b)—that it is very difficult to focus on the effects of one compound without considering the others. In addition, identification and quantification of individual cyanide species is not commonly practiced in WWTPs. Rather, bulk measurements such as "total cyanide" are usually employed.

Very few studies have been conducted to investigate the formation of cyanide in WWTPs through a comprehensive cyanide-monitoring program. In a study with a major municipal utility district in the State of California, elevated concentrations of cyanide (>10 μg/L) in the chlorinated final effluent of the wastewater treatment plant were detected, which was inconsistent with the low cyanide concentrations found in the influent (Environ, Inc., 1996). A review of the monitoring data revealed that thiocyanate was measured in the influent from 0.7 mg/L to below detection limit (i.e., < 0.01 mg/L) with an average concentration of 0.268 mg/L. It was also found that the average thiocyanate concentration in the secondary effluent decreased by approximately 60% to 0.161 mg/L (Environ, Inc., 1996). However, a positive correlation between the cyanide increase and the thiocyanate decrease could not be established. No single factor was identified as responsible for the cyanide formation (Environ, Inc., 1996).

The City of Sunnyvale, California, reported periodic episodes of cyanide occurrence in the chlorinated effluent from its wastewater treatment plant at levels that exceeded the daily average discharge limit of 5 µg/L specified in its NPDES permit. A study was performed by Malcolm Pirnie, Inc. (1997) to determine the cause for the occurrence of cyanide including the measurement of total cyanide and several potential cyanide precursors at several sampling points located from the secondary effluent to the final effluent. The causative agent or reaction sequence could not be identified.

Weinberg et al. (2000) conducted an assessment of trace cyanide concentrations in municipal WWTP discharges in North Carolina, which indicated that most cyanide exceedances were associated with those plants practicing terminal chlorination.

An intensive WWTP monitoring program was implemented at six full-scale WWTPs in the United States to investigate the fate and formation of cyanide in wastewater treatment processes, with a focus on chlorination and dechlorination processes. A review of historical monitoring data for cyanide species in these WWTPs was also conducted. The aim of the WWTP monitoring program was to obtain a better understanding of the cyanide formation phenomenon as it occurs in the field. The primary objectives of the monitoring program were as follows:

- To monitor the removal or production of cyanide species in chlorination/dechlorination processes of a WWTP over a one-year cycle of seasons; and
- To monitor intensively the fate of cyanide in the chlorination process for periods when postchlorination effluent cyanide concentrations were elevated.

4.2 Methods

4.2.1 WWTP Participants

The participating WWTPs are listed in Table 4-1. The characteristics of WWTPs are indicated in the table. These plants, ranging from a 12-mgd full-scale tertiary treatment facility to a 380-mgd secondary treatment plant, provided a range of wastewater compositions, process conditions, and cyanide inputs.

4.2.2 Selection of Cyanide Analytical Techniques

A comprehensive cyanide analytical method evaluation was performed, as a part of Task 1 of this study, before the commencement of the monitoring program (see Chapters 2.0 and 3.0). The criteria for selection of methods employed in WWTP monitoring were as follows: 1) method performance with wastewater samples including percent recovery and method detection limit; 2) cost of implementation; and 3) ease of implementation. On the basis of the study, three primary methods were chosen to perform cyanide measurements routinely in WWTPs: total cyanide (APHA/AWWA/WEF, 1998 – Method 4500-CN-C) WAD cyanide (APHA/AWWA/WEF, 1998 – Method 4500-CN-I) and free cyanide by microdiffusion (ASTM, 1998a – Method D4282-95). Other cyanide speciation measurements including automated analysis of total cyanide with in-line UV digestion (U.S. EPA, 1983 – Method 335.2) automated analysis of total cyanide by thin film distillation (ASTM, 1998d – Method 4374-93) were employed in the monitoring program at selected facilities. The analytical methods used for cyanide species analyses are listed in Table 4-2.

Table 4-1. WWTP Cyanide Monitoring Study Participants.

WWTP	Facility	Plant type	Chlorination	Dechlorination
MWRDGC,[a] Chicago	Hanover Park	Tertiary, 12 mgd	Yes	Yes
GCA,[b] Pasadena, TX	Washburn Tunnel Facility	Secondary, 60 mgd	No	No
MWRA,[c] Boston	Deer Island Treatment Plant	Secondary, 380 mgd	Yes	Yes
EBMUD,[d] Oakland, CA	Oakland Water Pollution Control Plant	Secondary, 168 mgd	Yes	Yes
City of Sunnyvale, CA	Sunnyvale Water Pollution Control Plant	Secondary, 29.5 mgd	Yes	Yes
City of Los Angeles,	Glendale Water Reclamation Plant	Secondary, 20 mgd	Yes	Yes

[a] Metropolitan Water Reclamation District of Greater Chicago
[b] Gulf Coast Waste Disposal Authority
[c] Massachusetts Water Resources Authority
[d] East Bay Municipal Utility District

Table 4-2. Standard Cyanide Analytical Methods for Wastewater.

Method	Species measured	Reference
Total cyanide after distillation	Total CN	SM 4500-CN-C
Total cyanide, spectrophotometric	Total CN	EPA Method 335.2
Free cyanide by microdiffusion	Free CN	ASTM D4282-95
WAD cyanide	WAD-CN	SM 4500-CN-I
Automated total cyanide by thin film distillation	Total CN	ASTM D 4374-93
Automated WAD cyanide by thin film distillation	WAD-CN	SM 4500-CN-I
Total cyanide by low-power UV digestion	Total CN	EPA Method 335.3

*Preservation, NaOH, pH >12.

4.2.3 WWTP Monitoring Schedule

A comprehensive WWTP monitoring workplan, focused on the chlorination process, was developed and can be found in Appendix C of this report. The monitoring program provided for periodic measurement of several cyanide species before and after the chlorination/dechlorination processes at the participating WWTPs. This monitoring program was planned to last for 12 months. Table 4-3 summarizes the sampling schedule in terms of sampling duration, frequency, sample type, and sample locations. Cost and resource availability were considered in a general way in establishing this schedule.

Table 4-3. Cyanide Sampling Schedule for WWTP Monitoring Program.

Monitoring type	Duration	Frequency	Sample type	Sampling locations[a,b]
Incident-detections	May 2000 – May 2001	1 day/week	24-hour composite, auto-sampler (every 3 hours)	1 location: Postchlorination
Incident-driven	5 days subsequent to an incident	5 days per incident	Four 6-hour composites, auto-sampler (every 1 hour)	2-4 locations: Prechlorination Postchlorination Pre-dechlorination (as applicable) Post-dechlorination (as applicable)

[a] Please refer to the diagram of each WWTP for detailed sampling locations.
[b] In the event that the residence time between the postchlorination and pre-dechlorination sample locations was insignificant, it was acceptable to disregard the postchlorination sample location and sample from only the prechlorination, pre-dechlorination, and post-dechlorination sample locations.

4.2.4 Incident-Detection Monitoring

Incident-detection monitoring was performed weekly using 24-hour flow rate-weighted composite sampling of postchlorination effluent. The composite samples were analyzed for total cyanide. An "incident" was defined as one of the following events:

- A total cyanide concentration at the postchlorination point greater than 5 μg/L; or
- An analytical result 50% greater than the average cyanide concentration (e.g., observation of 15 μg/L versus a typical effluent cyanide concentration of 10 μg/L); or
- An analytical result 100% greater than the method detection limit.

WWTPs used their discretion in identifying cyanide formation incidents based on their knowledge of the utility's treatment process and previous cyanide formation history. Conventional composite automatic samplers were employed. These devices typically take individual samples for eight, three-hour periods, and where feasible these samples were subsequently composited with proportioning according to flow during the sampling periods. Each individual container was prepared with NaOH for cyanide preservation and $Na_2S_2O_3$ was added to dechlorinate the postchlorination samples. All samples were collected and stored in brown plastic or amber glass containers. If an incident was considered in progress based on the criteria listed previously, an incident-driven monitoring sequence would be initiated as soon as possible.

4.2.5 Incident-Driven Monitoring

Once a utility identified an excursion, the plan was for the utility to implement an intensive five-day monitoring program. Separate samples would be collected during the incident-driven monitoring for cyanide analysis and for general characterization of the wastewater.

Wastewater samples for cyanide species and characterization analysis were collected as four, six-hour composite samples over 24 hours, with sampling of one gallon (3,785 mL) every six hours at each sampling location. The samples were collected according to a detailed procedure, which can be found in the Appendix C of this report. To mitigate the impact of volatilization during holding times, extraction, and analysis for cyanide species were performed within nine days of sample collection.

4.3 Results

Results from the WWTP monitoring program as well as the historical cyanide monitoring data review are provided on a plant-by-plant basis. The purpose for reviewing WWTP historical cyanide monitoring data was to gain a better understanding as to the occurrence of cyanide in the WWTPs, and the extent to which the historical cyanide formation was repeated in the monitoring activity associated with this study. Here, historical data are defined as regular monitoring data at participating wastewater plants before May 2000 when the intensive WWTP monitoring program commenced. The 2000 and 2001 data obtained as the result of the WWTP monitoring program are compared to the relevant historical data. The complete data set for WWTP monitoring can be found in a project database.

4.3.1 East Bay Municipal Utility District (EBMUD)

EBMUD is a public utility in the East San Francisco Bay Area that provides full secondary wastewater treatment for over 610,000 residential and industrial customers. Wastewater collected by interceptors flows to EBMUD's wastewater treatment plant in Oakland near the entrance of the San Francisco-Oakland Bay Bridge. Primary treatment removes floating materials such as oil and grease, as well as materials such as sand, silt, and organic solids, which are heavy enough to settle in water (Figure 4-1). Secondary aerobic biotreatment removes most of the suspended and dissolved organic compounds.

The wastewater treatment steps include prechlorination (for odor control), screening (to remove large objects), grit removal, primary sedimentation, secondary treatment using high-purity oxygen-activated sludge, final clarification, and disinfection via chlorination and dechlorination. The treated effluent is then discharged one mile off the East Bay shore through a deepwater outfall into San Francisco Bay. EBMUD provides secondary treatment for a maximum flow of 168 mgd. Primary treatment can be provided for up to 320 mgd. Storage basins provide plant capacity for a short-term hydraulic peak of 415 mgd. The average annual flow is currently 80 mgd.

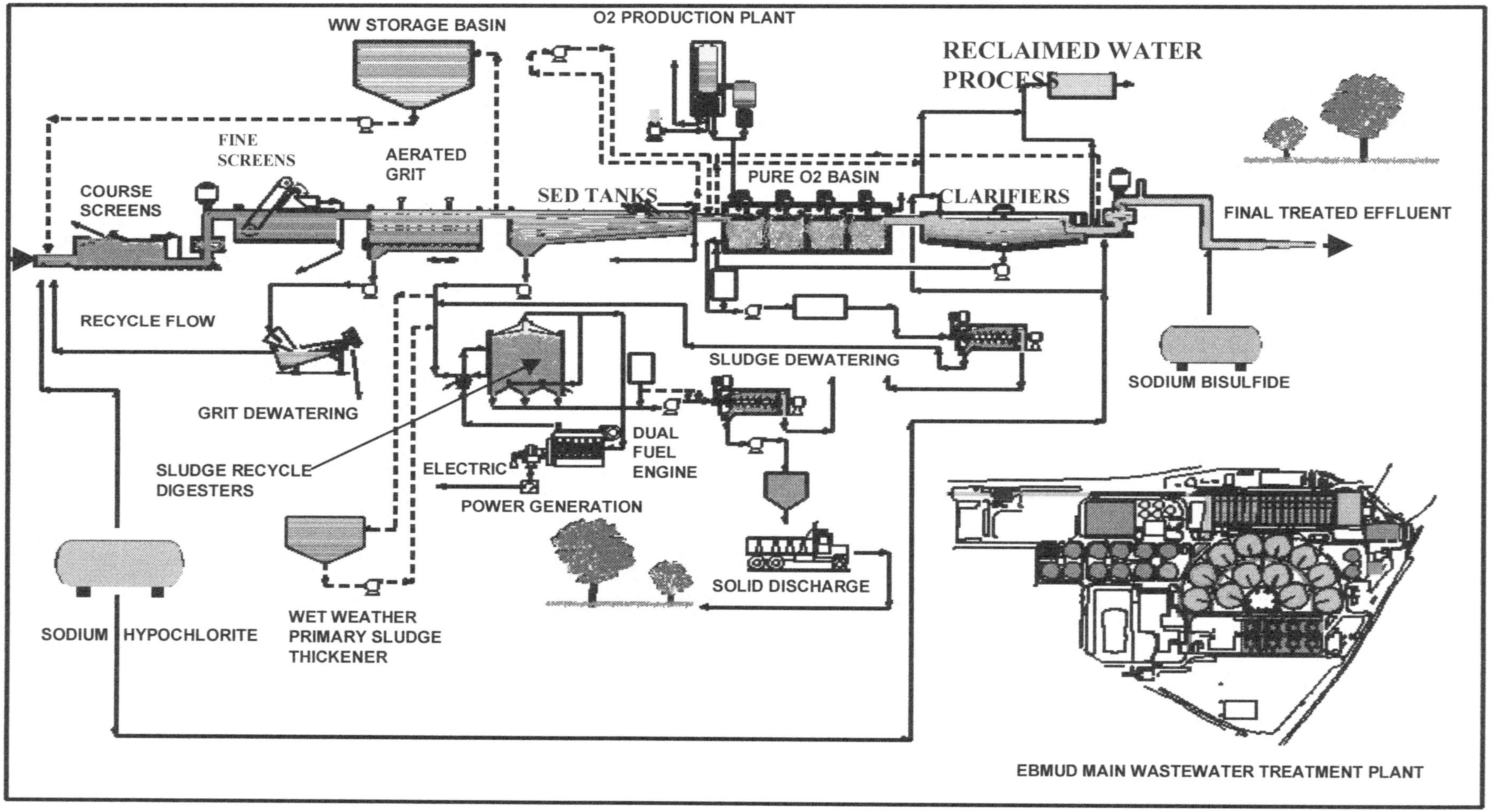

Figure 4- 1. Process Flow Diagram of Oakland Water Pollution Control Plant, EBMUD.

4.3.1.1 Cyanide Historical Data Review

The occurrence of cyanide formation in the EBMUD plant has been described by Ellgas (1999) who stated that "elevated concentrations of cyanide have infrequently appeared for many years in EBMUD's postchlorination, secondary effluent samples collected and analyzed for the NPDES permit". From September 1995 to February 1996, EBMUD experienced four exceedances of its cyanide NPDES permit limit (10 μg/L as a maximum daily average of eight grab samples collected over 24 hours and composited). Only data from 1997 to 1999 for total cyanide at the chlorinated effluent and dechlorinated effluent were available.

In 1997 there were 54 monitoring events, with both chlorinated effluent and dechlorinated effluent cyanide concentrations recorded in each event (Figure 4-2). Among the 54 events, the chlorinated effluent total cyanide concentration exceeded 10 μg/L in nine events. However, in only one out of those nine events did the dechlorinated effluent cyanide concentration exceed 10 μg/L. The maximum concentration of total cyanide in a single day occurred on March 3, 1997, with 48 μg/L in chlorinated effluent and 46 μg/L in dechlorinated effluent. The average total cyanide concentration in chlorinated and dechlorinated effluents was 7 μg/L and 6 μg/L, respectively. There were four occurrences in which the total cyanide concentrations in the dechlorinated effluent were greater than those in chlorinated effluent.

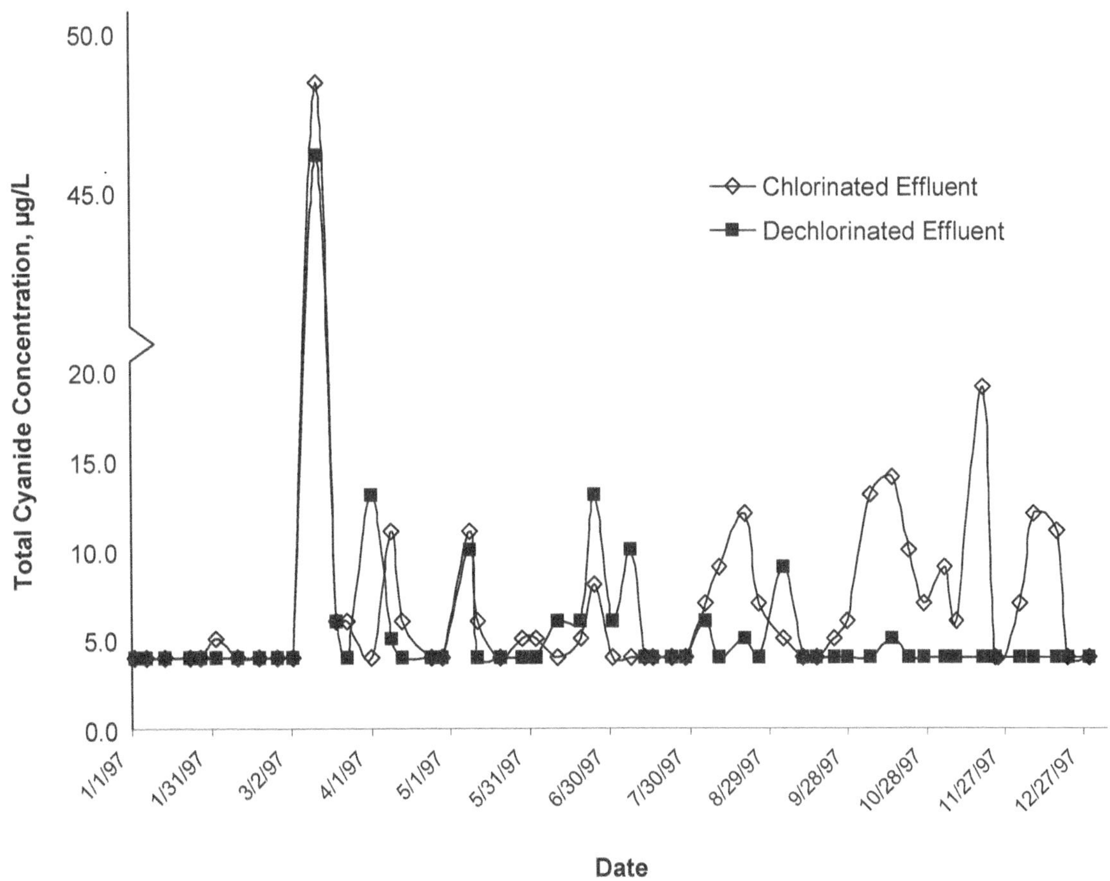

Figure 4-2. EBMUD Total Cyanide Monitoring Data for 1997.

In 1998, 52 monitoring events were recorded (Figure 4-3). Among these, the chlorinated effluent total cyanide concentration exceeded 10 μg/L in three events, but in none of those three events did dechlorinated effluent cyanide concentration exceed 8 μg/L. The maximum concentration of total cyanide in a single day occurred on December 3, 1998, with 16 μg/L in chlorinated effluent and 8 μg/L in dechlorinated effluent. There were three occurrences in which the total cyanide concentrations in the dechlorinated effluent were greater than those in chlorinated effluent.

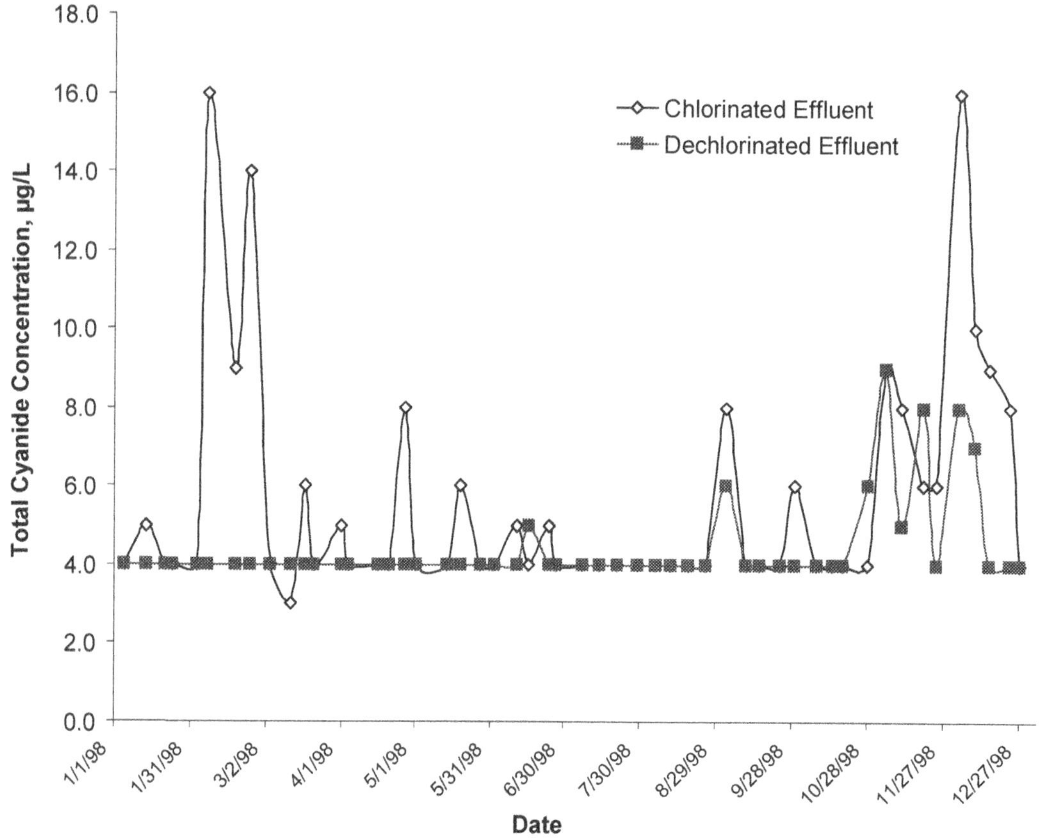

Figure 4-3. EBMUD Total Cyanide Monitoring Data for 1998.

In 1999, there were 52 monitoring events recorded (Figure 4-4). In three events the chlorinated effluent total cyanide concentration exceeded 10 μg/L, but in none of those three events did the dechlorianted effluent cyanide concentration exceed 8 μg/L. The maximum concentration of total cyanide in a single day occurred on January 1, 1999 with 17 μg/L in chlorinated effluent and 4 μg/L in dechlorinated effluent. There were five occurrences in which the total cyanide concentrations in the dechlorinated effluent were greater than those in chlorinated effluent.

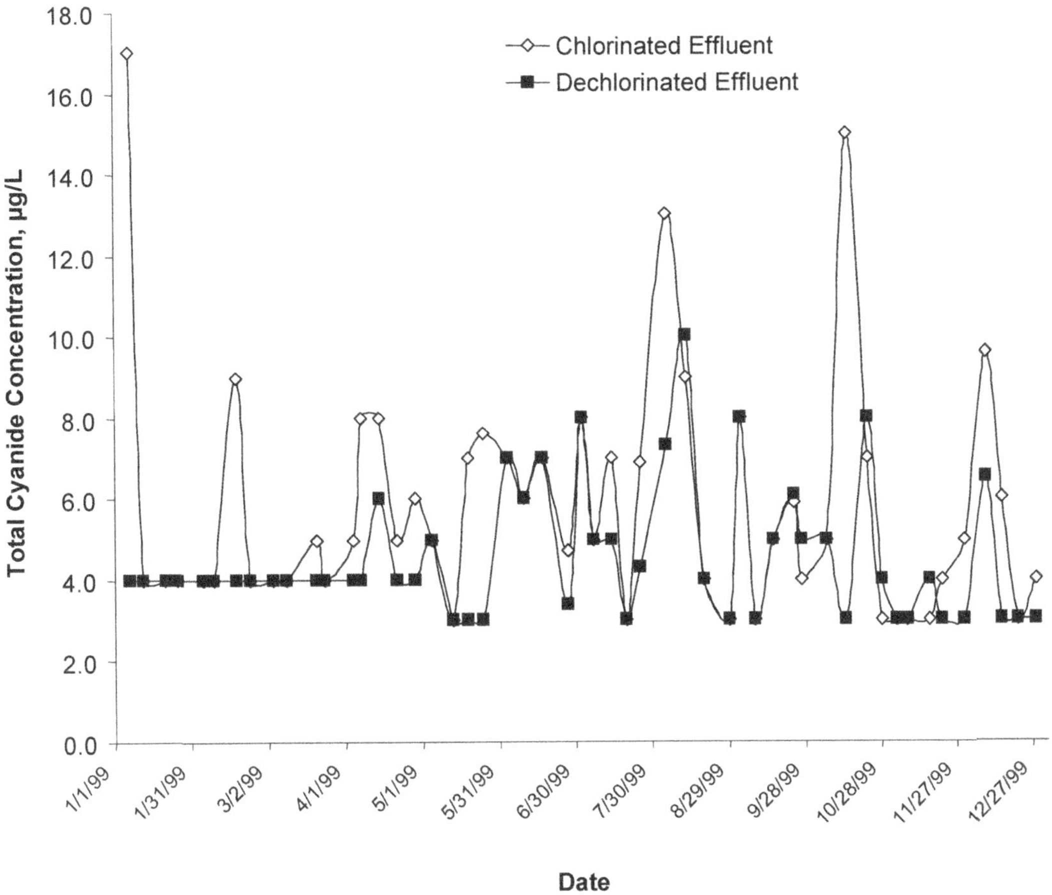

Figure 4-4. EBMUD Total Cyanide Monitoring Data for 1999.

Overall, relatively high cyanide concentrations (above 10 µg/L) were commonly detected in the chlorinated effluent and less commonly in the dechlorinated effluent. Although there were no monitoring data available for the primary influent cyanide concentration, cyanide levels in the influent were either very low or nondetectable according to Ellgas (1999). Therefore, high concentrations of cyanide in chlorinated and dechlorinated effluents cannot be attributed to cyanide sources for the influent.

4.3.1.2 WWTP Monitoring in 2000–2001

EBMUD sampled weekly for the incident-detection monitoring beginning August 4, 2000. From August 4, 2000 to April 16, 2001, grab samples were collected at the postchlorination point each Sunday beginning at 8:00 a.m., then every 3 hr thereafter through Monday morning at 5:00 a.m. The resulting eight grab samples were flow-composited to a single sample that was analyzed for total cyanide using U.S. EPA Method 335.2 (U.S. EPA, 1983). From April 25 to May 30, 2001, a slightly modified incident-detection monitoring program was implemented. Weekly grab samples were collected beginning on Tuesday at 9:00 a.m. and then every three hours until the following Wednesday at 5:00 a.m. Three locations were sampled including prechlorination, postchlorination, and post-dechlorination. The eight grab samples for each location were subsequently flow-composited to a single sample that was analyzed for approximately 38 analytes including total, WAD, and free cyanide.

The average total cyanide value of the postchlorination effluent for 47 consecutive weeks was 5.6 µg/L (Figure 4-5). A maximum value of 11 µg/L occurred on August 14, 2000. There were 16 events where the total cyanide from the incident-detection sampling exceeded the incident-driven trigger specified in the workplan (i.e., >5 µg/L of total cyanide). However, it was not possible to implement a full five-day incident-driven sampling sequence because of staffing constraints in the WWTP. A simplified incident-driven sampling sequence was conducted on September 24-25, 2000, and results from four sampling events completed in this period are shown in Figure 4-6. Three locations were sampled including prechlorination, postchlorination, and post-dechlorination. Samples were then analyzed for total, WAD, and free cyanide, respectively. Because of inconsistencies with the total cyanide analyses results, only the WAD and free cyanide values were used for comparison.

Among the four events, in three cases the chlorinated and dechlorinated effluent cyanide (WAD) concentrations exceeded those in the unchlorinated effluent. Moreover, all the WAD cyanide detected (from 7-9 µg/L) in the chlorinated effluent was in the form of free cyanide as confirmed by the free cyanide analysis. However, no free cyanide was detected in either prechlorination or post-dechlorination WWTP effluents, indicating that only free cyanide (HCN and CN^-) was formed in the chlorination process.

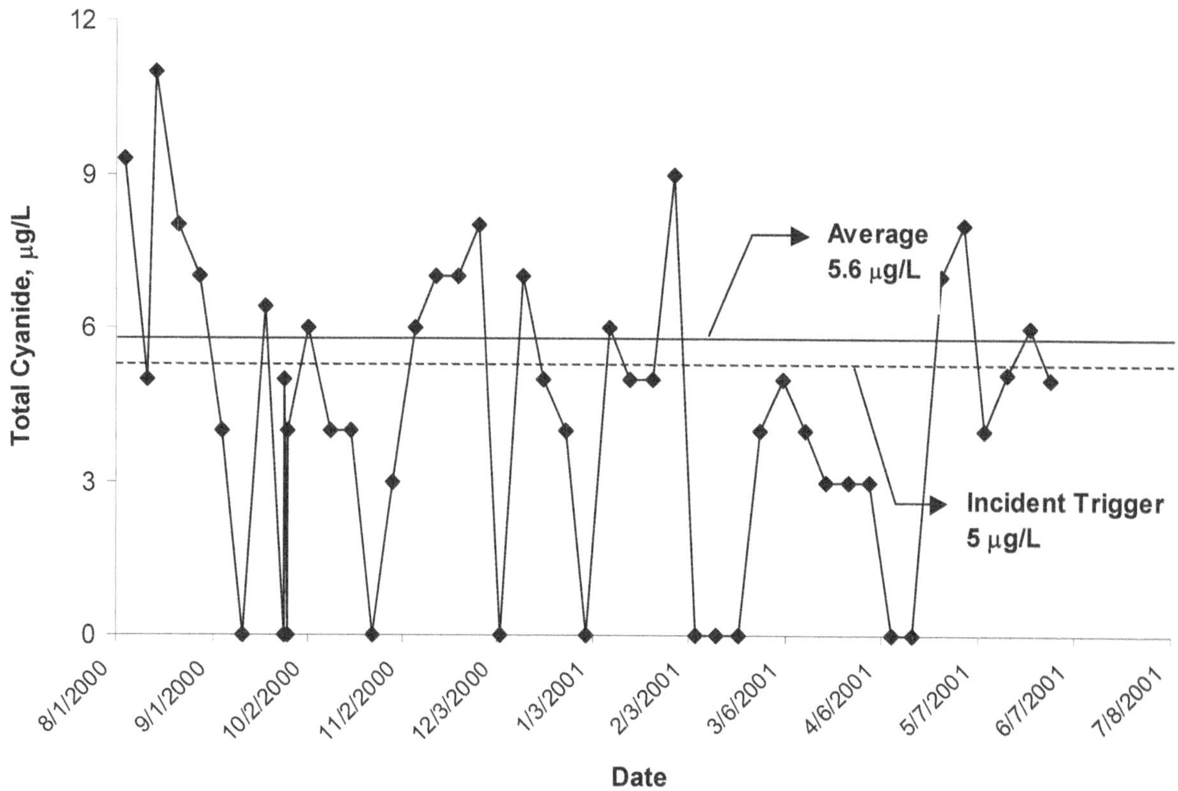

Figure 4-5. EBMUD Total Cyanide Monitoring Data of Postchlorination Effluent (August 2000 – June 2001).

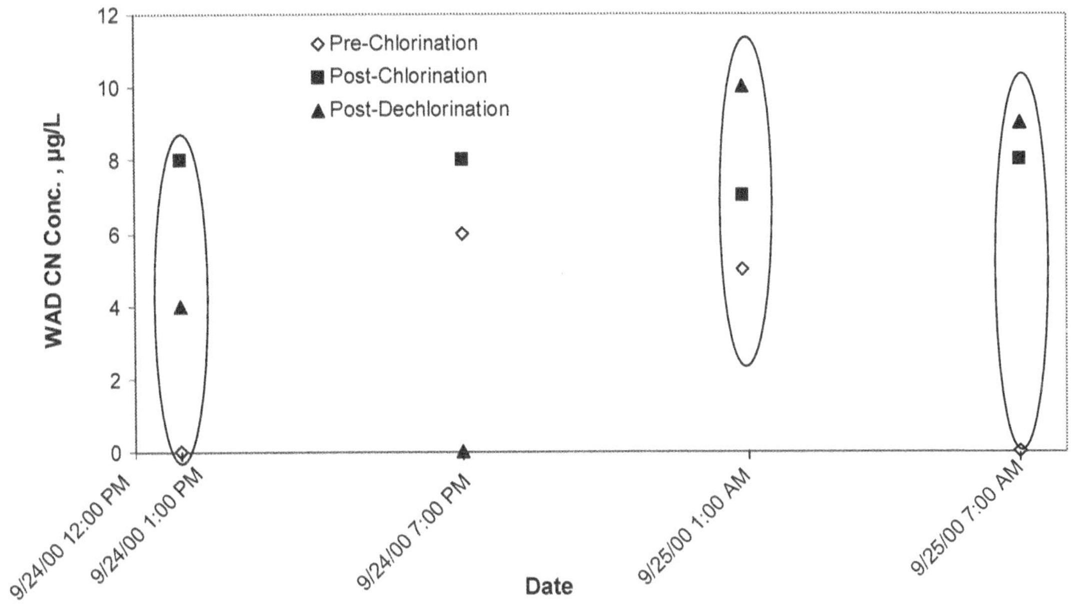

Figure 4-6. WAD Cyanide Monitoring Data from a Simplified Incident-Driven Sampling Conducted at Three Locations of EBMUD (September 24–25, 2000).

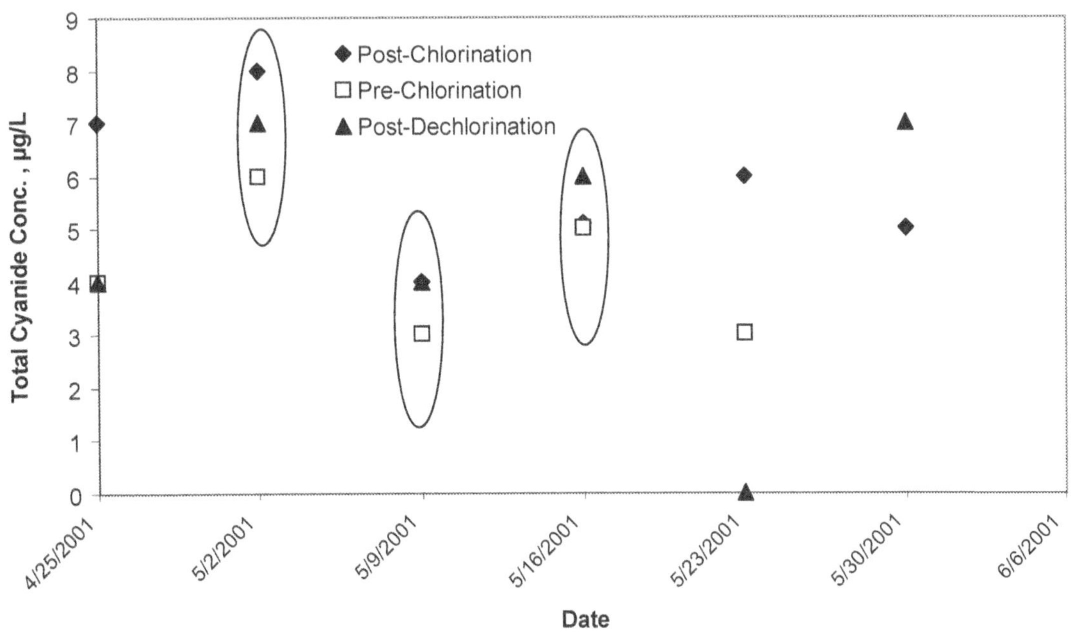

Figure 4-7. Total Cyanide Monitoring Data at Three Locations of EBMUD (April - July 2001).

Results for the monitoring events from April 2001 to July 2001 in which three locations were sampled weekly are given in Figure 4-7. Among the six events, in five the chlorinated and dechlorinated effluent total cyanide concentrations exceeded those in the unchlorinated effluent. The cyanides in the prechlorination effluent were all complexed cyanides, whereas the cyanides in postchlorination and post-dechlorination effluents were almost all in free cyanide form at elevated concentrations.

4.3.2 Gulf Coast Waste Disposal Authority (GCA)

GCA, based in Pasadena, Texas, is a nontax-supported unit of local government dedicated to waste management activities. GCA owns and/or operates four industrial wastewater treatment facilities that process liquid wastes from more than 60 manufacturing (largely petrochemical) plants. Among the four plants, the Washburn Tunnel Wastewater Treatment Facility (WTF), the participating WWTP, is a regional collection and treatment facility that treats the waste from one municipality, nine industrial plants via pipeline, two industrial plants via barge, and the WTF's trucked-in waste receiving station known as the Vince Bayou Receiving Station (VBRS). The VBRS primarily receives septic and chemical toilet waste. However, Class I and Class II nonhazardous industrial wastewater is also accepted for treatment.

The WTF plant began operations in May 1973 and was recognized as a first in the United States to mix industrial waste streams from different industries in a single biological joint wastewater treatment process. This 60-mgd facility employs a modified activated sludge process and, because of the composition of the waste streams, requires the use of pH control chemicals and nutrient addition (Figure 4-8). The processes effectively remove > 93-98% of the BOD and total suspended solids (TSS). WTF is the only WWTP participating in this monitoring program that does not chlorinate the secondary effluent as other WWTPs do.

The waste streams from the plants and VBRS are pumped through force mains to the head of the system for primary treatment which consists of pH control, coarse screening, grit removal, primary clarification, secondary treatment, and final clarification (Figure 4-8). The combined waste stream passes through 3" coarse bar racks, mechanically cleaned bar screens and two grit chambers. The effluent from the primary clarifiers becomes the influent to the secondary system after the addition of micronutrients. Secondary treatment consists of a cooling cycle (as required), an activated sludge treatment system with sludge recycle, clarification, and final discharge into the Houston Ship Channel. The cooled wastewater is treated by biological stabilization in three aeration basins having a detention time of 14 hours each under normal flow conditions. The oxygen demand is met in the aeration basins by 24 surface aerators, which aerate and mix the waste and biomass. The mixed liquor flows into six secondary clarifiers. The settled biosolids are either returned to the aeration basins or wasted for ultimate disposal. The effluent is discharged via two 40" lines to the Houston Ship Channel, Segment No. 1007, of the San Jacinto River Basin. Laboratory analyses are conducted routinely for organic, inorganic, and heavy metal constituents to ensure compliance with regulatory effluent standard.

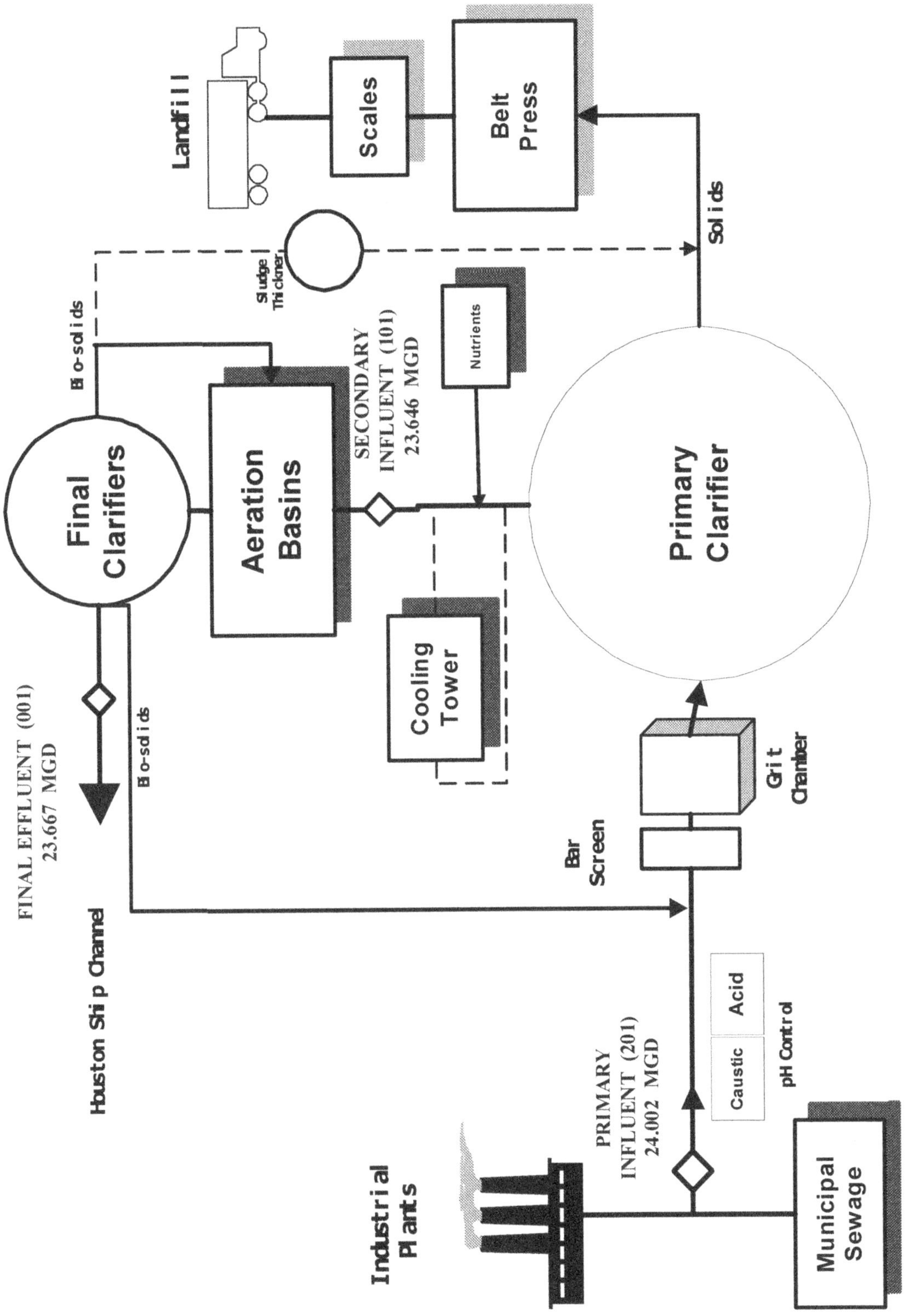

Figure 4-8. Process Flow Diagram of Washburn Tunnel Wastewater Treatment Facility (WTF), GCA.

4.3.2.1 Cyanide Historical Data Review

The monitoring locations within WTF were primary influent (labeled 201 in Figure 4-8), secondary influent (i.e., primary clarifier effluent, labeled 101) and final effluent (labeled 001). Generally, the cyanide concentrations in the primary influent were 0.1-1.3 mg/L. The cyanide concentrations were reduced to 0.01-0.7 mg/L in the secondary influent, and further reduced to < 0.034 mg/L in the final effluent. A summary of total cyanide monitoring data for each month of 1999 is presented in Table 4-4. There clearly was removal of cyanide in the plant, especially in the final effluent because monitoring data consistently show that cyanide concentrations in the final effluent were significantly lower than those in the primary treatment influent and secondary treatment influent. However, there were a few days when the total cyanide concentration in the secondary treatment influent did exceed that in the primary treatment influent. For example, among the 156 monitoring events in 1999 in which total cyanide concentration in all three locations were recorded, in seven events (about 4.5%) the secondary influent total cyanide concentration exceeded that in the primary influent (Table 4-5).

Because WTF discharges the biosolids to the headwork of the treatment plant (a common practice for some WWTPs) as indicated in Figure 4-8, it is possible that cyanide contained in the sludge. If the sludge did contain cyanide, that may account for the elevated cyanide concentration of the primary clarifier effluent. No data were available for the cyanide concentration in the biosolids. It should be noted that the elevated total cyanide in secondary influent was largely caused by the increase of free cyanide concentration. That is, free cyanide was either generated during the primary treatment process or liberated from the sludge discharged to the headwork. Investigation into the monitoring data didn't reveal any potential cause of cyanide formation in these particular seven days in 1999.

4.3.2.2 WWTP Monitoring in 2000–2001

Because GCA does not chlorinate secondary treated wastewater, a modification of the workplan was implemented. Samples were collected daily as grab samples from primary influent (labeled 201 in Figure 4-8), secondary influent (labeled 101) and final effluent (labeled 001), and analyzed for total cyanide, free cyanide (only in final effluent) and a broad range of chemical parameters such as BOD, TOC, NH_3-N, sulfide, phenols, and some heavy metals. The monitoring data from January 2000 to June 2001 showed no significant differences with data obtained in the past (see historical data review) in terms of the apparent cyanide removal pattern in the final effluent (Table 4-6). Generally, the total cyanide concentrations in the primary influent were from nondetectable to 1.63 mg/L. Cyanide concentrations were reduced to around nondetectable to 1.15 mg/L in the secondary influent, and further reduced to nondetectable to 0.132 mg/L in the final effluent. However, there were some few days when the total cyanide concentration in the secondary influent did exceed that in the primary influent indicating cyanide formation in primary treatment (clarification). Among the 280 monitoring events recorded, in 14 events (about 5%) the secondary influent total cyanide concentration and mass exceeded those in the primary influent (Table 4-7). Investigation into the monitoring data didn't reveal any potential cause of cyanide formation in these particular days.

Table 4-4. Summary of Total Cyanide Monthly Monitoring Data in WTF of GCA in 1999[a] (units: mg/L).

	Primary influent				Secondary influent				Final effluent			
Month	No.	Min.	Max.	Avg.	No.	Min.	Max.	Avg.	No.	Min.	Max.	Avg.
Jan.	13	0.105	0.737	0.348	13	0.028	0.628	0.205	13	0.008	0.072	0.029
Feb.	12	0.115	0.738	0.413	12	0.047	0.189	0.119	12	0.015	0.077	0.036
Mar.	14	0.084	0.606	0.304	14	0.042	0.536	0.199	14	0.007	0.047	0.025
Apr.	12	0.021	0.203	0.064	12	0.020	0.142	0.050	12	0.006	0.050	0.018
May	14	0.061	0.876	0.470	14	0.048	0.707	0.274	14	0.012	0.035	0.021
Jun.	13	0.296	1.090	0.762	13	0.134	0.702	0.405	13	0.011	0.062	0.029
Jul.	13	0.230	0.666	0.487	13	0.017	0.354	0.139	13	0.018	0.038	0.029
Aug.	13	0.118	0.746	0.412	13	0.049	0.531	0.187	13	0.006	0.065	0.040
Sept.	13	0.107	0.819	0.479	13	0.116	0.504	0.265	13	0.024	0.073	0.043
Oct.	13	0.141	0.782	0.442	13	0.048	0.206	0.117	13	0.020	0.068	0.041
Nov.	13	0.151	1.090	0.687	13	0.053	0.741	0.303	13	0.021	0.065	0.044
Dec.	13	0.304	1.310	0.824	13	0.195	1.670	0.710	13	0.035	0.080	0.050

[a] No. = the number of samples collected in each month.
Min. = the minimum value of total cyanide concentration measured.
Max. = the maximum value of total cyanide concentration measured.
Avg. = the average value of total cyanide concentration measured.

Table 4-5. Cyanide Species Monitoring Data in WTF of GCA in 1999 (Total Cyanide in Secondary Influent Exceeded Total Cyanide in Primary Influent, Both in Concentration and Mass).

	Primary influent						Secondary influent					
	Total CN		Complex-CN		Free CN		Total CN		Complex-CN		Free CN	
Date	Conc. (mg/L)	Mass (lb)	Conc. (mg/L)	Mass (lb)	Conc. (mg/L)	Mass (lb)	Conc. (mg/L)	Mass (lb)	Conc. (mg/L)	Mass (lb)	Conc. (mg/L)	Mass (lb)
2/17/99	0.505	140	0.169	47	0.336	93	0.533	149	0.174	49	0.359	100
2/20/99	0.481	107	0.271	60	0.21	47	0.494	109	0.153	34	0.341	76
4/24/99	0.044	7	0.045	8	ND[a]	0	0.054	9	0.065	11	ND	0
4/26/99	0.025	5	0.039	8	ND	0	0.037	7	0.035	7	ND	0
9/25/99	0.107	20	NA[b]	NA	NA	NA	0.248	44	NA	NA	NA	NA
12/13/99	0.805	152	NA	NA	NA	NA	1.67	309	NA	NA	NA	NA
12/15/99	0.686	110	NA	NA	NA	NA	0.747	117	NA	NA	NA	NA

[a] ND = Not detected.
[b] NA = Not analyzed.

Table 4-6. Summary of Total Cyanide Monthly Monitoring Data in WTF of GCA, January 2000–June 2001 (units: mg/L).

Month	Primary Influent				Secondary Influent				Final Effluent			
	No.	Min.	Max.	Avg.	No.	Min.	Max.	Avg.	No.	Min.	Max.	Avg.
Jan. 00	14	0.240	1.16	0.910	14	0.318	0.980	0.700	14	0.028	0.097	0.058
Feb. 00	12	0.290	1.40	0.880	12	0.249	1.15	0.606	12	0.020	0.070	0.047
Mar. 00	12	0.367	1.41	0.645	13	0.096	0.595	0.237	13	0.014	0.051	0.033
Apr. 00	13	0.354	1.31	0.843	13	0.127	0.788	0.447	13	0.034	0.079	0.055
May 00	14	0.260	1.63	0.708	14	0.035	0.628	0.252	14	0.012	0.068	0.039
Jun. 00	12	0.175	1.05	0.658	12	0.043	0.681	0.333	13	0.023	0.063	0.039
Jul. 00	14	0.634	0.983	0.754	14	0.136	0.586	0.371	14	0.022	0.053	0.038
Aug. 00	13	0.242	1.09	0.857	13	0.075	0.492	0.288	13	0.009	0.072	0.049
Sept. 00	13	0.322	1.28	0.878	13	0.086	0.632	0.352	13	0.017	0.064	0.050
Oct. 00	13	0.344	1.77	0.909	13	0.258	1.02	0.612	13	0.029	0.068	0.046
Nov. 00	13	0.139	0.817	0.462	13	0.120	0.504	0.314	13	0.018	0.048	0.034
Dec. 00	13	0.270	1.60	0.689	13	0.177	0.972	0.517	13	0.025	0.114	0.068
Jan. 01	14	0.212	1.09	0.606	14	0.214	0.905	0.489	14	0.026	0.132	0.077
Feb. 01	7	0.148	0.764	0.591	7	0.145	0.493	0.311	7	0.043	0.081	0.062
Mar. 01	30	0.053	0.679	0.427	30	0.078	0.533	0.325	13	0.026	0.067	0.040
Apr. 01	30	0.043	0.470	0.178	30	0.033	0.468	0.118	13	0.012	0.036	0.023
May 01	31	0.076	0.522	0.269	31	0.045	0.151	0.097	14	0.022	0.050	0.039
Jun. 01	12	ND	0.122	0.032	13	ND	0.079	0.032	13	ND	0.045	0.011

No. = the number of samples collected in each month.
Min. = the minimum value of total cyanide concentration measured.
Max. = the maximum value of total cyanide concentration measured.
Avg. = the average value of total cyanide concentration measured.
ND = Not detected.

Table 4-7. Cyanide Monitoring Data in WTF of GCA from January 2000 to June 2001 (Total Cyanide in Secondary Influent Exceeded that in Primary Influent, Both in Concentration and Mass).

	Primary influent		Secondary influent		Final effluent	
Date	Conc. (mg/L)	Mass(lb)	Conc. (mg/L)	Mass (lb)	Conc. (mg/L)	Mass (lb)
1/26/2000	0.24	40	0.98	151	0.028	4.25
6/26/2000	0.355	70	0.667	126	0.049	9.23
9/9/2000	0.322	67	0.365	78	0.058	12.5
11/25/2000	0.139	27	0.188	38	0.044	8.87
12/2/2000	0.27	52	0.429	81	0.092	17.5
12/4/2000	0.293	56	0.435	82	0.053	9.86
1/10/2001	0.334	77	0.399	91	0.040	9.20
3/30/2001	0.104	25	0.119	29	NA[a]	NA
3/31/2001	0.053	11	0.078	15	0.036	6.95
6/2/2001	ND[b]	0	0.032	5	0.012	1.93
6/4/2001	0.025	4	0.035	6	ND	0
6/18/2001	0.039	8	0.061	12	0.016	3.10
6/20/2001	0.039	8	0.067	14	0.018	3.72
6/25/2001	ND	0	0.027	5	0.026	5.09

[a] NA = Not analyzed.
[b] ND = Not detected.

Overall, there clearly was removal of cyanide in the WTF plant by secondary (biological) treatment. There were some occasions (about 5% of sampling events) in which cyanide concentration increased in the primary clarifier effluent indicating either cyanide formation in primary treatment or cyanide regeneration from biosolids discharged to the headwork of the treatment plant.

4.3.3 The Los Angeles-Glendale Water Reclamation Plant (LAGWRP)

The LAGWRP is located at the southwest junction of the Los Angeles River Flood Control Channel and Colorado Boulevard. LAGWRP is designed to treat an average dry weather flow of 20 mgd. The plant is normally operated on a constant flow basis, providing full secondary treatment using a conventional plant configuration that includes the standard waste activated sludge process (Figure 4-9). The secondary effluent receives advanced treatment through coagulation and filtration along with chlorination and dechlorination. The advanced secondary treated effluent from the plant is either reclaimed and/or discharged to the Los Angeles River.

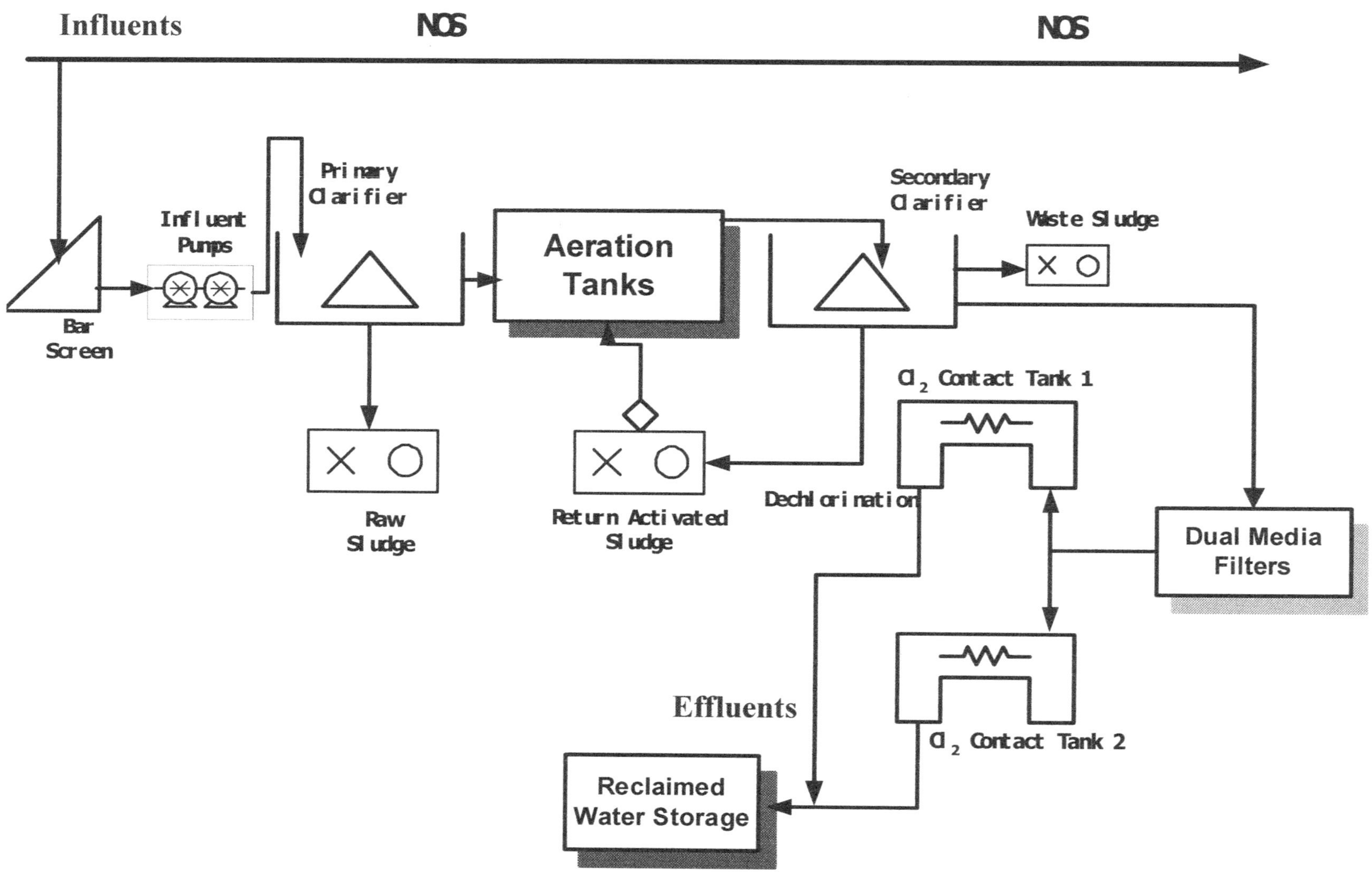

Figure 4-9. Process Flow Diagram of the Los Angeles-Glendale Water Reclamation Plant (LAGWRP).

4.3.3.1 Cyanide Historical Data Review

Historical data from 1996 to 1999 for cyanide species monitored at the influent and effluent of the plant were analyzed. In 1996, there were 12 sampling events with either influent or effluent total cyanide concentration recorded in each event. The influent and effluent flow rates were also documented. Among the 12 events, only two events had both influent and effluent cyanide data available. Both events clearly showed that cyanide was formed during the wastewater treatment processes in the plant as effluent cyanide concentrations unambiguously exceeded influent concentrations (Figure 4-10). The calculation of cyanide mass based on cyanide concentrations and flow rates of the influent and effluent, showed a net gain of cyanide following wastewater treatment (see the insert in Figure 4-10). On February 2, 1996, the cyanide mass in the raw wastewater was 0.766 lb/day whereas the mass in the effluent was 1.868 lb/day. On August 1, 1996, the cyanide mass in the influent was 0.339 lb/day compared to 0.746 lb/day in the effluent. The net gain of the cyanide mass in these two days was 144% and 120%, respectively.

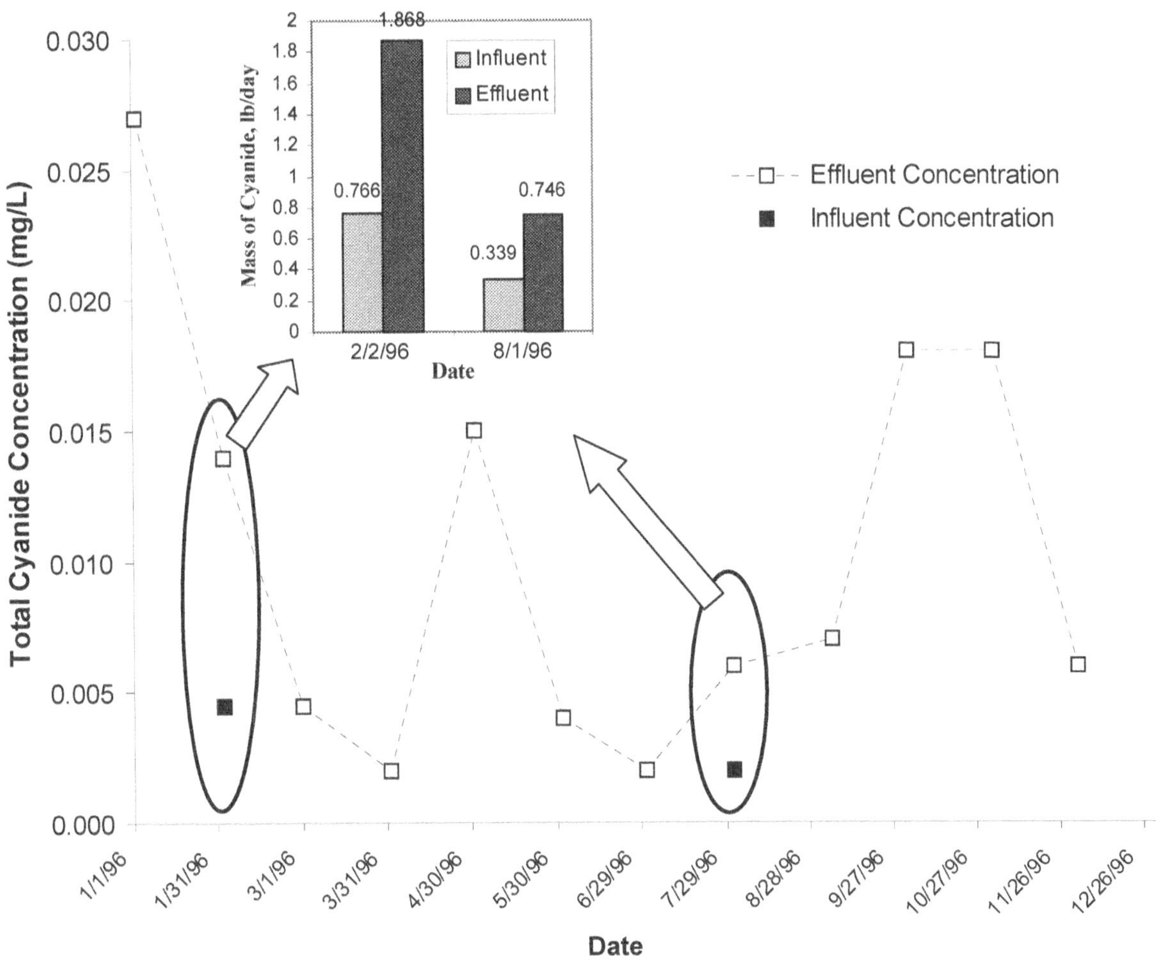

Figure 4-10. Total Cyanide Monitoring Data for LAGWRP in 1996.

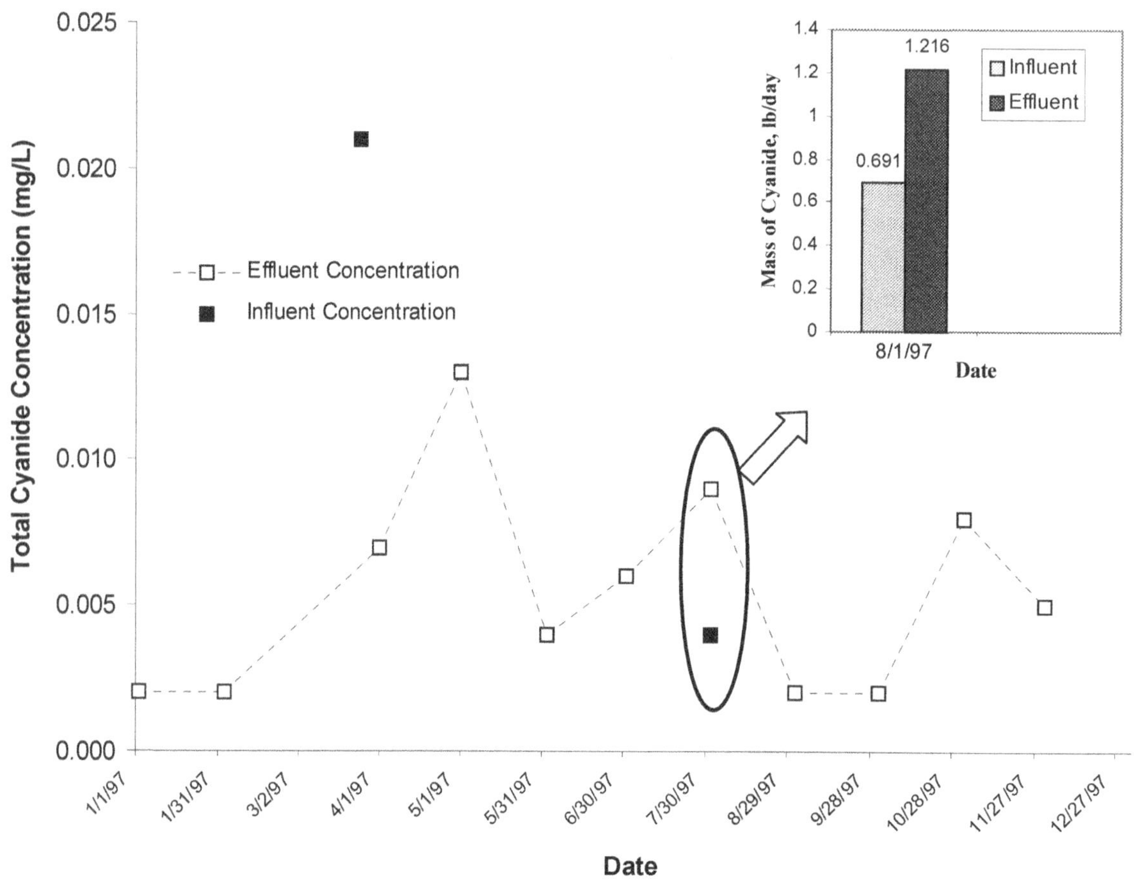

Figure 4-11. Total Cyanide Monitoring Data for LAGWRP in 1997.

In 1997, there were also 12 sampling events with either influent or effluent total cyanide concentrations recorded in each event. The influent and effluent flow rates were also documented. Among the 12 events, only one event had both influent and effluent total cyanide data available, which clearly showed cyanide was formed during wastewater treatment in the plant (Figure 4-11). The calculation of cyanide mass showed a 76% net gain of cyanide wastewater treatment (see the insert in Figure 4-11). On August 1, 1997, the cyanide mass in the influent was 0.691 lb/day compared to 1.216 lb/day in the effluent.

In 1998, there were 12 sampling events with either influent or effluent total cyanide concentrations recorded in each event. The influent and effluent flow rates were also documented. Four out of the 12 events had both influent and effluent cyanide data available. Among those four, there were two events showing that the total cyanide concentration increased from the influent to the effluent (Figure 4-12). However, only one set of the data appeared to have significant differences between influent and effluent cyanide concentrations. On May 2, 1998, the cyanide concentration increased significantly from 0.005 mg/L in the influent to 0.012 mg/L in the effluent. The cyanide mass on that day increased accordingly from 0.771 lb/day in the influent to 1.66 lb/day in the effluent, or about 115% net gain. On August 1, 1998, the cyanide concentration increased from 0.001 mg/L in the influent to 0.002 mg/L in the effluent. Both concentrations were close to the method detection limit. However, in terms of cyanide mass, the increase of cyanide was a little more significant, 0.171 lb/day in the influent vs. 0.284 lb/day in the effluent, or about 66% net gain.

In 1999, there were 45 sampling events with either influent or effluent total cyanide concentrations recorded in each event (Figures 4-13 and 4-14). The influent and effluent flow rates were also documented. There were 29 events with both influent and effluent total cyanide data available. Among those, seven events showed that the total cyanide concentration increased from the influent to the effluent. There is one set of the data where the difference is not significant enough to be considered as a cyanide increase. The biggest difference occurred on August 31, 1999, when the cyanide concentration increased dramatically from 0.002 mg/L in the influent to 0.033 mg/L in the effluent. The cyanide mass on that day increased accordingly from 0.32 lb/day in the influent to 4.13 lb/day in the effluent, or about 1174% net gain. On the other five days, the cyanide mass gain was 49-139%.

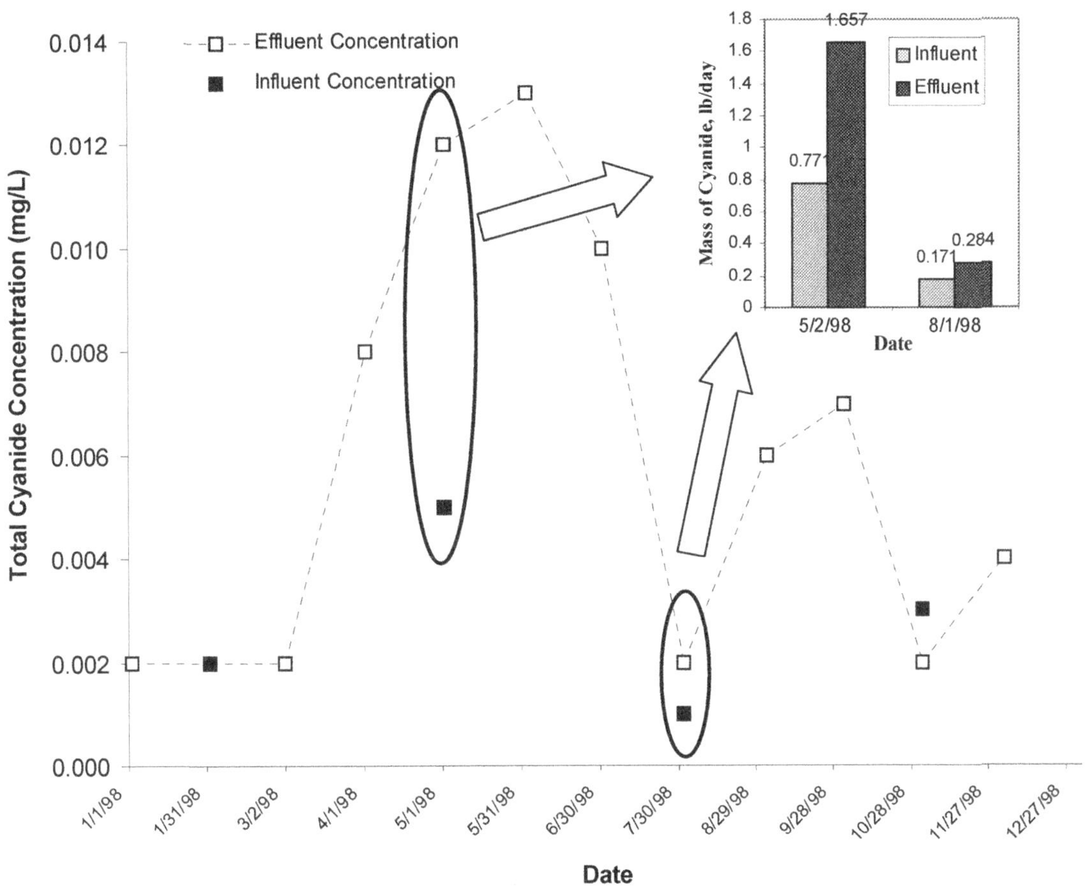

Figure 4-12. Total Cyanide Monitoring Data for LAGWRP in 1998.

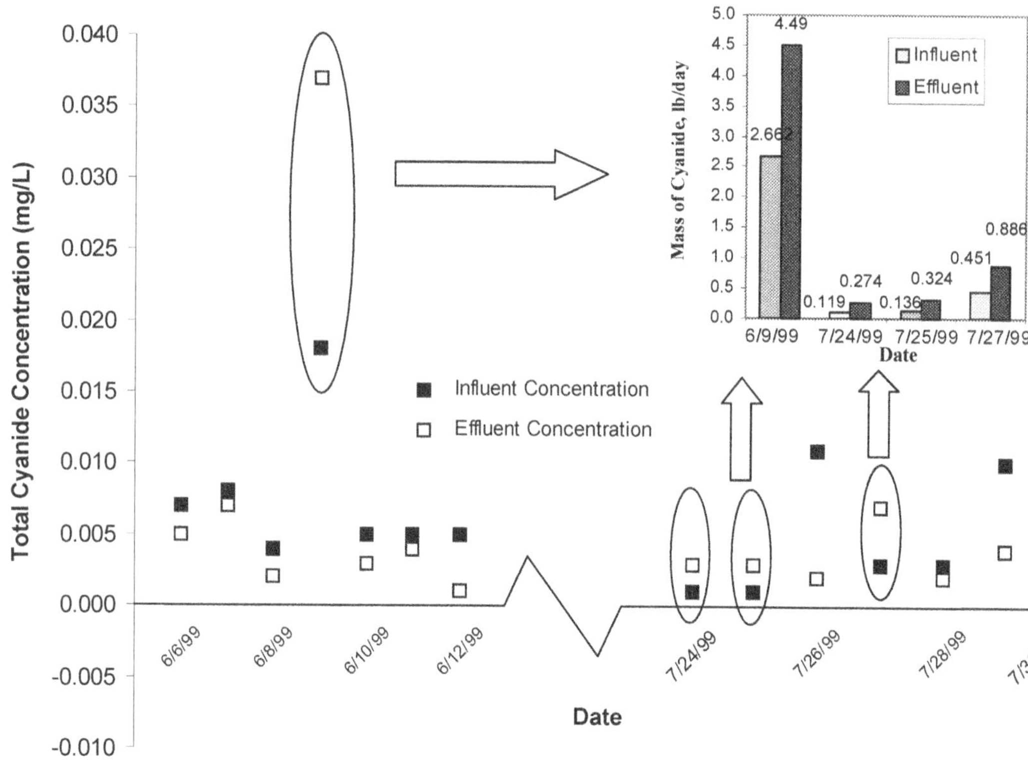

Figure 4-13. Total Cyanide Monitoring Data for LAGWRP from June to July 1999.

From the above analyses, it is evident that cyanide formation occurred periodically at LAGWRP. Most of the cyanide spikes, in terms of both concentration and mass, occurred in the summer time, especially in August. The cause of cyanide formation is unclear because of limited complementary data for other chemical compounds as well as plant operational parameters.

4.3.3.2 WWTP Monitoring for May 2000–May 2001

The LAGWRP conducted the monitoring program from May 2000 to May 2001. Incident-detection sampling was performed with 24-hour compositing once per week. The composite samples for plant influent, postchlorination and post-dechlorination were analyzed for total cyanide. No events that would trigger incident-driven sampling were observed during the one-year period. Figures 4-15 and 4-16 show the total cyanide results for plant influent, postchlorination, post-dechlorination, monthly average limit, and daily maximum limit for 2000 and 2001, respectively. Total cyanide concentrations in the plant influent were 2–16 μg/L, whereas total cyanide in postchlorination and post-dechlorination effluents was 2– 4 μg/L except for a few days. Specifically, there were occasional high spikes of effluent total cyanide concentrations on days such as August 22, 2000, December 19, 2000, and April 17, 2001, etc. Among the 48 monitoring events, only one effluent (post-dechlorination) sample exceeded the NPDES permit daily maximum cyanide limit of 22.0 μg/L with a value of 23.0 μg/L (April 17, 2001). Only one other sample exceeded the NPDES permit cyanide monthly average limitation of 5.2 μg/L with a value of 8.0 μg/L (July 6, 2000).

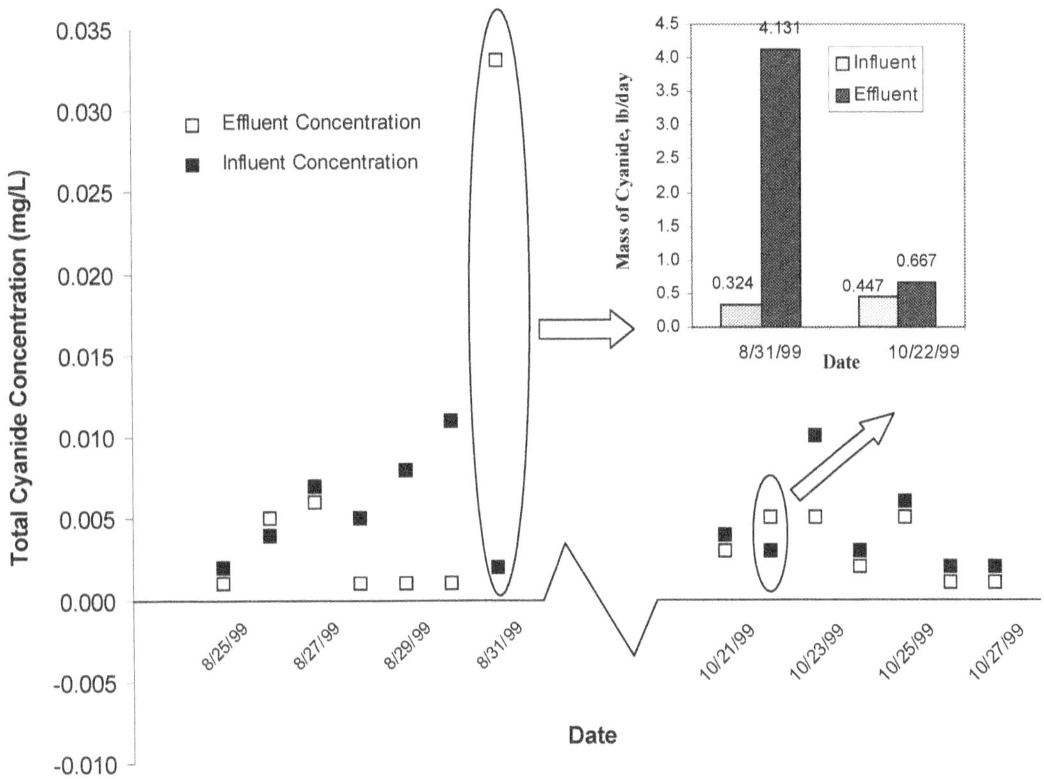

Figure 4-14. Total Cyanide Monitoring Data for LAGWRP from August to October 1999.

It should be noted that nitrification is normally carried out at LAGWRP, and that the plant did not experience high levels of nitrification in the summer during the monitoring period, which may account for the lack of elevated cyanide observations. The presence of nitrite can lead to cyanide formation under some conditions, and can cause false positives in cyanide analysis (see Chapter 7.0). It is also not clear whether the high effluent cyanide concentration reflects cyanide formation during wastewater treatment or other factors such as interference during total cyanide analysis.

4.3.4 The City of Sunnyvale Water Pollution Control Plant (WPCP)

WPCP is an advanced WWTP treating wastewater for either reclamation or discharge into the Guadalupe slough. The plant has a hydraulic capacity of 29.5 mgd. The treatment processes at the plant consist of aerated grit removal, primary sedimentation, oxidation ponds, fixed growth reactors (FGR), dissolved air flotation tanks for polymer assisted algae removal, dual media filters, chlorination with chlorine gas addition, and dechlorination with sulfur dioxide gas addition. Figure 4-17 depicts the tertiary process at the WPCP. Two sampling locations used for regular monitoring are the secondary effluent (E-3) and final effluent (T-3).

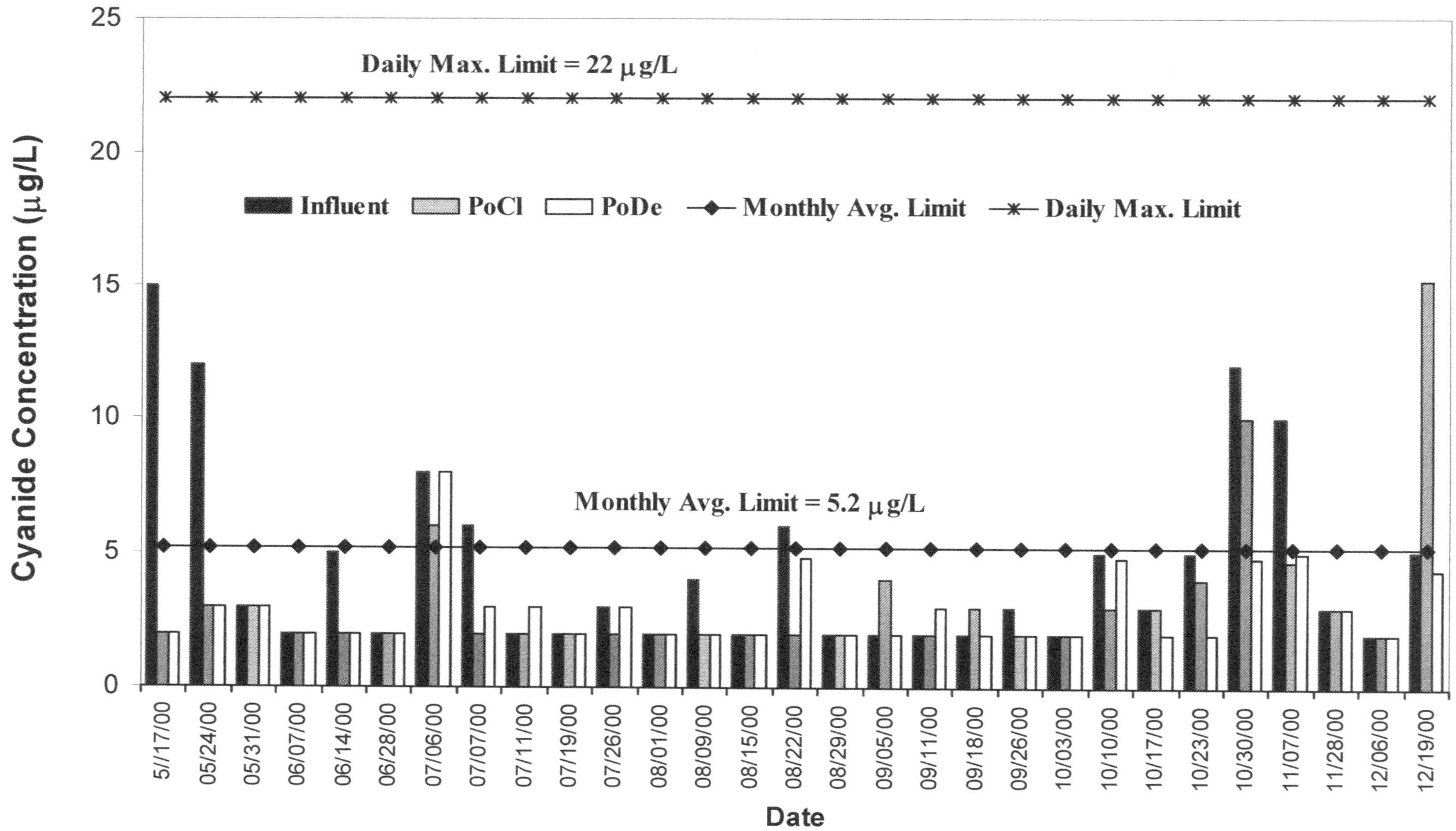

Figure 4-15. Total Cyanide Monitoring Data for LAGWRP in 2000.

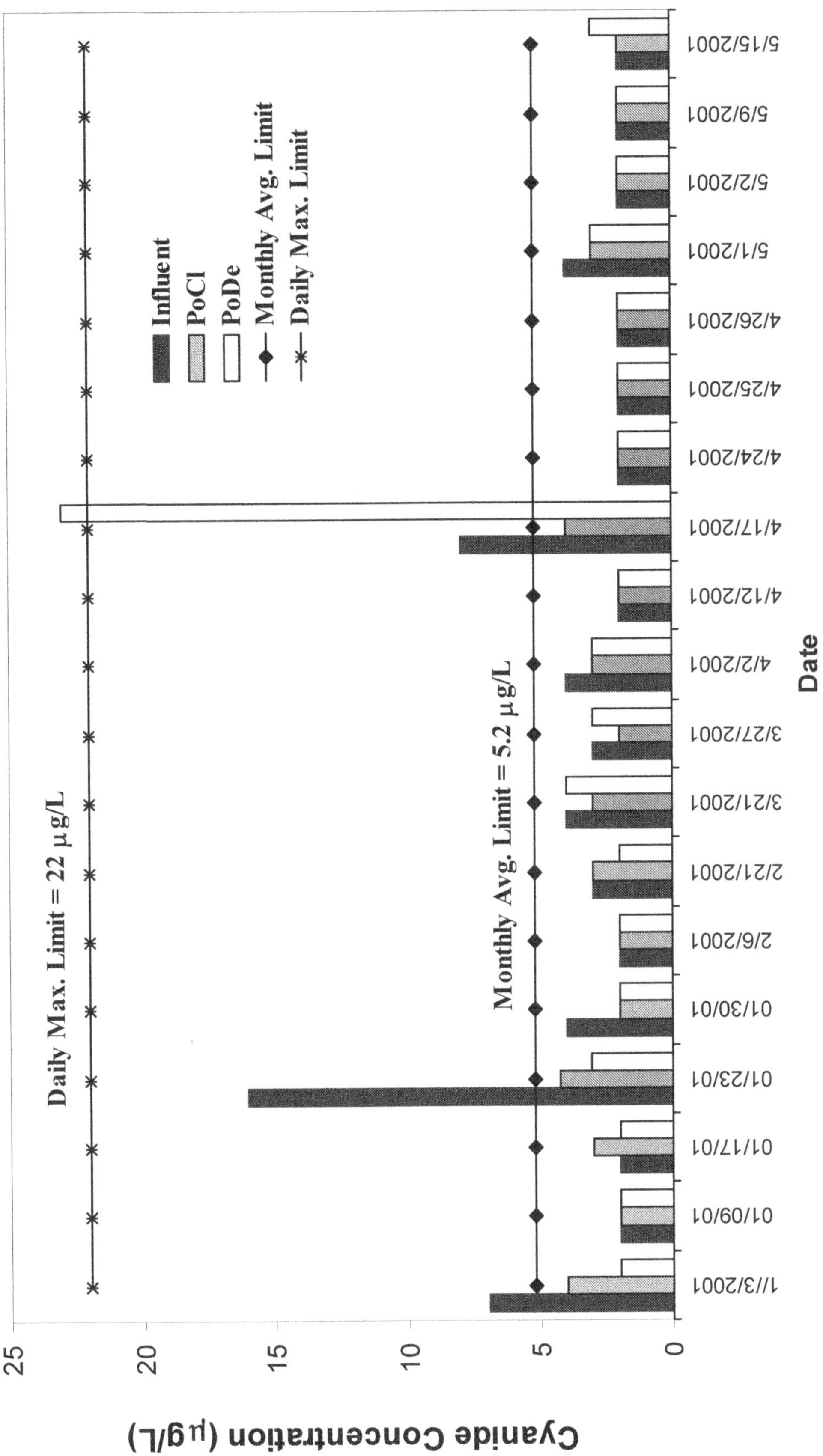

Figure 4-16. Cyanide Monitoring Data for LAGWRP in 2001 (as of May 15, 2001).

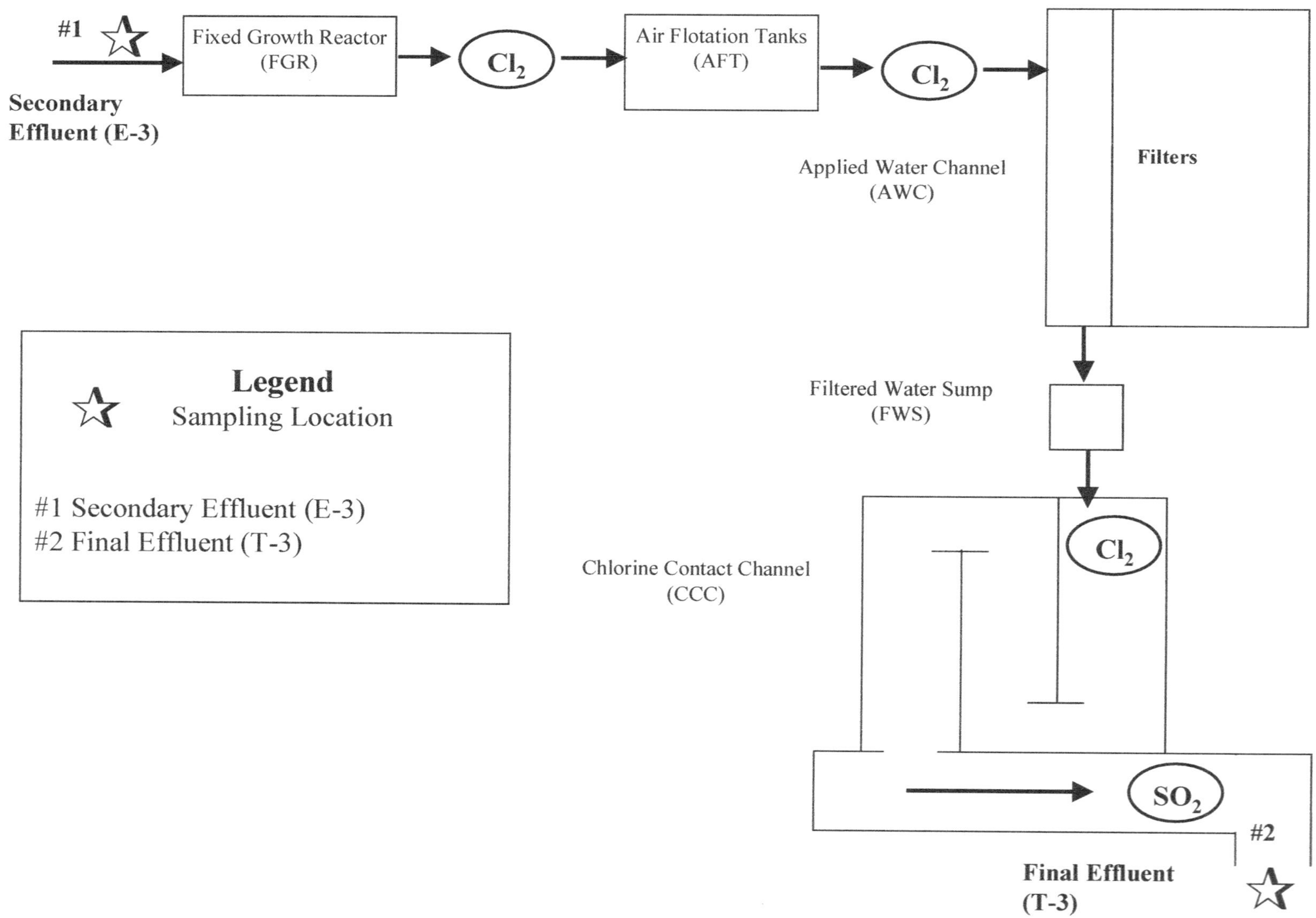

Figure 4-17. Tertiary Process Diagram of the City of Sunnyvale Water Pollution Control Plant (WPCP).

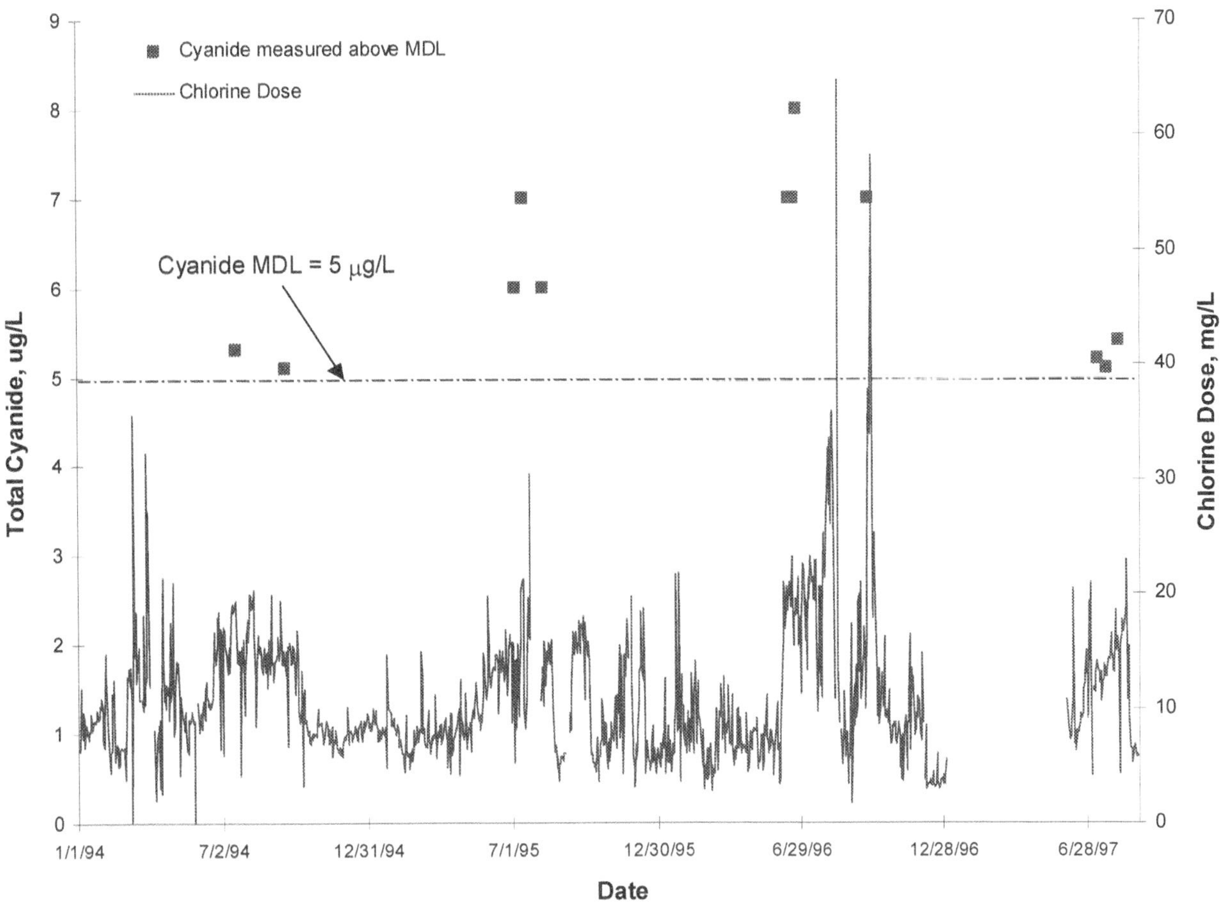

Figure 4-18. Cyanide Monitoring Data for the City of Sunnyvale Water Pollution Control Plant (WPCP), 1994-1997.

4.3.4.1 Cyanide Historical Data Review

WPCP has reported periodic episodes of cyanide occurrence in the chlorinated effluent at levels that exceed the daily average discharge limit of 5 μg/L specified in the NPDES permit. Detectable levels of cyanide have been observed only during the summer in the tertiary processes at the plant. A detailed investigation was carried out by Malcolm Pirnie, Inc. (1997). The researchers concluded that analytical interferences and internal sources were the likely candidates responsible for the cyanide occurrence. It was also stated in the report that "because cyanide formation only occurs when the plant is fully nitrifying during the summer months, it appeared that the distribution of nitrogen species plays a role in cyanide occurrence" (p. 1-4).

Figure 4-18 shows the total cyanide measured above MDL (5 μg/L) from composite T-3 samples (final effluent) and corresponding chlorine dose versus time from January 1, 1994, to August 31,1997. There were 12 cyanide occurrences detected in the final effluent over 1,338 days, with total cyanide of 5.1– 8.0 μg/L with a mean of 6.18 μg/L. The date and concentration of cyanide occurrence as well as the corresponding chlorine dose are presented in Table 4-8 and Figure 4-19.

As illustrated in Table 4-8 and Figure 4-19, most of the cyanide occurrences were in summer and all of the cyanide occurrences happened when the chlorine dose exceeded the average dosage (10.8 mg/L). However, there were also numerous monitoring data for the summer periods demonstrating no cyanide occurrence under high chlorine dosage. Therefore, it is inconclusive that chlorination is solely responsible for cyanide formation. Other parameters of potential interest are nitrate and nitrite concentrations. It was found that in all cyanide occurrence days, the nitrite concentration always decreased from 0.1 mg/L to 1.98 mg/L in the secondary effluent to < 0.01 mg/L (MDL for nitrite) in the final effluent. This implies that oxidized nitrogen species may have played a role in the cyanide formation process.

Overall, the monitoring data clearly showed that cyanide formation occurred mostly in the summer when complete nitrification was occurring in the secondary biotreatment process.

Table 4-8. Cyanide Concentration Detected at the Final Effluent (T-3) of WPCP, City of Sunnyvale, Above MDL from 1/1/1994 to 8/31/1997.

No.	Date	Cyanide concentration, μg/L	Chlorine dose, mg/L
1	7/19/1994	5.3	14.7
2	9/19/1994	5.1	12.9
3	7/5/1995	6.0	14.3
4	7/14/1995	7.0	21.2
5	8/9/1995	6.0	15.8
6	6/13/1996	7.0	21.0
7	6/18/1996	7.0	15.5
8	6/23/1996	8.0	18.2
9	9/22/1996	7.0	37.8
10	7/10/1997	5.2	13.3
11	7/23/1997	5.1	13.3
12	8/6/1997	5.4	14.6

4.3.4.2 WWTP Monitoring for 2000-2001

WPCP initiated the cyanide-monitoring program in June 2000 with weekly sampling at three locations: FGRs, filtered water sump (FWS), and the end of chlorine contact channel (CCC) (see Figure 4-17). However, because of the lack of detection of cyanide formation events at the final effluent sample location, WPCP changed the weekly sampling program to a monthly sampling program (24-hour composites) for total cyanide in December 2000 because it was their experience over the years that cyanide is not detected at the treatment plant during the winter months. In addition to low ambient temperatures during the winter, the lack of cyanide formation in the plant was also attributed to the lack of significant nitrification. WPCP changed back to biweekly sampling from May 2001 to July 2001. During the one-year monitoring period, only one cyanide formation event was observed. On May 21, 2001, total cyanide was nondetectable in the FGR combined sump (located after the FGRs), and was 7.5 μg/L in FWS (located after the filters and before the chlorination) and 5.7 μg/L at the end of CCC (located after the chlorination). It should be noted that all conditions on that particular day were almost the same as those on other days except the temperature (23.5°C) was 1.5-3.5°C higher than other late

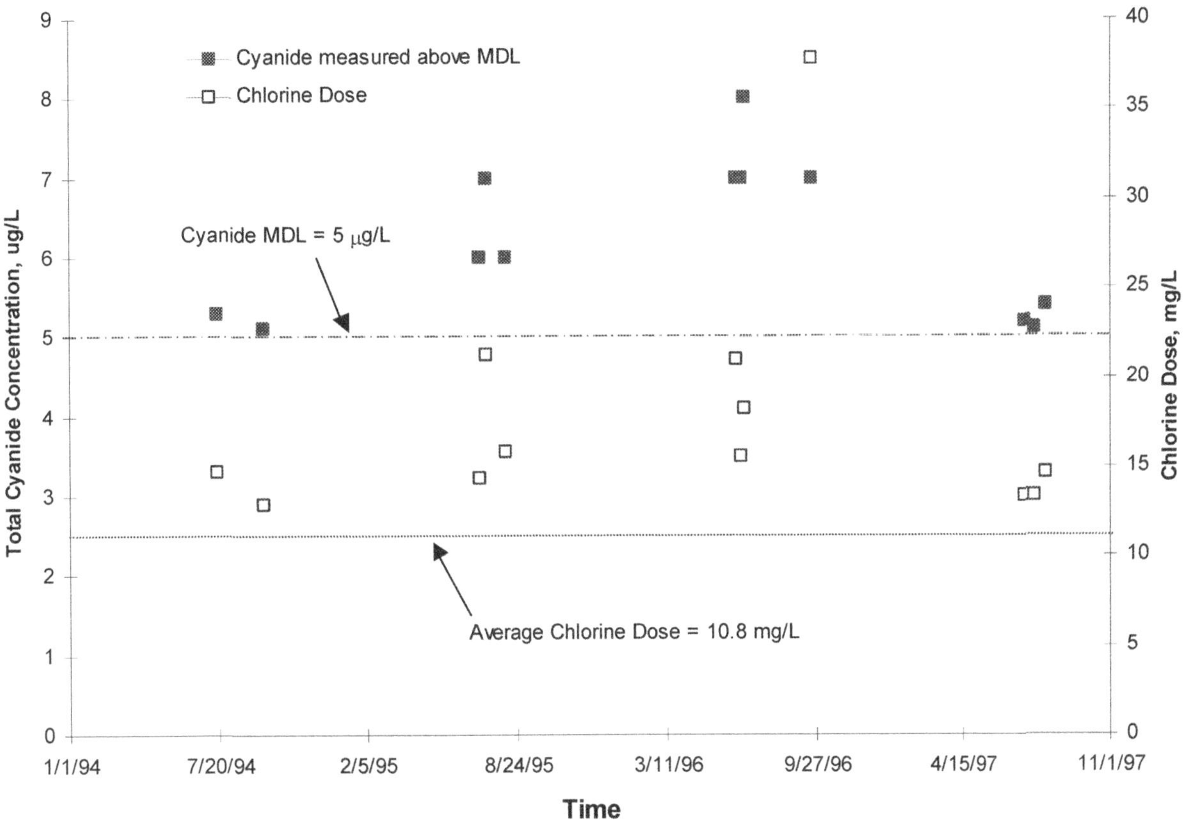

Figure 4-19. Cyanide Occurrence Data in the City of Sunnyvale Water Pollution Control Plant (WPCP), 1994–1997.

spring/summer days, and 9-12°C higher than winter days. Additionally, the cyanide detected was close to the detection limit of 5 µg/L, indicating a possibility of false positive in the cyanide analytical method. No incident-driven sampling was conducted following that single event.

4.3.5 Metropolitan Water Reclamation District of Greater Chicago (MWRDGC)

MWRDGC is a government agency created in 1889 to protect the water quality of Lake Michigan, the major drinking water supply for the Chicago area. The Hanover Park Water Reclamation Plant (WRP) of MWRDGC was the nation's first major tertiary wastewater treatment facility. It has a design capacity of 12 mgd. The flow diagram of Hanover Park WRP is provided in Figure 4-20. It consists of four major sections. The pretreatment section includes coarse screens, fine screens, and grit chambers. Primary treatment is conducted in four batteries of preliminary settling tanks followed by secondary treatment, achieved by four batteries of aeration tanks and final settling tanks. Tertiary treatment is performed using six filter beds. Sodium hypochlorite is added to the head of the tertiary filters to control the water quality flowing to the filters. Sodium bisulfite is added to the end of the filters as dechlorinating agents.

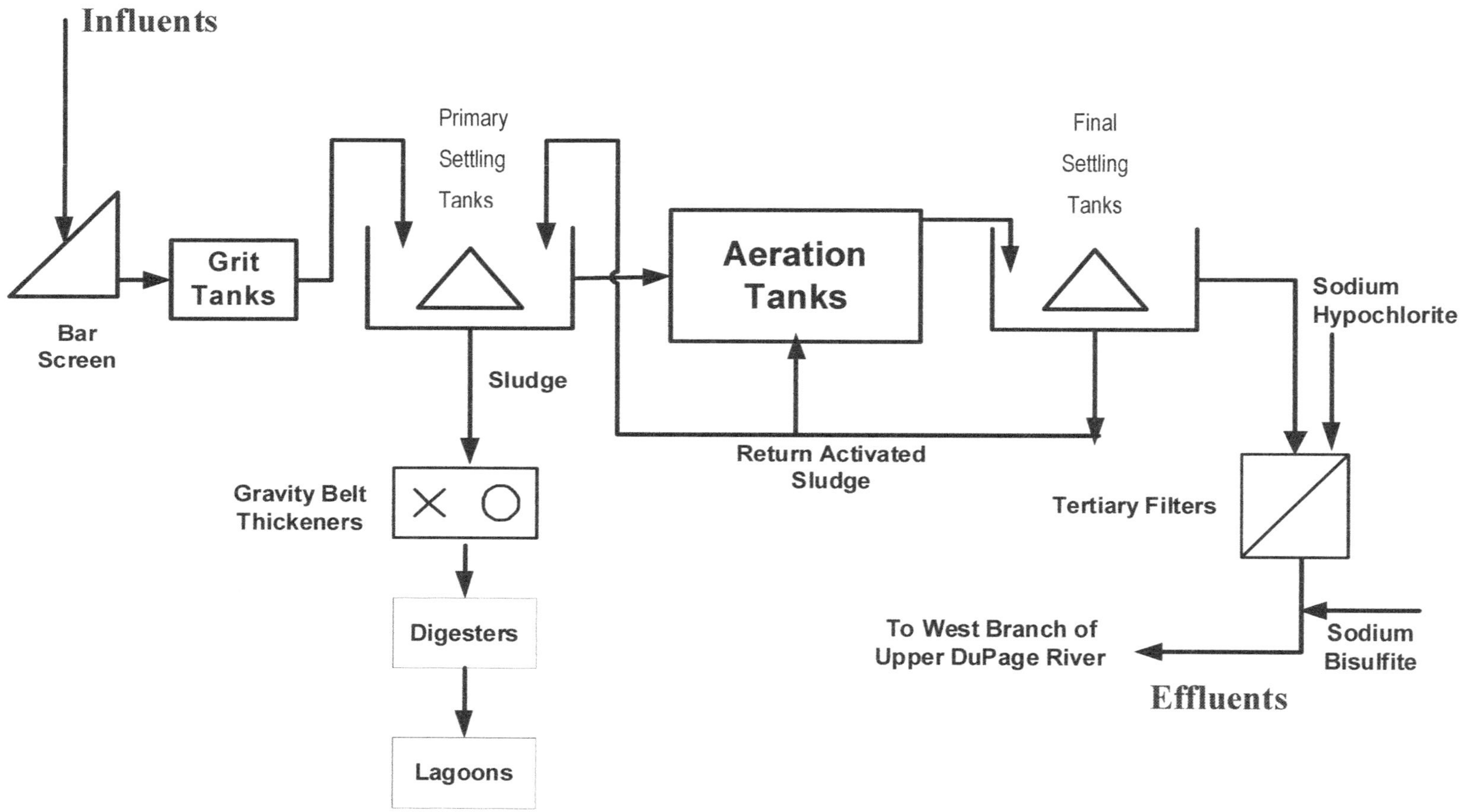

Figure 4-20. Process Flow Diagram of Hanover Park Water Reclamation Plant, MWRDGC.

4.3.5.1 Cyanide Historical Data Review

Around 1995, an increase in cyanide concentration in final effluent was observed during the summer at MWRDGC when chlorination/dechlorination was practiced at certain facilities. Extensive monitoring data had verified that chlorination was causing an increase WAD cyanide. Table 4-9 presents a statistical summary of annual WAD cyanide concentrations in the influent and effluent of the Hanover Park WRP (1996-1999). From the table, it is clear that the mean WAD cyanide concentration in the effluent exceeded that in the influent for most of the years. Maximum WAD cyanide concentrations in the effluent were always higher than those in the influent. On the basis of the data of monthly averages of raw sewage and final effluent WAD cyanide at the Hanover Park WRP, Sawyer et al. (1996) observed that when the chlorination/dechlorination system was operational, the monthly average final effluent WAD cyanide concentrations increased to the 4.0- to 6.0-μg/L level although raw sewage WAD cyanide concentrations remained at either a 1.0- or 2.0-μg/L level. They also found that a similar pattern occurred at the James C. Kirie WRP where effluent WAD cyanide concentrations averaged either 1.0 μg/L or 2.0 μg/L from November through April when chlorination/dechlorination was inoperative versus monthly averages of 3.0 μg/L and 4.0 μg/L from May through October when chlorination/dechlorination was practiced (Sawyer et al., 1996).

Table 4-9. Summary of Annual WAD Cyanide Concentrations in Hanover Park WRP, MWRDGC, 1994–2000 (units: mg/L).

	Influent			Effluent		
Year	Min	Mean	Max	Min	Mean	Max
1994	0.000	0.001	0.003	0.000	0.000	0.011
1995	NA[a]	NA	NA	0.000	0.003	0.011
1996	0.000	0.001	0.003	0.001	0.003	0.016
1997	0.000	0.001	0.003	0.000	0.003	0.010
1998	0.000	0.001	0.003	0.000	0.003	0.010
1999	0.000	0.001	0.005	0.000	0.003	0.012
2000	0.000	0.001	0.011	0.000	0.003	0.011

[a] NA = Not available.

4.3.5.2 WWTP Monitoring for 2000

MWRDGC initiated a cyanide incident-detection sampling event in June 2000, and continued monitoring for total cyanide and WAD cyanide in their unchlorinated secondary effluent, chlorinated secondary effluent, and dechlorinated tertiary effluent at the Hanover Park WRP until November 2000 when the chlorination process was ceased for the winter and spring as a regular practice. Because MWRDGC observed elevated total cyanide concentrations in chlorinated secondary effluent almost continuously during the monitoring period, they conducted several intensive cyanide incident-driven monitoring events, once in September 2000 and once in October 2000. Samples were collected for four consecutive days and analyzed for the entire range of chemical parameters specified in the workplan.

The evaluation of the MWRDGC monitoring data indicates that thiocyanate was being generated in the chlorination process. The total cyanide analysis technique employed in the MWRDGC laboratory (UV irradiation with thin film distillation) can also detect thiocyanate in addition to free and metal complexed forms of cyanide. In the conventional distillation method for total cyanide (APHA/AWWA/WEF, 1998 – Method 4500-CN-C), SCN^- is not detected. The MWRDGC monitoring data include two total cyanide values, one is actually free cyanide plus metal-complexed cyanide plus SCN, reported as total cyanide (designated herein as TOTCN+SCN), and the other is total cyanide less thiocyanate (designated herein as TOTCN). That is, total cyanide similar to what one gets by the manual distillation technique that does not detect SCN. Figure 4-21 shows the September 2000 monitoring data of TOTCN+SCN and TOTCN versus time. From the graph, it is obvious that the TOTCN+SCN was higher by about a factor of two in the chlorinated and dechlorinated samples compared to unchlorinated samples, but the TOTCN remained at the same level in unchlorinated, chlorinated, and dechlorinated samples (very close to the detection limit). The SCN^- concentrations in the samples can be obtained by subtracting TOTCN values from the TOTCN+SCN values [which is the ASTM Method D4374-93 (1998d) to determine SCN]. Figure 4-22 gives the calculated SCN^- values versus time. It is obvious that SCN^- was being created by the chlorination process. The analyses of data from other months (August and October 2000) showed the same pattern.

An independent laboratory performed direct analysis of SCN^- for all MWRDGC samples because the MWRDGC laboratory does not perform SCN^- as a routine analysis. SCN^- was reported as below 0.1 mg/L, the method detection limit for their procedure (APHA/AWWA/WEF, 1998 – Method 4500-CN-M). The calculated SCN^- results (usually 10-15 μg/L) obtained from the subtraction of the TOTCN from the TOTCN+SCN are much less than the MDL for the procedure. Currently MWRDGC doesn't have any QA/QC data showing complete recovery of SCN^- from CN+SCN analysis and nonrecovery of SCN^- for TOTCN analysis. Therefore, there is still a possibility that the calculated concentration of thiocyanate obtained from subtracting TOTCN from TOTCN+SCN^- is a false positive. However, this finding poses a big question: What is the main reason for potential SCN^- formation in the chlorination process? The thiocyanate is perhaps derived from oxidative decomposition of an organic compound, but this requires further investigation.

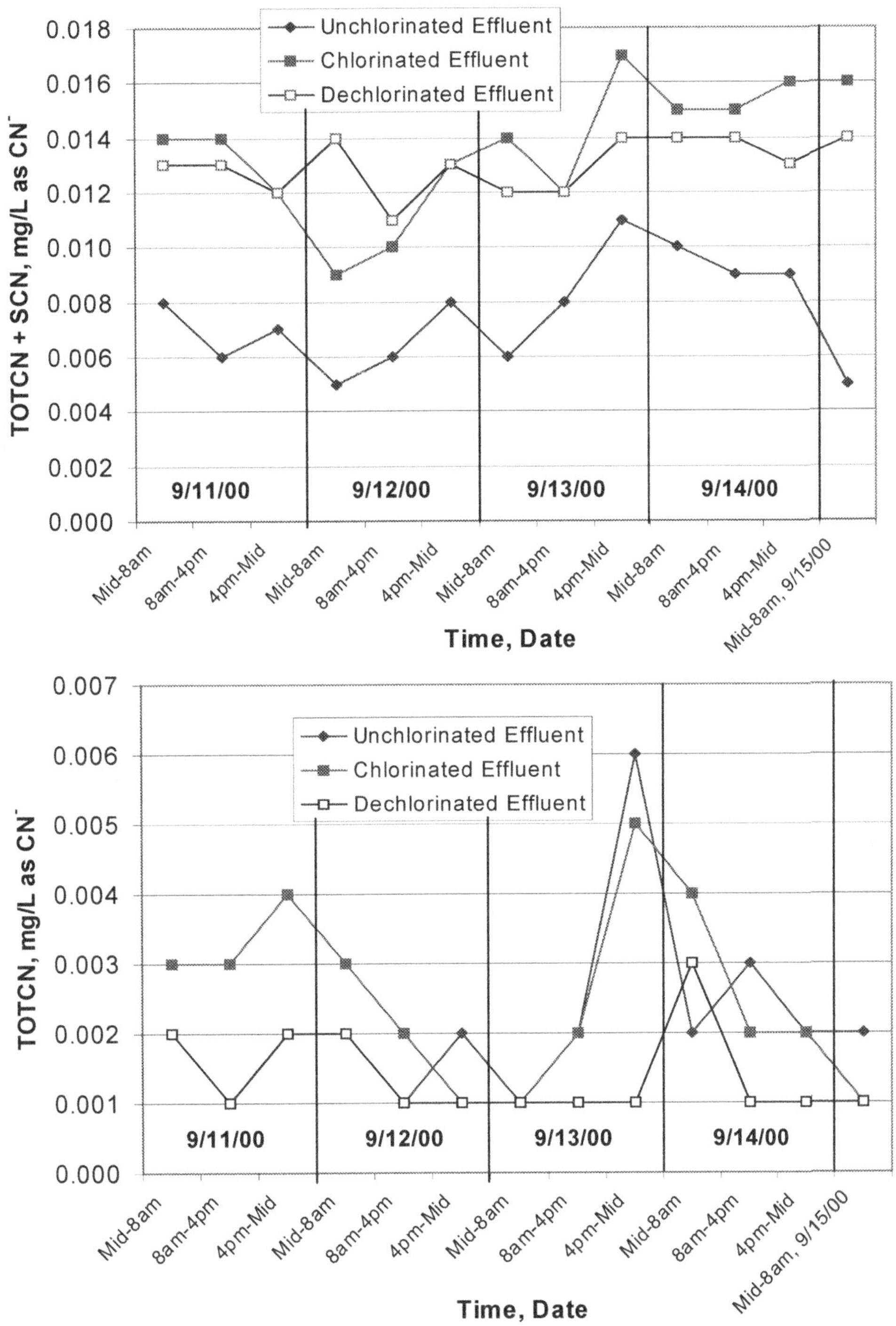

Figure 4-21. Monitoring Data for Cyanide Species in Hanover Park WRP, MWRDGC (September 2000).

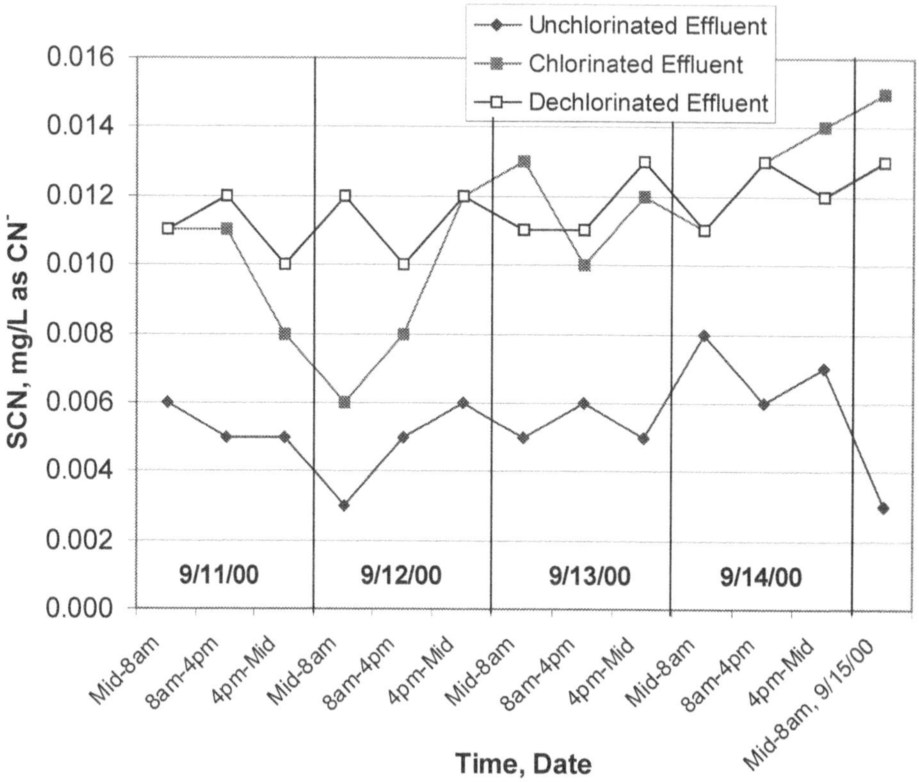

Figure 4-22. Calculated SCN^- Concentrations in Hanover Park WRP, MWRDGC (September 2000).

The WAD cyanide data showed a similar pattern as occurred previously in the Hanover Park WRP (see previous section). The average WAD cyanide concentration in the unchlorinated secondary effluent was usually very low, from nondetectable (< 0.4 μg/L) to 0.7 μg/L, and the WAD cyanide increased to a higher level of 0.48–1.7 μg/L in the chlorinated secondary effluent and to an even higher level of 1.2–2.2 μg/L in the dechlorinated secondary effluent (Table 4-10). The increase in WAD cyanide concentration, which does not measure SCN^-, was at most 1.3 μg/L, and this does not seem to account for all the increase in total cyanide concentration described before, which is hypothesized to be the effect caused by the increase of SCN^-.

Table 4-10. Summary of Monthly WAD Cyanide Mean Concentrations in Secondary Effluents at Hanover Park WRP, MWRDGC, from June to October 2000 (units: μg/L).

Month	Unchlorinated	Chlorinated	Dechlorinated
June	0.13	0.48	1.3
July	0.67	1.1	1.2
August	0.60	1.7	1.7
September	0.54	1.4	1.6
October	0.092	0.96	2.2

Consultation with MWRDGC revealed that TOTCN+SCN is reported for four of the District's treatment plants. WAD cyanide, determined using an automated version of *Standard Methods* 4500-CN-I (APHA/AWWA/WEF, 1998), is reported for the other three treatment plants including the Hanover Park WRP.

Overall, data from MWRDGC indicate that thiocyanate is formed in the chlorination process, as verified by the total cyanide plus thiocyanate concentration measured in prechlorination, postchlorination, and post-dechlorination effluents. As shown in the laboratory studies (see Chapters 5.0-7.0), chlorine can react with and decompose SCN^- to yield free cyanide. For MWRDGC, no increase in free cyanide upon chlorination was observed, indicating that the chlorine dosing was sufficiently high to destroy any free cyanide produced.

4.3.6 Massachusetts Water Resources Authority (MWRA)

MWRA is a Massachusetts public authority established in 1984 to provide wholesale water and sewer services to 2.5 million people and more than 5,500 large industrial users in 61 metropolitan Boston communities. The Deer Island Treatment Plant (DITP) serves more than two million people and is the second largest sewage treatment plant in the United States. The plant's peak capacity is 1.27 billion gallons of wastewater per day. Average daily flow is about 380 mgd. Disinfection with chlorine is implemented in the treatment plant (Figure 4-23). After passing through primary and secondary treatment, wastewater is disinfected with sodium hypochlorite. There are two disinfection basins, each approximately 500' with a capacity of four million gallons, in which the effluent is mixed with sodium hypochlorite. In September 2000, DITP put a 9.5-mile ocean outfall into operation. It also began dechlorinating its final effluent with sodium bisulfite. After disinfection and dechlorination, the effluent is discharged to the marine environment.

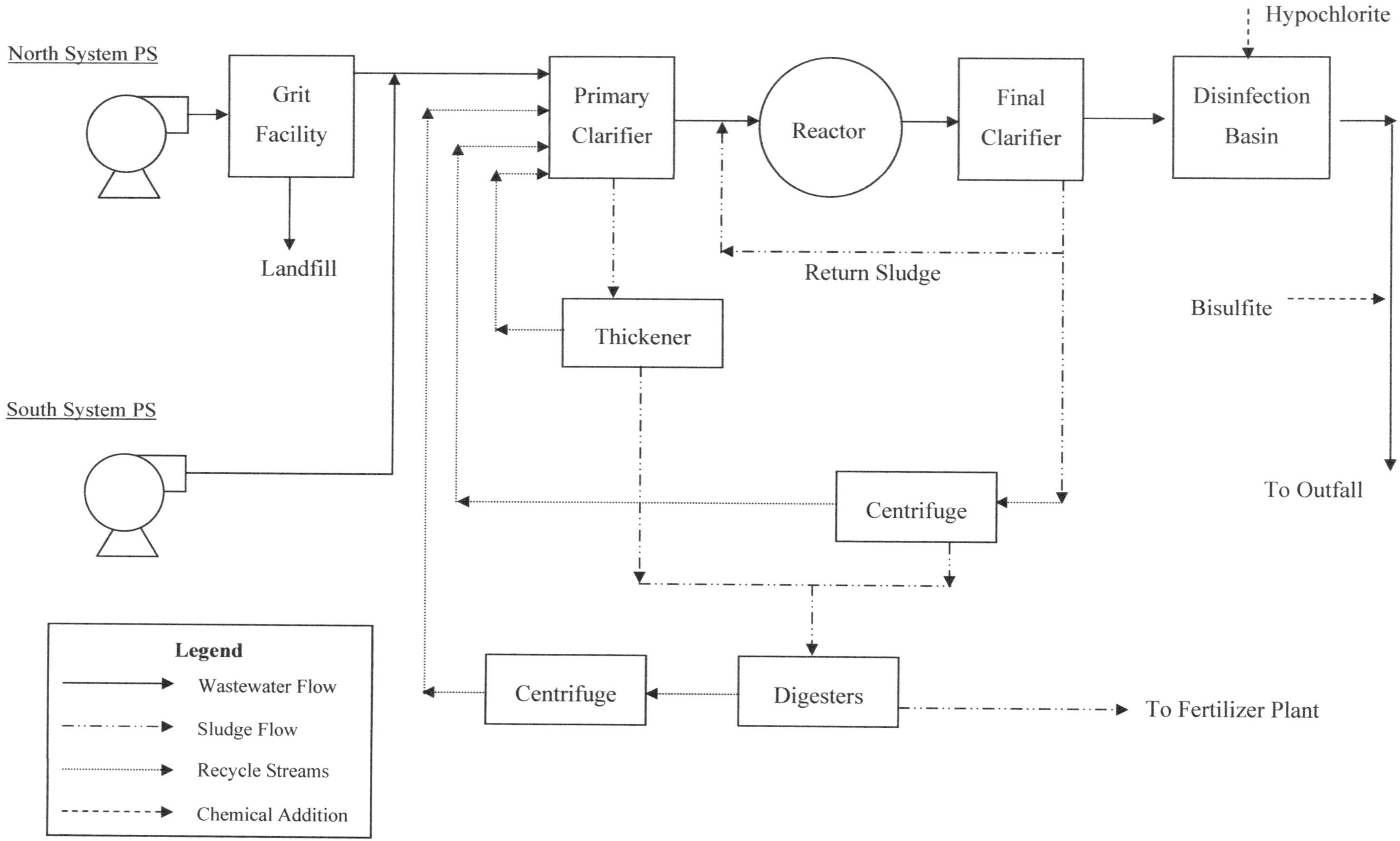

Figure 4-23. Process Flow Diagram of Deer Island Treatment Plant, MWRA.

4.3.6.1 Cyanide Historical Data Review

MWRA began to detect what appeared to be cyanide at low levels in chlorinated primary effluent (and prechlorinated influent) in 1995 at the time when the MWRA Central Laboratory went into operation at the DITP. Cyanide formation was first reported by Delaney et al. (1997) who stated "At times, low levels of cyanide have been detected in the plant's chlorinated primary effluent. Cyanide is rarely observed in the plant's influent." It was unclear whether this was a sampling or analysis artifact, or whether cyanide was actually formed during the chlorination process. This report also concluded that cyanide appeared to form from thiocyanate when WWTP wastewater was chlorinated with NaOCl.

4.3.6.2 WWTP Monitoring for 2001

MWRA began collecting and testing samples for the WWTP monitoring program in March 2001. A change in the monitoring plan was made to only collect weekly samples until the end of June and to test them for a wider range of parameters than was intended for incident-driven sampling. Samples were collected at three locations including prechlorination secondary effluent, postchlorination effluent, and post-dechlorination final effluent. At each of these three locations two composite samplers were used to collect 24-hour composite samples. One of the two composite samples was pre-preserved with sodium hydroxide. Following the collection of the composite sample, aliquots of each composite were prepared and any additional sample preservation was performed. For each of these three weekly samples, analyses were performed to test total cyanide (by online UV digestion/distillation), thiocyanate, nitrite, nitrate, TSS, total organic carbon, nine metals by ICP, and total residual chlorine.

The evaluation of the MWRA monitoring data indicates that total cyanide was generated consistently in the chlorination process, probably from the chlorination of thiocyanate. Figure 4-24 shows the total cyanide concentration in prechlorination, postchlorination, and post-dechlorination effluents. The postchlorination and post-dechlorination samples were dechlorinated with two agents, ascorbic acid (AsA) and sodium thiosulfate. Results show that total cyanide was nondetectable (<5 μg/L) in prechlorination samples whereas significant cyanide concentrations (up to 14 μg/L) were observed in dechlorinated postchlorination and post-dechlorination samples. Generally, samples dechlorinated with ascorbic acid yielded more total cyanide concentration than samples with sodium thiosulfate, reinforcing a concern about using ascorbic acid as an analytical dechlorinating agent for total cyanide analysis (Carr et al., 1997).

Additionally, MWRA analyzed concentrations of thiocyanate, a typical cyanide-related compound encountered in most WWTPs, in all samples collected, which provided valuable insight on the formation of cyanide in the chlorination process. Thiocyanate is normally not analyzed in most WWTPs because it is far less toxic than free cyanide and thus is not a regulated pollutant. Moreover, most of the total cyanide analytical methods employed in WWTPs do not measure thiocyanate (SCN^-) as a cyanide species to be counted in total cyanide concentration, the aggregate of all compounds containing the cyanide group. Table 4-11 provides monitoring data for thiocyanate and total cyanide in three locations. The results show that thiocyanate was present in the prechlorination effluent at a concentration of 13.7–21.8 μg/L and decreased to a concentration from nondetectable to 8.63 μg/L in postchlorination samples. The thiocyanate was almost nondetectable in post-dechlorination samples except for one case where a concentration of 5.14 μg/L SCN^- was reported, which was just beyond the detection limit of 5 μg/L. It is hypothesized that chlorination of thiocyanate caused the formation of cyanide because the relationship between the disappearance of SCN^- and formation of CN^- in the chlorination process was clear. This hypothesis is supported by laboratory results (see Chapter 5.0) indicating formation of cyanide in thiocyanate solutions after chlorination in which the chlorine dose was not sufficient to destroy SCN^- completely thus leaving no chlorine residual to destroy any CN^- produced from reaction of chlorine with SCN^-.

Table 4-11. Summary of Weekly Monitoring Data of Thiocyanate and Total Cyanide in Three Locations at Deer Island Treatment Plant, MWRA, April–June 2001 (units: μg/L).

Date	Prechlorination		Postchlorination[a]			Post-dechlorination[a]		
			CN			CN		
	CN	SCN	AsA	$Na_2S_2O_3$	SCN	AsA	$Na_2S_2O_3$	SCN
04/19/2001	ND[b]	15.2	ND	ND	ND	ND	ND	ND
04/26/2001	ND	20.0	7.33	6.13	5.60	5.35	5.44	ND
05/03/2001	ND	17.7	9.67	6.10	7.20	5.24	5.72	ND
05/10/2001	ND	21.8	13.2	8.75	8.00	7.62	7.9	3.7
05/17/2001	ND	15.7	7.84	ND	7.14	5.15	ND	ND
05/24/2001	ND	17.2	14.0	8.03	8.63	6.72	9.09	ND
05/31/2001	ND	16.7	8.18	ND	7.47	6.73	7.51	ND
06/07/2001	ND	13.7	11.5	7.78	7.43	ND	ND	5.14

[a] Postchlorination and post-dechlorination samples for cyanide analysis were dechlorinated with two reagents, ascorbic acid (AsA) and sodium thiosulfate ($Na_2S_2O_3$), before cyanide analysis.
[b] ND = Not detected.

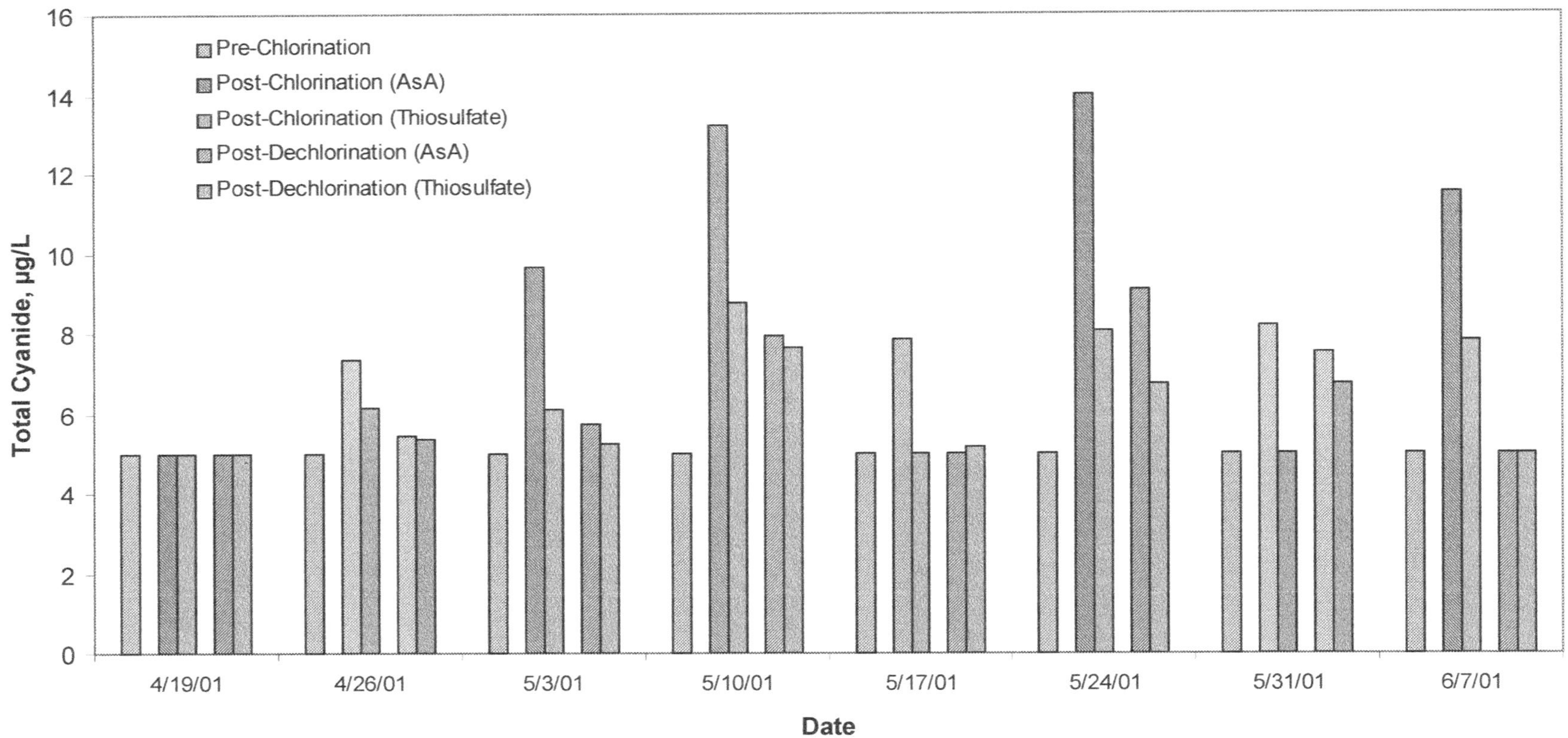

Figure 4-24. Monitoring Data for Total Cyanide Concentrations at Three Locations in Deer Island Treatment Plant, MWRA from April–June 2001.

4.4 Discussion and Conclusions

The monitoring data obtained from this one-year WWTP monitoring program provided a database for the investigation of cyanide formation in WWTP wastewater secondary treatment (Table 4-12). Data from all participating WWTPs showed evidence of cyanide formation in this one-year monitoring effort, although the cyanide formation pattern varied significantly from one plant to another, and from one season to another. Along with the WWTP monitoring program, a laboratory study of mechanisms of formation of cyanide during wastewater treatment was performed at Carnegie Mellon University (CMU). This study investigated several chemical reactions (or reaction groups) that may lead to the formation of cyanide upon secondary treatment. Secondary effluents from several WWTPs were used in the CMU studies. Some reaction mechanisms studied in the laboratory seemed very relevant to explaining the monitoring data, whereas other mechanisms examined did not appear to explain the cyanide formation observed at the WWTPs. Among all mechanisms considered, the chlorination of thiocyanate seems to be the most important mechanism for the formation of cyanide in wastewater treatment processes, especially in chlorination/dechlorination as observed both in the laboratory study and in the MWRA and MWRDGC monitoring events.

Table 4-12. Summary of Formation of Cyanide in WWTPs.

Plant [a]	Chlor.	Observation of cyanide formation	CFF, % [b]	Cyanide formation mechanism
EBMUD	Yes	WAD-CN in chlorinated and dechlorinated effluents frequently exceeded those in unchlorinated effluent.	79	Unknown
GCA	No	Occasionally total cyanide concentration in the secondary effluent exceeded those in the primary effluent.	5	Unknown
LAGWRP	Yes	Occasionally total cyanide concentration in the chlorinated and dechlorinated effluent exceeded those in the influent.	8	Unknown
Sunnyvale WPCP	Yes	Very limited cyanide formation was observed due to a lack of significant nitrification in the plant.	5	Unknown
MWRDGC	Yes	Significant total cyanide increase after chlorination process, due to the inclusion of thiocyanate in the total cyanide analysis. Evidence for production of SCN^- in chlorination. Slight increase of WAD-CN in the chlorination process.	98	The formation of SCN^- in chlorination process may be attributable to the formation of CN^-, which then reacted with sulfur compounds present in the wastewater to form SCN.
MWRA	Yes	Total cyanide increased and thiocyanate decreased in the chlorination process.	88	SCN^- reacts with chlorine to produce CN^-. If chlorine dose is less than equivalent SCN^- concentration, there is no residual chlorine to react with CN^- and free cyanide residual may exist.

[a] Chlor., Chlorination; CFF, Cyanide formation frequency; EBMUD = East Bay Municipal Utility District; GCA = Gulf Coast Waste Disposal Authority; LAGWRP = Glendale Water Reclamation Plant, City of Los Angeles; WPCP = Water Pollution Control Plant, City of Sunnyvale; MWRDGC = Metropolitan Water Reclamation District of Greater Chicago; MWRA = Massachusetts Water Resources Authority.

[b] Based on the available monitoring data.

Thiocyanate (SCN^-), a linear electronegative polyatomic ion, is a cyanogen that is commonly encountered in municipal wastewater (Hung and Pavlostathis, 1997). Thiocyanate is used or generated in various industrial processes such as photo finishing, coke gasification, herbicide and insecticide production, ore mining process, dyeing and electroplating, etc. Wastewaters from those operations can contain thiocyanate at a concentration of 5 mg/L or more (Csikai and Barnard, 1983), and WWTPs receiving these kinds of industrial wastewaters will have thiocyanate in their influents. Another source of SCN^- is from plants of the *Brassicaceae* family such as cauliflower, turnips, and cabbage where SCN^- occurs naturally (Brown and Morra, 1993). Thiocyanate is generally considered far less toxic than free cyanide; that is, HCN and CN^- (APHA/AWWA/WEF, 1998; Boening and Chew, 1999). Most of the total cyanide analytical techniques employed in WWTPs do not measure SCN^- as a cyanide species to be counted in total cyanide for compliance purposes. Additionally, most WWTPs do not routinely monitor SCN^- concentration in the influent or effluent because SCN^- is not usually a regulated pollutant, and it is readily biodegraded in secondary treatment (Luthy and Jones, 1980). Therefore, the fate of SCN^- during WWTPs treatment processes is not well understood, probably due to the complicated cyanide–thiocyanate interactions that have not been adequately studied.

Simultaneous with the WWTP monitoring program in this study, effects of thiocyanate on the formation of cyanide during chlorination and UV disinfection were investigated in a laboratory study with synthetic solutions and WWTP secondary effluents. Results indicated that free cyanide was formed in thiocyanate solutions upon chlorination in which the chlorine dose was not sufficient to destroy SCN^- completely.

This result has been observed in the monitoring events at the Deer Island Treatment Plant, MWRA, which is one of the two WWTPs that intensively monitored SCN^-. The other WWTP, the Hanover Park facility of the MWRDGC, observed a total cyanide plus thiocyanate formation pattern, which further analysis revealed was caused by thiocyanate formation in the chlorination process. This thiocyanate increase was not accompanied by a free cyanide increase, most likely because chlorine dosing was sufficient to destroy any free cyanide produced. It was also found in the laboratory study that thiocyanate can be broken down to yield cyanide by UV irradiation, although this was not observed in this WWTP monitoring program because no participating WWTPs used UV irradiation as a replacement of chlorine disinfection. There is also a possibility that ozonation can convert thiocyanate to cyanide under some conditions (Jensen and Tuan, 1993; Soto et al., 1995).

4.5 Recommendations

Results from this WWTP monitoring program as well as the related laboratory research efforts in the project indicate a positive correlation between cyanide increases and thiocyanate decreases during chlorination. This formation pattern has not been adequately addressed or studied by WWTPs receiving significant thiocyanate concentrations in the influent, probably due to the lack of accurate monitoring data for thiocyanate, which is not routinely measured in most WWTPs in the United States. This may be the main reason why many questions associated with this issue are still unanswered. Therefore, the frequent measurement of thiocyanate (SCN^-) in wastewater influent and effluent is strongly recommended for WWTPs that have encountered the cyanide formation problem in the past or WWTPs that receive industrial discharges containing thiocyanate in the influent.

CHAPTER 5.0

THE EFFECTS OF THIOCYANATE ON FREE CYANIDE FORMATION DURING CHLORINATION AND UV DISINFECTION OF WWTP SECONDARY EFFLUENT

5.1 Introduction

Thiocyanate (SCN^-), a linear, electronegative, polyatomic ion, is a cyanogen compound commonly encountered in municipal wastewater (Hung and Pavlostathis, 1997). Thiocyanate is used or generated in various industrial processes such as photofinishing, coke gasification, herbicide and insecticide production, ore mining process, dyeing and electroplating, etc. Wastewaters from those operations can contain thiocyanate at a concentration of 5 mg/L or more (Csikai and Barnard, 1983) and WWTPs receiving these kinds of industrial wastewaters will have thiocyanate in their influents. Another source of SCN^- is from plants of the *Brassicaceae* family such as cauliflower, turnips, and cabbage where SCN^- is released naturally from the hydrolysis of glucosinolates (thioglucosides) (Newman, 1975; Brown and Morra, 1993).

Thiocyanate is generally considered far less toxic than free cyanide—that is, HCN and CN^- (APHA/AWWA/WEF, 1998; Boening and Chew, 1999). As intended, most of the total cyanide analytical techniques employed in WWTPs do not measure SCN^- as a cyanide species to be counted in total cyanide concentration for compliance purposes. Additionally, most WWTPs do not routinely monitor SCN^- concentration in the influent or effluent because SCN^- is not usually a regulated pollutant.

Although the biological degradation of thiocyanate has been studied extensively (Neufeld and Valiknac, 1979; Luthy and Jones, 1980; Landreth, 1989; Paruchuri et al., 1990; Banerjee, 1996; Hung and Pavlostathis, 1997), these studies primarily focused on the treatment of thiocyanate-laden industrial wastewaters. Overall, the fate of SCN^- in WWTP treatment processes is still poorly understood, especially in oxidative processes such as nitrification, chlorine disinfection, and UV irradiation for disinfection.

No studies have been carried out to investigate the reaction pathways for production of cyanide from thiocyanate in WWTP effluent. This pathway may be in part responsible for the production of free cyanide in WWTPs because the fission of the S-CN group may take place in some WWTP oxidation processes such as chlorination or UV disinfection.

The goal of the study was to address the possibility of elevated cyanide levels in WWTP secondary effluents from reactions involving thiocyanate. A laboratory study was conducted to investigate the potential for cyanide formation in biotreated wastewater containing thiocyanate during chlorination or UV irradiation for disinfection. The overall objective of the study was to gain a better understanding of interactions between cyanide and thiocyanate in WWTPs as well as the potential formation of cyanide during WWTP treatment processes.

5.2 Potential Reaction Processes for Cyanide Formation from Thiocyanate Decomposition

5.2.1 Chlorination

Oxidative decomposition of thiocyanate by chlorine can liberate free cyanide, which may remain in solution if there is no residual chlorine in solution. Chlorine (e.g., in the form of hypochlorite ion, OCl^-) reacts readily with the cyanide anion (CN^-) to form cyanate (CNO^-), based on the following well-known reaction scheme, which has been exploited extensively in industry for the treatment of cyanide-bearing wastewaters (Eilbeck and Mattock, 1987).

$$CN^- + OCl^- \rightarrow CNO^- + Cl^- \qquad (5\text{-}1)$$

In general, thiocyanate is oxidized to sulfate and either cyanide or cyanate depending on the pH of solution (Newman, 1975). Thus, a prime candidate for a chemical compound that can be potentially converted to cyanide upon chlorination in WWTPs is thiocyanate (SCN^-). Generally, the chlorination of thiocyanate is described as leading to the formation of cyanate (CNO^-) or cyanogen chloride (CNCl) depending on the pH (Allen et al., 1948; Newman, 1975).

$$SCN^- + 4\,Cl_2 + 5OH^- \rightarrow SO_4^{2-} + CNO^- + 5H^+ + 8Cl^- \qquad (5\text{-}2)$$

$$SCN^- + 4\,Cl_2 + 4H_2O \rightarrow HSO_4^- + CNCl + 7H^+ + 7Cl^- \qquad (5\text{-}3)$$

It has been suggested that solutions containing thiocyanate can form free hydrogen cyanide in acidic media (U.S. EPA, 1979), though no data in support of this statement have been provided.

5.2.2 UV Irradiation

UV irradiation has gained in popularity as an alternative method of disinfecting wastewater because of the concern over the potential toxicity and carcinogenicity of chlorination byproducts. Additionally, UV irradiation has been applied to the decomposition of metal cyanide complexes in treatment processes (Theis et al., 1997) and in some analytical methods for total cyanide analysis (ASTM, 1998d). Irradiation can readily break down strong metal cyanide complexes, even iron cyanide and cobalt cyanide complexes (Moggi et al., 1966).

Although there is no direct information regarding the potential formation of cyanide during UV irradiation process of wastewater, the possibility warrants investigation because of the potential for UV to decompose cyanide-bearing compounds such as thiocyanate or organocyanide compounds.

5.2.3 Artifacts of the Total Cyanide Analysis Technique

The most commonly used total cyanide analysis technique employs a strong acidic distillation and colorimetric determination (APHA/AWWA/WEF, 1998 – Method 4500-CN-C).

Hydrogen cyanide is liberated from an acidified sample by distillation and purging with air. HCN in the distillate is collected by passing the distillate gas through an NaOH scrubbing solution. Cyanide concentration in the NaOH solution is determined by a colorimetric procedure in which cyanides are converted to cyanogen chloride by reaction with chloramine-T, which subsequently reacts with pyridine and barbituric acid to give a red-colored complex. The intensity of red color is correlated with cyanide concentration and measured by light absorbance at 578 nm. Theoretically, the distillation step can be omitted if only free cyanides (i.e., CN^- and HCN) are present in the sample, as is the case when the standard curve for colorimetric determination of CN^- is developed.

Thiocyanate can cause a significant positive error in the colorimetric determination of cyanide (without distillation) in that thiocyanate, as well as free cyanide, reacts with chloramine-T to produce cyanogen chloride (CNCl) and gives a red-colored complex upon reaction with pyridine-barbituric acid. However, for the total cyanide analysis employing strong acidic distillation, the SCN^- interference can be effectively eliminated if SCN^- is present without oxidizing agents such as nitrate (NO_3^-) because SCN^- is not distilled. It has been the project team's experience that a control sample consisting of a standard 10 mg/L SCN^- solution yielded nondetectable cyanide in the total cyanide analysis. Thiocyanate mixed with nitrate had been reported to be a significant interference in the total cyanide analysis (with distillation) (Csikai and Barnard, 1983). It seems that the highly acidic condition in the analysis procedure for total cyanide in wastewaters favors converting thiocyanate to cyanide if oxidants such as nitrate are present. This analytical artifact is recognized in *Standard Methods* (APHA/AWWA/WEF, 1998 – Method 4500-CN-B).

5.3 Methods

5.3.1 WWTP Wastewater Used in the Study

Wastewater effluents from two WWTPs were used in this study. The primary wastewater used was collected at the postaeration tank (after sand filters) of a secondary treatment plant of the Metropolitan Water Reclamation District of Greater Chicago (MWRDGC), Cicero, Illinois. Two batches of unchlorinated secondary effluent were collected at different times and designated as Effluent A and Effluent B. The other wastewater used was unchlorinated secondary effluent collected from the Allegheny County Sanitary Authority WWTP, Pittsburgh, Pennsylvania, and designated as Effluent C. The characteristics of three WWTP secondary effluents (chemical entities related to cyanide species) are listed in Table 5-1.

5.3.2 Laboratory Studies

A series of laboratory experiments were conducted to investigate the reaction of thiocyanate in WWTP wastewater during chlorination and UV irradiation. Experiments were conducted with solutions prepared with reagent (deionized) water as well as with WWTP secondary effluent, with and without spiking of SCN^-.

Table 5-1. Chemical Characteristics of Three Unchlorinated Secondary Effluents.

Description	Effluent A	Effluent B	Effluent C
Utility / WWTP[a]	MWRDGC	MWRDGC	ALCOSAN
Date sampled	11/29/2000	2/28/2001	9/18/2000
pH	7.01	7.12	6.81
Total CN, mg/L as CN	ND[b]	ND	ND
WAD – CN, mg/L as CN	ND	ND	ND
Free CN, mg/L as CN	ND	ND	ND
SCN, mg/L as CN	ND	ND	ND
CNCl, mg/L as CN	ND	ND	ND
NO_3 – N, mg/L as NO_3	63.5	57.3	NA[c]
NO_2 – N, mg/L as NO_2	ND	ND	NA

[a] MWRDGC = Metropolitan Water Reclamation District of Greater Chicago; ALCOSAN = Allegheny County Sanitary Authority.
[b] ND = Not detected. Detection limits were 0.010 mg/L SCN^- (as CN), 0.002 mg/L CN^- (as CN), and 0.005 mg/L CNCl (as CN), 0.5 mg/L for NO_3-N as NO_3^- and 0.5 mg/L for NO_2-N as NO_2^-.
[c] NA = Not analyzed.

5.3.2.1 Thiocyanate Analytical Procedure

A method for the analysis of thiocyanate was developed, primarily based on a procedure adopted from Meeussen et al. (1989) with some minor changes in colorimetric procedures in which an aldehyde is added first to convert any cyanide to cyanohydrin. 10 milliliters (10.0 mL) of sample solution containing thiocyanate at pH 12.0 was placed in a dried volumetric flask (50 mL). A volume of 0.05 mL of 3.7% formaldehyde solution was added to the flask and mixed thoroughly. The flask was then placed in a 50°C water bath for 0.5 hour and allowed to cool down for 10 minutes before analysis. Acetate buffer (1 mL, pH = 4.5) and 2 mL of 1% chloramine-T solution were added, and the flask was shaken gently. The solution was allowed to stand for exactly two minutes. Five milliliters (5 mL) of pyridine-barbituric acid reagent were then added, the solution was diluted with distilled water to 50 mL, and the flask was stoppered and mixed thoroughly. The resulting solution was allowed to stand for eight minutes. The absorbance at 578 nm was then measured against a reagent blank with a 10-mm cell. Cyanogen chloride must be removed from the sample in situations where pH is not above 12 because it reacts directly with pyridine-barbituric acid to give a red color.

5.3.2.2 Chlorination Study

Chlorination experiments were conducted first in reagent deionized water spiked with different amounts of thiocyanate under various chlorination conditions. Analytical reagent grade chemicals (Fisher Scientific, Pittsburgh, PA) and reagent water (ASTM, 1998g) were used in all operations for the experimental study. The typical thiocyanate concentrations spiked in reagent water were 1 mg/L and 5 mg/L in the form of KSCN. In some cases, a lower concentration (0.1 mg/L) was used to reflect concentrations commonly encountered in WWTPs. In some experiments, nitrate (as $NaNO_3$) was spiked as well. Laboratory chlorination was carried out in 1-liter glass amber bottles (catalog no. 02-911-6, Fisher Scientific, Pittsburgh, PA) without headspace. The appropriate amount of sodium hypochlorite solution was added from a prepared 5 mg/mL stock solution of NaOCl. The pH of the chlorinated water sample was adjusted with 1N HCl or 1N NaOH to the desired pH, 3 or 11. For solutions with targeted pH 6.5, a buffer solution made of potassium dihydrogen phosphate and sodium hydroxide was used. The chlorinated solution was incubated in the dark at ambient temperature for a pre-determined period before analysis. Total cyanide (Method 4500-CN-C), WAD cyanide (Method 4500-CN-I), and cyanogen chloride (Method 4500-CN-J) analyses were performed according to *Standard Methods* (APHA/AWWA/WEF, 1998), whereas free cyanide by microdiffusion analysis followed the ASTM method (ASTM, 1998a). Analysis of thiocyanate was based on the procedure described above. Total residual chlorine (TRC) was measured by the DPD colorimetric method (APHA/AWWA/WEF, 1998 – Method 4500-Cl-G). In cases where TRC was detected, an appropriate amount of dechlorinating agent (either sodium thiosulfate or ascorbic acid) was applied to the solution before analyses based on the recommendation in *Standard Methods* (APHA/AWWA/WEF, 1998 – Method 4500-CN-B).

5.3.2.3 UV Irradiation Study

This part of the study explored the potential for cyanide formation upon UV irradiation of wastewater containing thiocyanate. UV irradiation experiments were conducted with a Rayonet photochemical reactor RPR-100 manufactured by The Southern New England Ultraviolet Company (Branford, CT). A schematic illustration of the reactor is shown in Figure 5-1. The reactor RPR-100 can hold up to 16 UV lamps around the outside rim, and a cylindrical quartz photochemical reaction vessel containing the sample solution was placed into the reactor. The vessel was exposed to UV irradiation for several predetermined periods of time. A ventilation fan inside the reactor prevented heat accumulation. Phosphor-coated low-pressure mercury lamps emitting monochromatic light at 253.7 nm were used in all experiments. In this research, two (2) lamps with a power consumption of 70 watts/lamp were employed. The light intensity at 2" from the lamps was about 16 mw/cm^2.

5.4 Results and Discussion

5.4.1 Thiocyanate Analytical Procedure

The technique to determine thiocyanate is very similar to the method for measuring total cyanide, except for the omission of the acidic high-temperature distillation step. The reason is that thiocyanate, as well as free cyanide, reacts with chloramine-T to produce an intermediate compound (cyanogen chloride) that subsequently reacts with pyridine-barbituric acid to produce a red-colored complex. Because free cyanide interferes with the thiocyanate test, 0.05 mL of 3.7% formaldehyde solution was added to the sample solution to mask the free cyanide (Meeussen et al., 1989). In practice, this dosage of formaldehyde is sufficient to cover concentrations of up to 250 μg/L of cyanide in 10 mL without any effect on the concentration of thiocyanate measured. It should be noted that cyanogen chloride itself, if present in the sample, is a method interference. This interference could be easily eliminated by adjusting the sample pH to 12 so that CNCl hydrolyzes to the cyanate ion, CNO^-, which is a common practice for preserving a sample for total cyanide analysis. This procedure was applied to the determination of thiocyanate by use of several mixed standard solutions containing thiocyanate, free cyanide, and cyanogen chloride. As Table 5-2 shows, accurate results for SCN^- were obtained in all cases, including cases where either CN^- or CNCl was present.

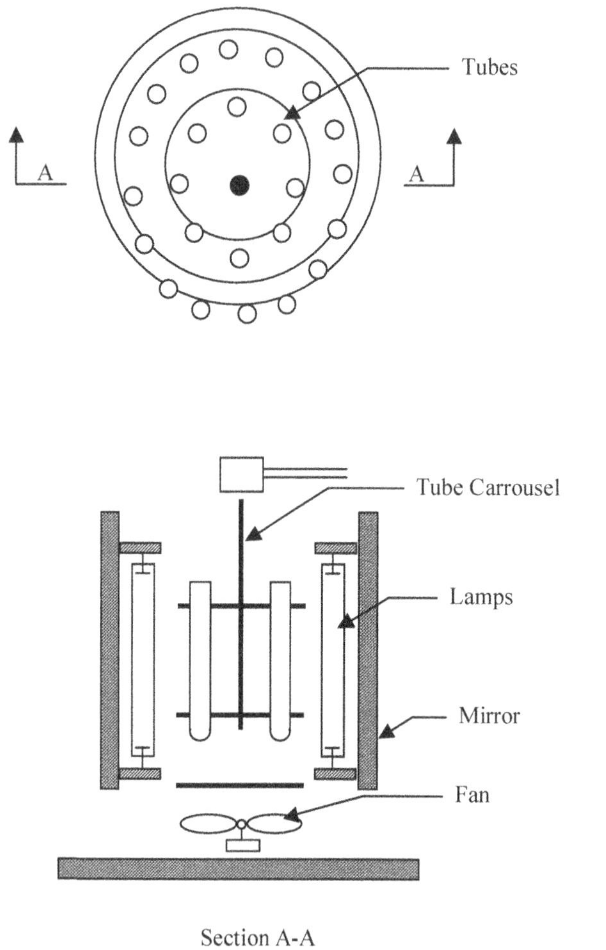

Figure 5-1. Schematic Illustration of the Photochemical Chamber Reactor RPR-100.

Table 5-2. Determination of Thiocyanate in Mixed Standard Solutions.

Exp. no.	Solution composition [a], mg/L as CN^-			SCN^- detected, (mg/L as CN^-)
	SCN^-	CN^-	CNCl	
1	0.020	0.0	0.0	0.020
2	0.050	0.0	0.0	0.050
3	0.100	0.0	0.0	0.099
4	0.200	0.200	0.0	0.202
5	0.100	0.000	0.100	0.104
6	0.0	0.100	0.0	ND [b]

[a] Solutions prepared in reagent (deionized) water with KSCN, NaCN, and preformed CNCl from reaction of NaCN with HOCl.
[b] ND = Not detected. Detection limit was 0.010 mg/L SCN^- (as CN).

5.4.2 Formation of Cyanide from Treatment Processes—Chlorination with Synthetic Thiocyanate Solutions

Figure 5-2 shows that the reaction of thiocyanate with chlorine at pH 3.0 for one hour produced some free cyanide when thiocyanate was not completely destroyed by chlorine. Five sets of the chlorination experiments with an initial SCN^- concentration of 1 mg/L and an OCl^- dosage of 2.5 mg/L (as Cl_2) produced 0.025–0.040 mg/L total cyanide (in the form of free cyanide) as well as 0.2 mg/L cyanogen chloride. Thiocyanate in the range of 0.7-0.8 mg/L remained in solution because of insufficient chlorine dosage. Analyses of thiocyanate and cyanide indicated that the reaction followed the stoichiometric equation listed below to form the major product, CNCl:

$$SCN^- + 4\,OCl^- + H^+ \rightarrow HSO_4^- + CNCl + 3Cl^- \qquad (5\text{-}4)$$

The low yield of CN^- from the SCN^- decomposition indicates that the cyanide ion, CN^-, is probably a reaction intermediate, which, once formed, is almost simultaneously attacked by the strong oxidant chlorine to form cyanogen chloride. The total CN group mass balance (including CN^-, SCN^-, and CNCl) for those experiments at pH 3.0 with initial SCN^- of 1 mg/L and chlorine dosage of 2.5 mg/L and 5.5 mg/L (as Cl_2) were from 80-100%, as presented in Figure 5-3.

The results in Table 5-3 show that the reaction of thiocyanate with chlorine at pH 6.5 also produced 0.02-0.05 mg/L free cyanide when the chlorine dose was not sufficient to destroy the SCN^- completely. For an initial SCN^- concentration of 1 mg/L, chlorination with an OCl^- dosage

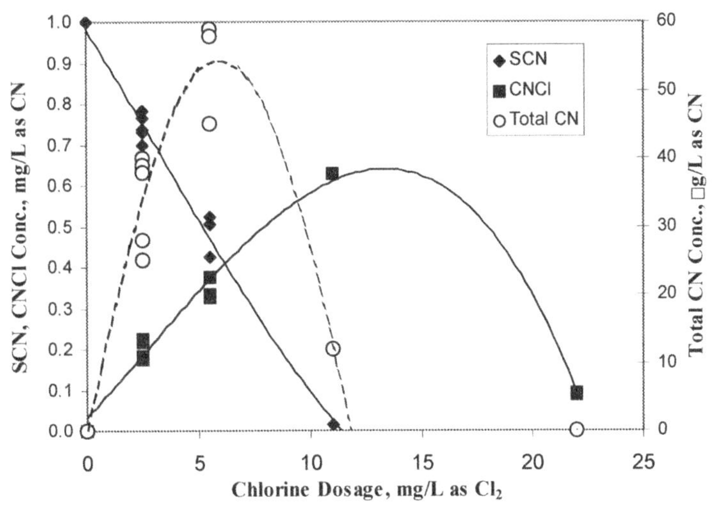

Figure 5-2. Results from Chlorination of 1.0 mg/L Thiocyanate Solution in Reagent Water at pH 3.0. (Chlorine was added as NaOCl. Reaction time is one hour).

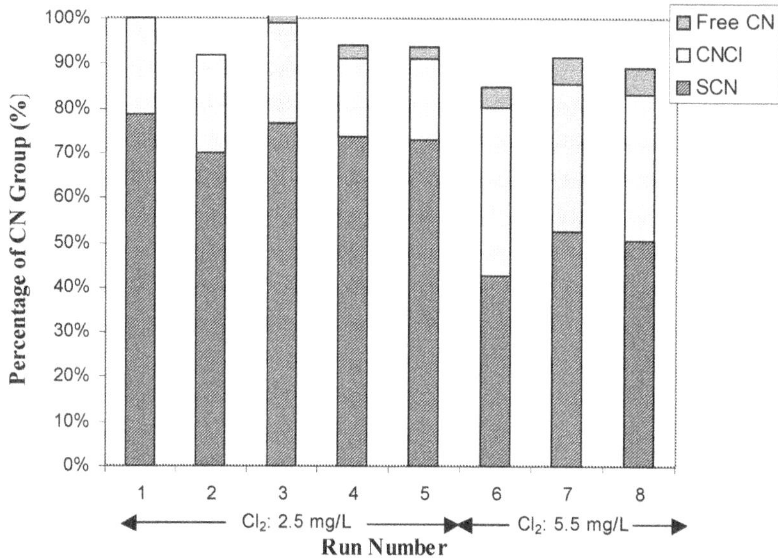

Figure 5-3. Total Mass Balance of CN Group for Chlorination of SCN^- with Initial Concentration of 1 mg/L (as CN) at pH 3.0. The total CN measurement was taken as the measure of free CN in the system.

of 5 mg/L (as Cl_2) resulted in 0.046 mg/L free CN, 0.402 mg/L CNCl, and about 0.512 mg/L of SCN^- remained in solution. The reaction seems to follow the stoichiometric equation as listed below, which is equivalent to Equation 5-5:

$$SCN^- + 4\,OCl^- \rightarrow SO_4^{2-} + CNCl + 3Cl^- \qquad (5\text{-}5)$$

Table 5-3. Results from Chlorination of Thiocyanate Solutions at pH 6.5.

	Initial condition		Chlorination products at time = 1 hr, mg/L as CN^-				
Exp. No.	SCN (mg/L as CN^-)	OCl^- (mg/L as Cl_2)	SCN	CNCl	Free CN[b] (meas. as Total CN)	WAD-CN	Free CN[c]
A1	1.0	5.0	0.512	0.402	0.046	0.027	0.029
A2	1.0	10.0	0.100	0.704	0.067	0.030	0.050
A3	1.0	20.0	ND[a]	0.086	ND	ND	ND
B1	0.1	0.5	0.049	0.030	0.022	0.019	0.022
B2	0.1	1.0	0.011	0.080	0.023	0.019	0.021
B3	0.1	2.5	ND	0.067	ND	ND	ND
B4	0.1	5.0	ND	0.053	ND	ND	ND
B5	0.1	10.0	ND	ND	ND	ND	ND

[a] ND = Not detected. Detection limits were 0.010 mg/L SCN^- (as CN), 0.002 mg/L CN^- (as CN), 0.002 mg/L WAD-CN (as CN), 0.002 mg/L Free CN (as CN), and 0.005 mg/L CNCl (as CN)
[b] Free cyanide was the sole cyanide form of the total cyanide produced by the chlorination, as measured in the total cyanide analytical method
[c] Free cyanide was measured using microdiffusion method, a species-specific analytical method

Table 5-4. Results from Chlorination of Thiocyanate Solutions at pH 11.0.

Exp. no.	Initial condition		Chlorination products at time = 1 hr, mg/L as CN^-				
	SCN (mg/L as CN^-)	OCl^- (mg/L as Cl_2)	SCN	CNCl	Free CN[b] meas. as Total CN	WAD-CN	Free CN[c]
C1	1.0	5.0	0.556	ND[a]	0.053	0.003	NA[d]
C2	1.0	5.0	0.539	ND	0.058	0.003	ND
C3	1.0	15.0	0.037	ND	ND	NA	NA
C4	1.0	20.0	0.025	ND	ND	NA	NA
C5	1.0	20.0	ND	ND	ND	ND	ND

[a] ND = Not detected. Detection limits were 0.010 mg/L SCN^- (as CN), 0.002 mg/L CN^- (as CN), 0.002 mg/L WAD-CN (as CN), 0.002 mg/L Free CN (as CN), and 0.005 mg/L CNCl (as CN)
[b] Free cyanide was the sole cyanide form of the total cyanide produced by the chlorination, as measured in the total cyanide analytical method
[c] Free cyanide was measured using microdiffusion method, a species-specific analytical method.
[d] NA = Not analyzed.

As shown in Table 5-4, the reaction of thiocyanate with chlorine at pH 11.0 also produced some free CN when thiocyanate was not completely consumed by chlorine. For an initial SCN^- concentration of 1 mg/L, chlorination with an equivalent Cl_2 dosage of 5 mg/L produced 0.06 mg/L free CN and some 0.55 mg/L of SCN^- remained in solution. The proposed reaction is given as Equation 5-6, in which the production of cyanate ion (CNO^-) is indicated.

$$SCN^- + 4\ OCl^- + 2\ OH^- \rightarrow SO_4^{2-} + CNO^- + 4Cl^- + H_2O \qquad (5\text{-}6)$$

Because of the difficulty for analyzing cyanate ion, no CNO^- was measured in the experiments. CNO^- is not captured by any analytical methods used for cyanide-related compounds in this work. From the SCN^- data, however, it seems that the results fit the equation very well. Theoretically, 5 mg/L of Cl_2 will react with 0.458 mg/L of SCN^- and leave 0.542 mg/L of SCN^- unreacted, which is very close to 0.547 mg/L detected experimentally. However, Equation 5.4.2-3 is not of great importance because the pH value is outside the range encountered in WWTPs.

The conclusion from these tests is that total cyanide is always detected in the thiocyanate solution under conditions in which the chlorine dose or the reaction time is not sufficient to destroy SCN^- completely. As noted earlier, residual chlorine will easily destroy free cyanide.

5.4.3 Chlorination with Thiocyanate—Spiked WWTP Effluents

To examine effects of the complexity of wastewater matrices, chlorination studies were also performed with several unchlorinated WWTP secondary effluents spiked with thiocyanate under various chlorination conditions.

It was found that cyanide was detected in wastewater solutions for which the chlorine dose was not sufficient to destroy the SCN^- completely (see Table 5-5).

Although in the clean water study it was found that cyanide formed directly from the decomposition of thiocyanate, in the study with wastewater effluent, some of the apparent cyanide formation may be attributed to artifacts of the total cyanide analysis in which reaction between thiocyanate and the nitrate ion present in the effluent at about 60 mg/L would be favored by the high-temperature acidic conditions (Csikai and Barnard, 1983). In the cases where total cyanide was detected, the mass balance of all cyanide groups were from 84-152%, with a mean of 116%. It is not clear whether the reaction between hypochlorite and thiocyanate under the reaction conditions or the reaction between nitrate and thiocyanate in the analysis dominated the results.

5.4.4 UV Irradiation with Synthetic Thiocyanate Solutions

Results of UV irradiation with thiocyanate-spiked synthetic solutions at different pH values are listed in Table 5-6. At pH of 6.5 and irradiation time of 10 min, 0.065 mg/L free cyanide (measured as total cyanide) was formed when initial $[SCN^-]$ = 1 mg/L, and 0.005 mg/L free cyanide (measured as total cyanide) was formed when initial $[SCN^-]$ = 0.1 mg/L. These values increased under test conditions at pH 3.0 and the same irradiation time of 10 min, with free cyanide (measured as total cyanide) concentration = 0.081 mg/L when initial SCN^- = 1 mg/L, and 0.007 mg/L when initial SCN^- = 0.1 mg/L. These results indicated that thiocyanate is broken down to yield free cyanide by UV irradiation, and that lower pH values favor the conversion of thiocyanate to cyanide. However, it is apparent that the conversion is not on an equimolar basis because the amount of free cyanide (in molar) formed was less than that of the initial thiocyanate, probably because of the relatively low UV intensity employed.

Table 5-5. Results from Chlorination of WWTP Effluents Spiked with SCN^-.

Exp. no.	WWTP effluent[a]	Initial condition		Chlorination products at time = 1 hr, mg/L as CN^-				
		SCN Spiked (mg/L as CN^-)	OCl^- (mg/L as Cl_2)	SCN	CNCl	Total CN	WAD-CN	Free CN
D1	A	0.1	0.0	0.104	ND	0.048	NA[b]	ND[c]
D2	A	0.1	0.5	0.064	0.023	0.035	NA	0.008
D3	A	0.1	1.0	0.035	0.055	0.042	NA	0.012
D4	A	0.1	2.0	0.022	0.076	ND	NA	ND
D5	A	0.1	4.0	0.021	0.065	ND	NA	ND
D6	B	0.1	0.0	0.102	ND	0.006	NA	NA
D7	B	0.1	0.5	0.075	0.024	0.002	NA	NA
D8	B	0.1	1.0	0.049	0.045	0.002	NA	NA
D9	B	0.1	2.0	0.031	0.070	ND	NA	NA
D10	B	0.1	4.0	0.016	0.023	ND	NA	NA
D11	C	0.1	0.0	0.080	ND	0.004	ND	NA
D12	C	0.1	0.5	0.068	0.018	0.020	0.008	NA
D13	C	0.1	1.0	0.047	0.032	0.033	0.016	NA
D14	C	0.1	2.5	ND	0.101	ND	0.003	NA
D15	C	0.121	0.0	0.121	ND	0.047	ND	ND
D16	C	0.121	0.5	0.083	0.034	0.027	0.007	0.008
D17	C	0.121	1.0	0.051	0.060	0.041	0.007	0.008
D18	C	0.121	2.5	ND	0.131	ND	0.002	0.002

[a] A = MWRDGC Effluent; B = MWRDGC Effluent; C = ALCOSAN Effluent, see Table 5-1.
[b] NA = Not analyzed.
[c] ND = Not detected. Detection limits were 0.010 mg/L SCN^- (as CN), 0.002 mg/L CN^- (as CN), 0.002 mg/L WAD-CN (as CN), 0.002 mg/L Free CN (as CN) and 0.005 mg/L CNCl (as CN).

Table 5-6. Results from UV Irradiation of Reagent Water Spiked with SCN^-.

	Initial condition at time = 0		UV irradiation products, mg/L as CN^-			
Exp. no.	SCN spiked (mg/L as CN^-)	pH	Time (min)	SCN	Total CN	WAD-CN
G1	1.00	3.0	5	0.989	0.045	0.046
G2	1.00	3.0	10	0.966	0.081	0.081
G3	0.10	3.0	5	0.090	0.002	0.002
G4	0.10	3.0	10	0.090	0.006	0.007
G5	0.10	3.0	15	0.090	0.010	0.011
G6	1.00	6.5	5	0.983	0.036	0.037
G7	1.00	6.5	10	0.955	0.064	0.065
G8	1.00	6.5	15	0.913	0.093	0.095
G9	0.10	6.5	5	0.092	ND[a]	ND
G10	0.10	6.5	10	0.090	0.004	0.005
G11	0.10	6.5	15	0.085	0.007	0.007

[a] ND = Not detected. Detection limits were 0.002 mg/L CN^- (as CN) and 0.002 mg/L WAD-CN (as CN).

5.4.5 UV Irradiation with Thiocyanate-Spiked WWTP Effluents

UV irradiation experiments were also performed with some thiocyanate-spiked WWTP effluents and results are presented in Table 5-7. Neither cyanide nor thiocyanate was detected in the original effluent sample (see Table 5-1). Total cyanide was detected upon UV irradiation of thiocyanate-spiked WWTP effluents. The comparison between the studies with synthetic thiocyanate solutions and thiocyanate-spiked WWTP effluents revealed that the percentage of thiocyanate conversion was much greater in WWTP effluent (about 12.3% for irradiation time of 10 min at pH 6.9) than in the reagent water matrix (about 6.4% for irradiation time of 10 minutes at pH 6.5). A reasonable explanation is that thiocyanate is also reacting with nitrate in the WWTP effluent, either during the UV irradiation process or during the total cyanide analytical process, to form cyanide (Csikai and Barnard, 1983).

Table 5-7. Results from UV Irradiation of WWTP Effluent C spiked with SCN^-.

	Initial condition at time = 0		UV irradiation products, mg/L as CN^-			
Exp. no.	SCN Spiked (mg/L as CN^-)	pH	Time min	SCN	Total CN	WAD-CN
H1	0.101	4.6	5	0.101	0.009	0.005
H2	0.101	4.6	10	0.090	0.019	0.014
H3	0.101	4.6	15	0.090	0.022	0.019
H4	0.106	6.9	5	0.106	0.008	0.004
H5	0.106	6.9	10	0.101	0.013	0.009
H6	0.106	6.9	15	0.099	0.016	0.013
H7	0.108	7.7	5	0.108	0.009	0.004
H8	0.108	7.7	10	0.101	0.011	0.008
H9	0.108	7.7	15	0.097	0.017	0.012

5.5 Formation of Cyanide in Total Cyanide Analysis Techniques

5.5.1 Synthetic Solutions Spiked with Thiocyanate and Nitrate

Experiments were performed to investigate the effects from the presence of nitrate in SCN^- solutions analyzed by conventional distillation methods. Results are given in Table 5-8. Table 5-8 shows that total cyanide was detected in the solutions containing SCN^- at a concentration of either 0.1 mg/L or 1 mg/L, mixed with NO_3^- at concentrations of 1.0-100 mg/L. For a solution consisting of 0.1 mg/L of SCN^- and 100 mg/L of NO_3^-, the formed free cyanide concentration could be as high as 0.086 mg/L. A typical range for nitrate in wastewater effluents is 15-20 mg/L as N (i.e., 66-89 mg/L as NO_3^-) (Metcalf and Eddy, Inc., 1991).

It should be noted that none of the solutions exhibited detectable WAD cyanide, which is measured in almost the same way as the total cyanide method except for a mildly acidic distillation at pH around 4.5. Therefore, it can be concluded that only the highly acidic condition favors the formation of cyanide from solutions containing thiocyanate and nitrate ion (i.e., this apparent interference only exists in the total cyanide analysis which is the most commonly used cyanide analytical method in WWTPs). Although this interference is addressed in *Standard Methods* (APHA/AWWA/WEF, 1998), no means to avoid the interference are discussed.

5.5.2 Thiocyanate-Spiked WWTP Effluents

To examine the effects of thiocyanate interference in total cyanide analysis of WWTP secondary effluent samples, the experiment was also performed in WWTP effluents A and B (see Table 5-9).

Table 5-9 shows that total cyanide was detected in the nitrate-bearing WWTP effluents spiked with SCN^-. For a WWTP effluent containing 0.1 mg/L of spiked SCN^- and 63.5 mg/L of background NO_3^-, total cyanide was formed at a concentration as high as 0.1 mg/L. This apparent interference could not be eliminated by adding sulfamic acid, a chemical reagent effective for removing NO_2^- interference in samples for total cyanide analysis.

Table 5-8. Thiocyanate Interference in Cyanide Analysis When Nitrate is Present in Reagent Water Matrix.

Exp. no.	Solution composition		Total CN analysis (mg/L as CN^-)	WAD-CN analysis (mg/L as CN^-)
	SCN^- (mg/L as CN^-)	NO_3^- (mg/L)		
E1	0.100	1.0	ND[a]	ND
E2	0.100	10.0	0.010	ND
E3	0.100	25.0	0.017	ND
E4	0.100	50.0	0.060	ND
E5	0.100	100	0.086	ND
E6	1.00	10.0	0.009	ND
E7	1.00	50.0	0.038	ND

[a] ND = Not detected. Detection limits were 0.002 mg/L CN^- (as CN) and 0.002 mg/L WAD-CN (as CN).

Table 5-9: Thiocyanate Interference in Total Cyanide Analysis of WWTP Effluent Samples When Nitrate is Present.

Exp. no.	WWTP effluent[a]	Solution composition				Total CN detected (mg/L as CN^-)
		Initial SCN^- (mg/L as CN^-)	Spiked SCN^- (mg/L as CN^-)	Initial NO_3^- (mg/L)	Spiked NO_3^- (mg/L)	
F1	A	ND[b]	0.100	63.5	0.0	0.100
F2	A	ND	0.100	63.5	50.3	0.096
F3	B	ND	0.100	57.3	0.0	0.025
F4	B	ND	0.100	57.3	50.0	0.060

[a] A and B were MWRDGC effluents, see Table 5-1 .

[b] ND = Not detected. Detection limit was 0.010 mg/L SCN^- (as CN).

5.6 Summary

Apparent formation of low concentrations (5-10 μg/L) of cyanide has been observed in WWTP treatment plants, in particular during chlorination. This work identifies thiocyanate (SCN^-) as a potential precursor to cyanide formation. A series of experiments with synthetic solutions and WWTP secondary effluents investigated pathways through which thiocyanate (SCN^-) can be converted to free cyanide (CN^-). Cyanide was detected in synthetic thiocyanate solutions undergoing incomplete oxidative destruction by chlorine because of nonstoichiometric chlorine dosing at all pH values tested (3.0, 6.5, and 11.0). For a SCN^- concentration of 1 mg/L, chlorination with OCl^- resulted in free CN at concentrations of 0.010-0.060 mg/L. Experiments performed with WWTP secondary effluents spiked with SCN^- supported this observation. Free residual chlorine can easily destroy free cyanide in the solution, so if chlorine dosing is more than the stoichiometric equivalent of SCN^-, it will destroy all SCN^- and any free cyanide liberated by its reaction with SCN^-. Thus, there is potential to avoid cyanide formation via thiocyanate by increasing the chlorine dose, without significantly increase the formation of disinfection byproducts.

The potential for UV irradiation to liberate free cyanide from SCN^- was also investigated. It was found that thiocyanate can indeed be decomposed by UV irradiation. Free cyanide (measured as total cyanide) was formed at concentrations of 0.065-0.081 mg/L upon UV irradiation of 10 min at a pH of either 6.5 or 3.0.

Thiocyanate can also cause a significant positive error in the conventional total cyanide analytical technique employing acidic distillation and spectrophotometric finish when nitrate is present in the sample. Thiocyanate present in a sample may bias the total cyanide test if nitrate is present.

CHAPTER 6.0

FORMATION OF FREE CYANIDE AND CYANOGEN CHLORIDE FROM CHLORINATION OF WWTP SECONDARY EFFLUENT CONTAINING RESIDUAL AMMONIA

6.1 Introduction

Because of widespread concern in the United States over the presence of organic disinfection by-products (DBPs) in chlorinated waters, many drinking water facilities are turning to chloramination and other alternative disinfection technologies as a total or partial replacement for chlorine (McGuire, 1989). Chloramines are also of interest in wastewater treatment because they can be produced in chlorination by hypochlorous acid reacting with ammonia and its complex derivatives in accordance with the following reactions:

$$NH_3 + HOCl \leftrightarrows H_2O + NH_2Cl \text{ (monochloramine)} \qquad (6\text{-}1)$$

$$NH_2Cl + HOCl \leftrightarrows H_2O + NHCl_2 \text{ (dichloramine)} \qquad (6\text{-}2)$$

$$NHCl_2 + HOCl \leftrightarrows H_2O + NCl_3 \text{ (nitrogen trichloride)} \qquad (6\text{-}3)$$

These reactions are governed primarily by pH and chlorine-to-nitrogen weight ratios. At pH levels commonly present at drinking water and WWTPs (7.0-9.5), monochloramine is the predominant species (Meyers, 1998). Because ammonia is commonly present in raw wastewater, and complete nitrification does not usually occur in WWTP biotreatment processes, the ammonia remaining upon nitrification can react with hypochlorite introduced during chlorination to produce a significant amount of chloramine.

One concern over the use of chloramines for disinfecting drinking water is the current level of uncertainty in the nature and health impacts of chloramine DBPs, among which cyanogen chloride (CNCl) has been identified and studied intensively (Krasner et al., 1989; Ferguson et al., 1990; Scully, 1990; Krasner et al., 1991; Pedersen et al., 1995). Cyanogen chloride, a highly toxic cyanide species, is often observed in chloraminated drinking waters at concentrations of several micrograms per liter (Krasner et al., 1989). Cyanogen chloride requires species-specific measurement and is not detected in conventional cyanide analysis (APHA/AWWA/WEF, 1998; ASTM, 1998a,b; U.S. EPA, 1983). In addition, cyanogen bromide (CNBr), another toxic cyanogen halide, may also be formed in the chloramination of waters containing bromine and/or bromide (Diehl et al., 2000).

The fate of CNCl in water is rather complicated. Its significance as a cyanide source depends highly on the reaction conditions such as pH and time. Generally, cyanogen chloride is an unstable compound whose half-life in water ranges from 1.75 min to 10 hr at 5°C (Boening and Chew, 1999). CNCl is always the first-step product of the chlorination of cyanide, which occurs instantaneously at all pH values (Gerritsen and Margerum, 1990).

$$CN^- + Cl_2 \rightarrow CNCl + Cl^- \quad (6\text{-}4)$$

In alkaline solutions with pH > 8.5, CNCl may continuously undergo hydrolysis to form cyanate (CNO^-). The rate of hydrolysis is accelerated by a slight excess of chlorine (Eden et al., 1950a; Chamberlin and Snyder, 1955).

$$CNCl + 2OH^- \rightarrow CNO^- + Cl^- + H_2O \quad (6\text{-}5)$$

Cyanate will undergo further hydrolysis and decomposes to HCO_3^- and NH_3 at neutral pH. With no excess chlorine present, CNO^- can persist for 24 hours at pH 9 (Pettet and Ware, 1955). When large residual chlorine is available, it may be further converted to nitrogen gas and bicarbonate (Eden et al., 1950b).

$$2CNO^- + 3OCl^- + 2OH^- \rightarrow N_2 + 3Cl^- + 2CO_3^{2-} + H_2O \quad (6\text{-}6)$$

No studies of the occurrence of cyanogen chloride in chlorinated WWTP effluent have been reported. Also, there are very few data pertaining to the potential transformation of CNCl to CN^- (Lewis, 1992).

The goal of this study was to address the possibility of formation of cyanogen chloride or cyanide from chloramination upon chlorination of WWTP secondary effluent containing residual ammonia.

6.2 Methods

6.2.1 WWTP Wastewater Used in the Study

This study used wastewater effluents from two WWTPs. The primary wastewater used was collected at the postaeration tank (after sand filters) of a secondary treatment plant of the Metropolitan Water Reclamation District of Greater Chicago in Cicero, Illinois. Two batches of unchlorinated secondary effluent were collected at different times and designated as Effluent A and Effluent B. The other wastewater used was unchlorinated secondary effluent collected from the Allegheny County Sanitary Authority WWTP (Pittsburgh, Pennsylvania), and designated as Effluent C. These wastewaters exhibited pH of 7.0 and 7.1, nondetectable nitrite, and nitrate concentrations of 63.5 and 57.3 mg/L. Additional characteristics of the three WWTP secondary effluents are provided in Table 5-1.

6.2.2 Laboratory Studies

Laboratory studies were conducted to investigate the formation of cyanogen chloride or cyanide in the chlorination of WWTP secondary effluent containing residual ammonia. Experiments were performed first on synthetic aqueous solutions containing some model organic compounds aimed at identifying cyanide as a potential product of chloramination. To assess the effects of chloramination in wastewater secondary effluent, chlorination experiments were conducted with some unchlorinated WWTP effluents.

6.2.2.1 Model Organic Compounds

Benzene, L-serine, catechin, and humic acid were chosen as representative of typical potentially reactive organics in WWTPs with serine as an amino acid, humic acid as a humic substance, benzene as a hydrocarbon, and catechin as a plant pigment. Structures for these compounds are given in Figure 6-1. All chemicals were reagent grade or better and were obtained from Fisher Scientific, Inc. (Pittsburgh, PA), Sigma Chemical Co. (St. Louis, MO), and Acros Organics (Fairlawn, NJ).

In the chloramination study, 0.1-mM solutions of the pure organic compounds were used to increase the detection probability of any cyanide or cyanogen chloride formed. All chloramination experiments were done in batch, in pH-buffered solutions (pH = 6.0 ± 0.5, phosphate buffer), with dosing of 3 mM of NaOCl and 2 mM of NH_4Cl (approximately 3:2 on a molar basis) to produce monochloramine almost entirely. Separate reaction batches were set up for samples corresponding to designated reaction times up to 125 hours. The batches of chloraminated solutions were kept in a constant temperature bath (25.0 ± 1.0°C) in the dark before analysis. At the end of the designated reaction time, the sample was removed and split, and half of the solution was treated with sodium thiosulfate with a dosage of 0.1 g/L for dechlorination as recommended in *Standard Methods* (APHA/AWWA/WEF, 1998 – Method 4500-CN-B). In some cases, ascorbic acid was also used as a dechlorinating agent for comparison with thiosulfate. The solutions were then analyzed for total cyanide, free cyanide, and cyanogen chloride.

6.2.2.2 Analytical Methods

Cyanogen chloride was measured by the *Standard Methods*, Method 4500-CN-J (APHA/AWWA/WEF, 1998). Total cyanide analysis was performed using *Standard Methods,* Method 4500-CN-C (APHA/AWWA/WEF, 1998) and free cyanide by microdiffusion was conducted in accordance with an ASTM method (ASTM, 1998a). Monochloramine, dichloramine, and total residual chlorine were measured using *Standard Methods*, Method 4500-Cl-G (APHA/AWWA/WEF, 1998).

6.3 Results and Discussion

6.3.1. Chloramine Formation from Chlorination of Ammonia-Spiked Solutions

In all experiments, ammonia was added in the form of ammonium chloride (NH_4Cl) and chlorine was added in the form of sodium hypochlorite (NaOCl). When dosing 3 mM of NaOCl and 2 mM of NH_4Cl to a 1-L glass bottle with reagent (deionized) water, approximately 124 mg/L (as Cl_2) monochloramine was measured. The total residual chlorine was measured at 165 mg/L (as Cl_2), which means monochloramine was the predominant species in the chlorination of ammonia-spiked solutions in this study. The other chlorine species formed was dichloramine as confirmed by measurement.

6.3.2 Formation of CNCl from Chloramination

6.3.2.1 Synthetic Solutions

In experiments with synthetic solutions of some model organics, cyanogen chloride was formed in significant concentrations by chloramination as indicated in Figure 6-1. It was found that cyanogen chloride was produced in the range of 12.6-241 μg/L after 24 hours of reaction for 0.1-mM model organic solutions, which were chloraminated but not dechlorinated. Among the four organics tested, L-serine exhibited the largest CNCl yield. This may be attributed to its simple structure and the lack of a benzene ring. Benzene, catechin, and humic acid appear to be resistant to breakdown by reaction with chloramines compared to L-serine. The yield of CNCl from the chloramination of catechin was the lowest of the four compounds studied. With the exception of catechin, which gave maximum CNCl yield at six hours, the CNCl yield reached a maximum at the reaction time between 24-72 hours, varying from one compound to the other.

From Figure 6-1, it is clear that the dechlorination of the sample solutions with sodium thiosulfate ($Na_2S_2O_3$) can destroy CNCl effectively. The final concentrations of CNCl after dechlorination were found to be very low (4-6 μg/L).

6.3.2.2 Effects of Chloramine Dosage

Effects of chloramine dosage on the production of CNCl were also examined. Figure 6-2 depicts the formation of CNCl from the chloramination of 0.1 mM of L-serine with two different chloramine doses, one with 3 mM of NaOCl and 2 mM of NH_4Cl (giving 2 mM of chloramine) and the other with 0.3 mM of NaOCl and 0.2 mM of NH_4Cl (giving 0.2 mM chloramine). As Figure 6-2 shows, the production of CNCl was approximately proportional to the amount of chloramine added to the solution given an initial concentration of organic compound greater than the chloramine dose.

6.3.2.3 WWTP Effluents

WWTP wastewaters contain numerous organic compounds including proteins (40-60% by weight), carbohydrates (25-50%), and fats and oils (10%) (Metcalf and Eddy, Inc., 1991). To assess the effects of chloramination in actual wastewater, experiments were conducted with WWTP secondary effluent B. Results from a 12-day chloramination study are presented in Figure 6-3. With the addition of 3 mM of NaOCl and 2 mM of NH_4Cl (giving 2 mM of chloramine), the WWTP effluent exhibited a cyanogen chloride maximum of 25 μg/L as CN^- after 96-120 hours of chloramination. The first detected CNCl formation was at 24 hours with a concentration of 10 μg/L.

Thus, cyanogen chloride can be formed in WWTP secondary effluent upon formation of chloramine. This implies that precursors of CNCl can exist in the WWTP secondary effluent, even with its reduced organic content from biotreatment.

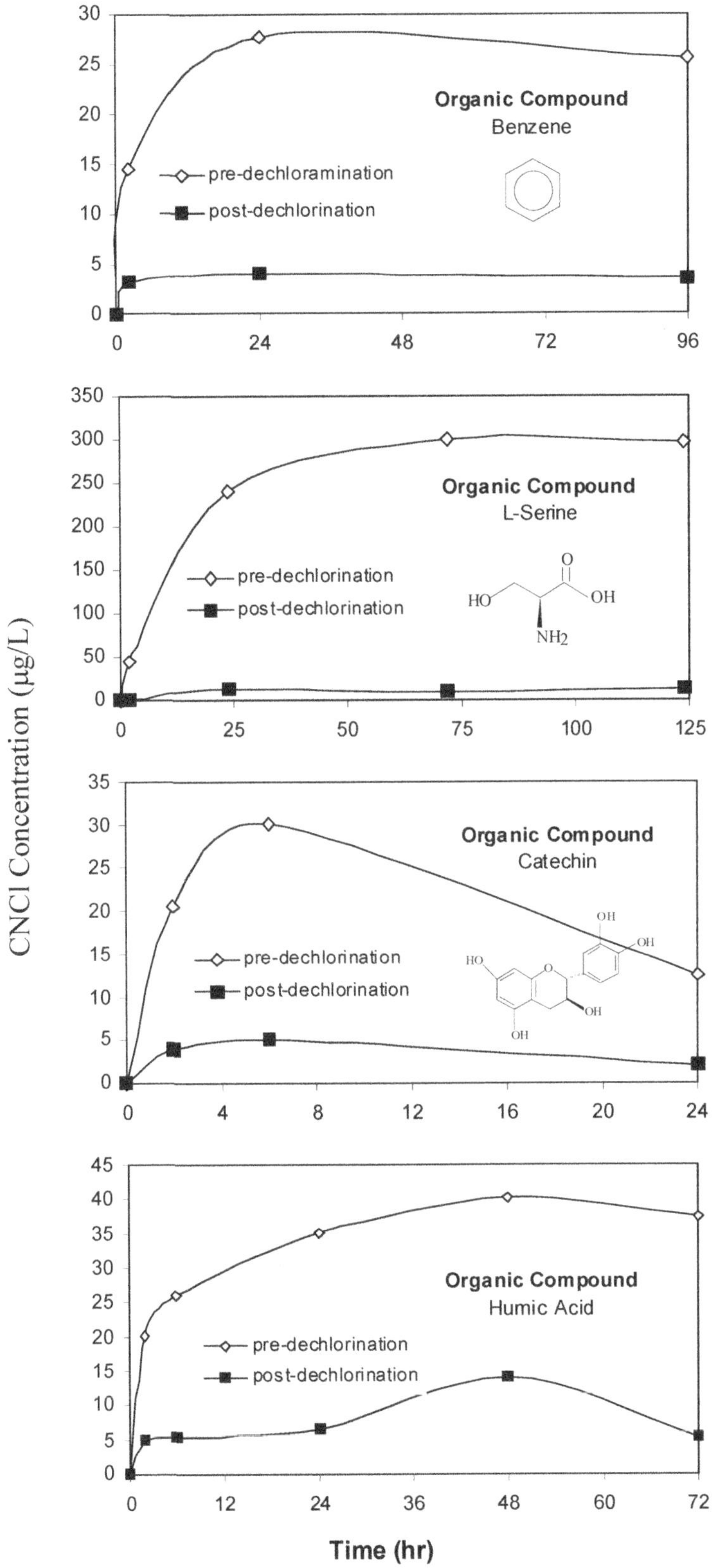

Figure 6-1. CNCl Formation from Chloramination (3 mM NaOCl + 2 mM NH_4Cl) of Organic Compounds (0.1 mM for each) at pH = 6.0 ± 0.5.

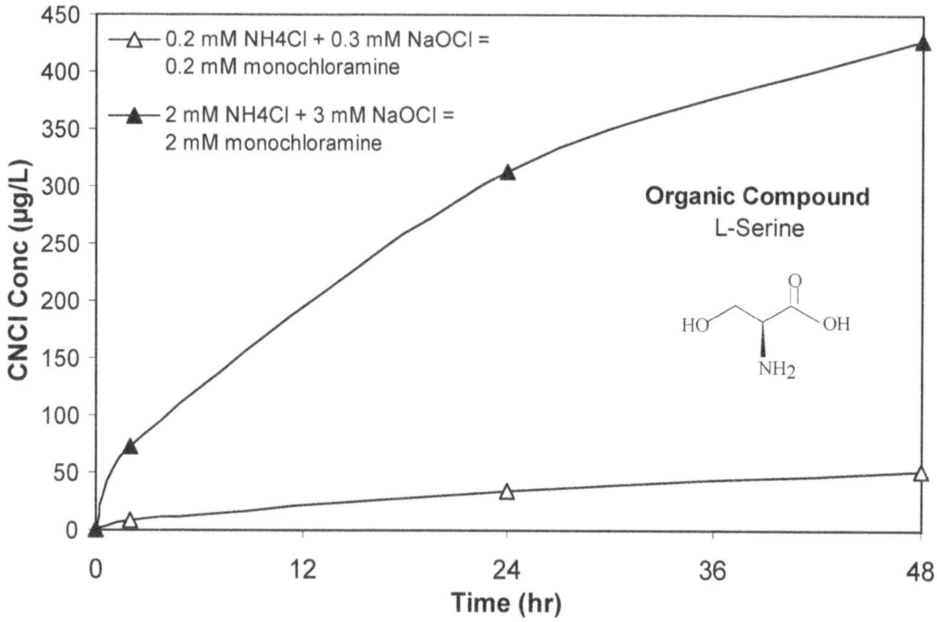

Figure 6-2. Effects of Chloramine Doses on the Formation of CNCl in Solution of 0.1 mM L-Serine at pH = 6.0 ± 0.5.

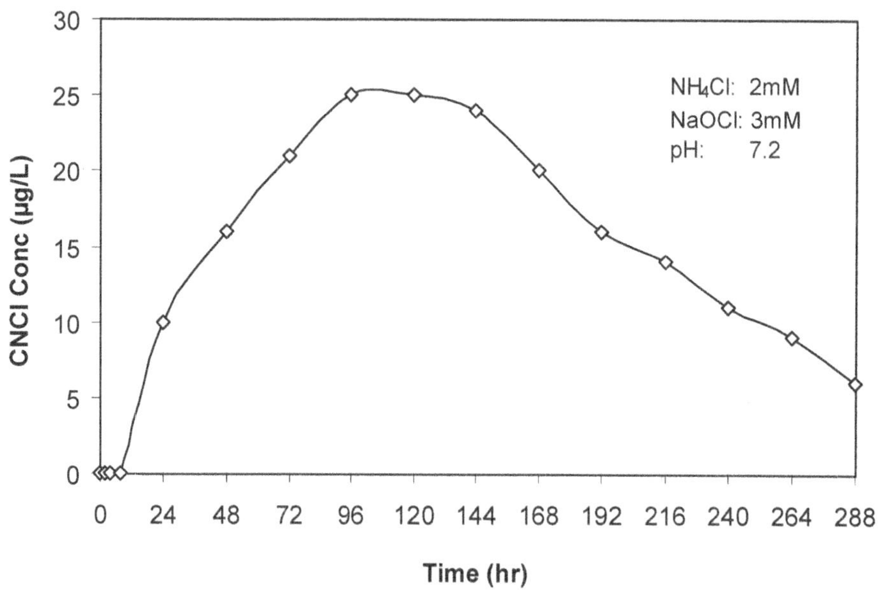

Figure 6-3. CNCl Formation from Chloramination of WWTP Effluent B.

6.3.2.4 WWTP Effluents Spiked with Model Organics

To examine effects of reactive organic compounds in the wastewater matrix on the formation of CNCl, the chloramination study was also performed with WWTP effluents spiked with 0.1 mM of L-serine. Figure 6-4 compares the CNCl production from chloramination of three WWTP effluents spiked with L-serine after correction by subtraction of CNCl concentration produced from raw effluents. The data show that different wastewater matrices yielded different amounts of CNCl formation from reaction of L-serine. Even the effluents from the same WWTP (Effluent A and B) sampled at different times yielded different amounts of CNCl formation. Results of this experiment indicate that CNCl formation is dependent on the specific organics in WWTP secondary effluents.

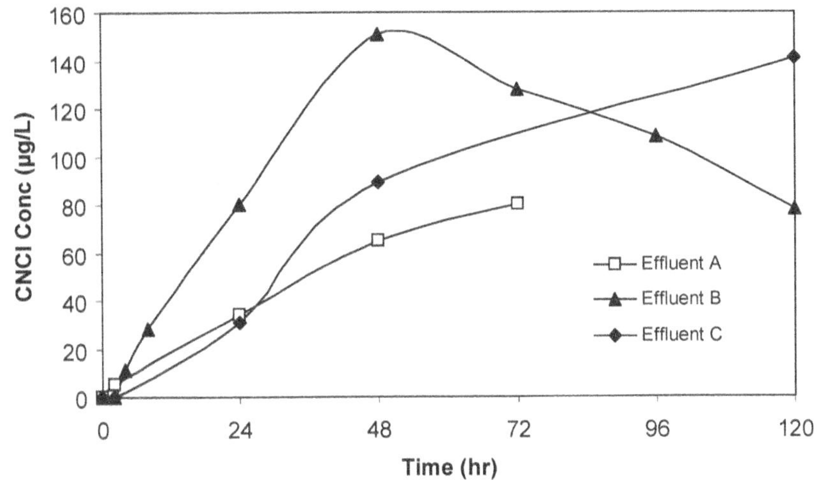

Figure 6-4. CNCl Formation from Chloramination (3 mM NaOCl + 2 mM NH_4Cl) of WWTP Secondary Effluents Spiked with 0.1 mM L-Serine. Data are Corrected for CNCl Production from Chloramination of Unspiked Effluent.

6.3.2.5 Comparison of Chloramination in Different Water Matrices and Effects of Dechlorinating Agents

CNCl was formed in much greater concentrations by chloramination of L-serine in wastewater than in deionized water as illustrated in Figure 6-5. Cyanogen chloride was produced at 34 µg/L after 24 hours of reaction for wastewater samples that were chloraminated but not dechlorinated, at 65 µg/L after 48 hours and at 80 µg/L after 72 hours of reaction. The results for the synthetic solutions in reagent water showed CNCl yield was less, giving 22 µg/L at 72 hours. For the dechlorination agents, thiocyanate not only effectively destroyed residual chlorine, but also removed cyanogen chloride, giving nondetectable CNCl for all reaction for both the WWTP wastewater and for the reagent water solutions. This indicates that $Na_2S_2O_3$ is not an appropriate dechlorinating agent for samples scheduled for CNCl analysis. It was also found that ascorbic acid did not react with CNCl. Thus, ascorbic acid is a preferred dechlorinating agent for samples to be analyzed for CNCl.

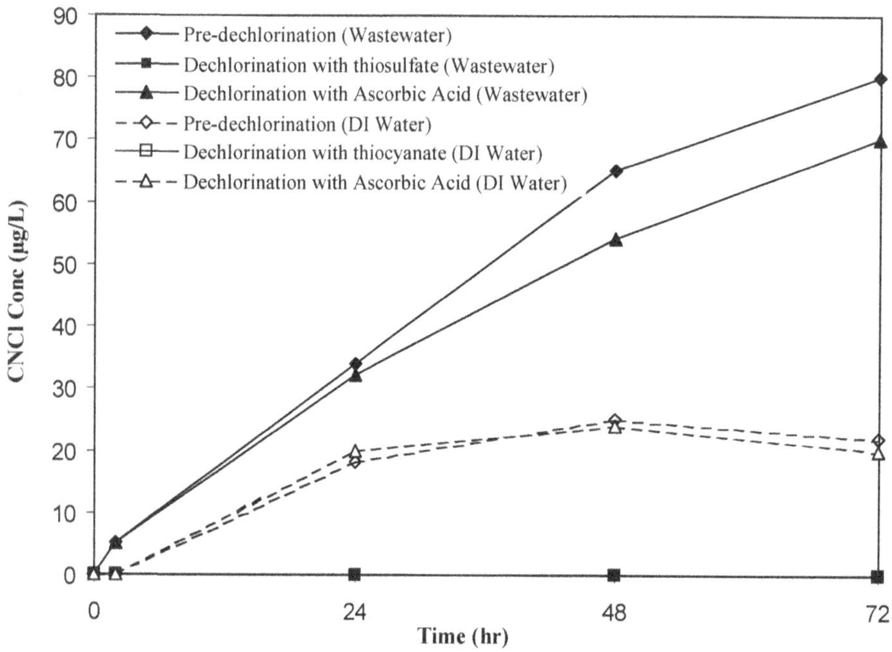

Figure 6-5. Comparison of Chloramination (3 mM NaOCl + 2 mM NH_4Cl) of 0.1 mM L-Serine in WWTP Effluent A and in Reagent (Deionized) Water.

6.3.3 Formation of Cyanide from Chloramination

6.3.3.1 Synthetic Solutions

The results discussed previously indicate that CNCl precursors occur in WWTP secondary effluents. Analyses of total cyanide and free cyanide were performed to determine whether other cyanide species are formed through chloramination of the model CNCl precursors. It was found that neither total cyanide nor free cyanide was detected in any of the chloraminated-but-not-dechlorinated synthetic solutions spiked with the model organic compounds. However, detectable cyanide, always at low levels (4.0-27 µg/L), was observed in chloraminated L-serine solutions that were also dechlorinated with sodium thiosulfate. In practice, total cyanide analysis is always performed with dechlorinated samples because chlorine as an oxidizing agent tends to destroy cyanide and cause false negatives.

Figure 6-6 presents the results from several runs of the chloramination of L-serine followed by dechlorination to remove residual chlorine and thus prevent destruction of any free cyanide produced. Because of slight changes of reaction conditions (temperature and pH), the total cyanide results exhibited variability. For example, total cyanide was usually detected around 10 µg/L at the reaction time of 24 hours, but the concentration was as high as 27 µg/L or as low as 4 µg/L in some cases.

Analysis of chloraminated and dechlorinated samples of 0.1 mM L-serine solution for CN^- also demonstrated the formation of free cyanide. As indicated in Figure 6-7, total cyanide was detected as 6.7 µg/L at 2-hour reaction time, with 4.8 µg/L as free cyanide. Total cyanide was measured at a concentration of 6.1 µg/L at 124 hours, whereas free cyanide decreased to 1.0 µg/L.

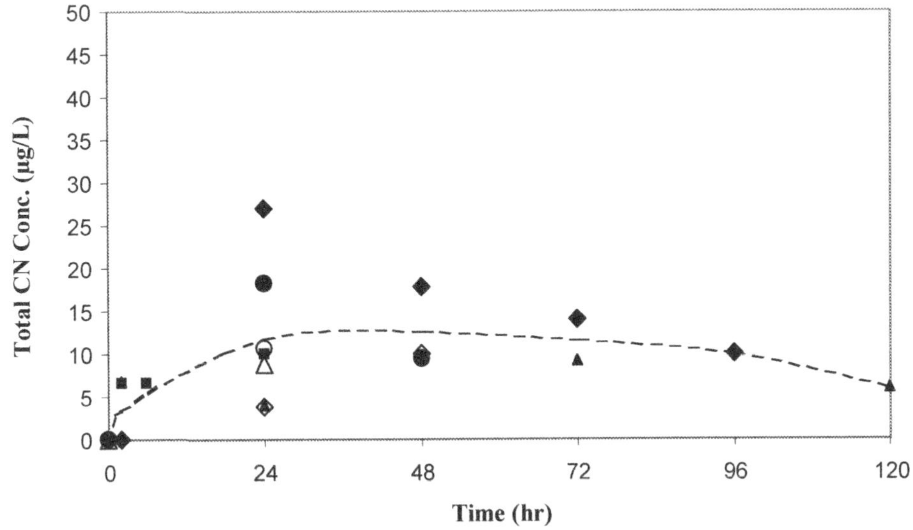

Figure 6-6. Detection of Total Cyanide from Chloramination (3 mM NaOCl + 2 mM NH_4Cl) of Synthetic L-Serine Solution (0.1 mM) at pH = 6.0 ± 0.5, Followed by Dechlorination with Sodium Thiosulfate (0.1 g/L).
The different symbols reflect different experimental trials with the same system.

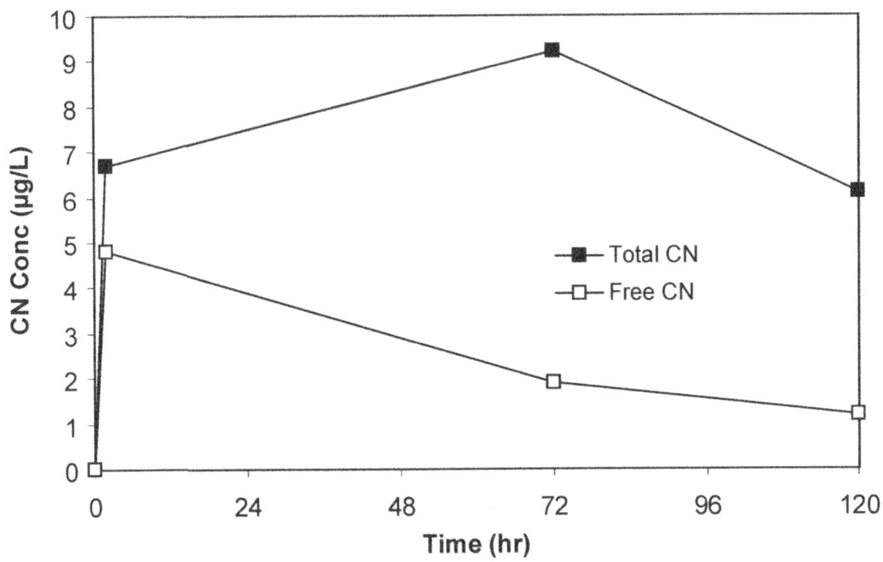

Figure 6-7. Detection of Total Cyanide and Free Cyanide from Chloramination (3 mM NaOCl + 2 mM NH_4Cl) of Synthetic L-Serine Solution (0.1 mM) at pH = 6.0 ± 0.5, Followed by Dechlorination with Sodium Thiosulfate (0.1 g/L).

6.3.3.2 WWTP Effluents

Similar experiments were conducted to investigate the potential formation of cyanide from chloramination of WWTP secondary effluents followed by dechlorination to prevent the destruction of any free cyanide produced. Sodium thiosulfate and ascorbic acid at a concentration of 0.1 g/L were used to test the effects of analytical dechlorinating agents. Results are given in

Figure 6-8. As shown there, total cyanide was detected in the chloraminated and dechlorinated wastewater effluents. The total cyanide concentration increased with time and became as high as

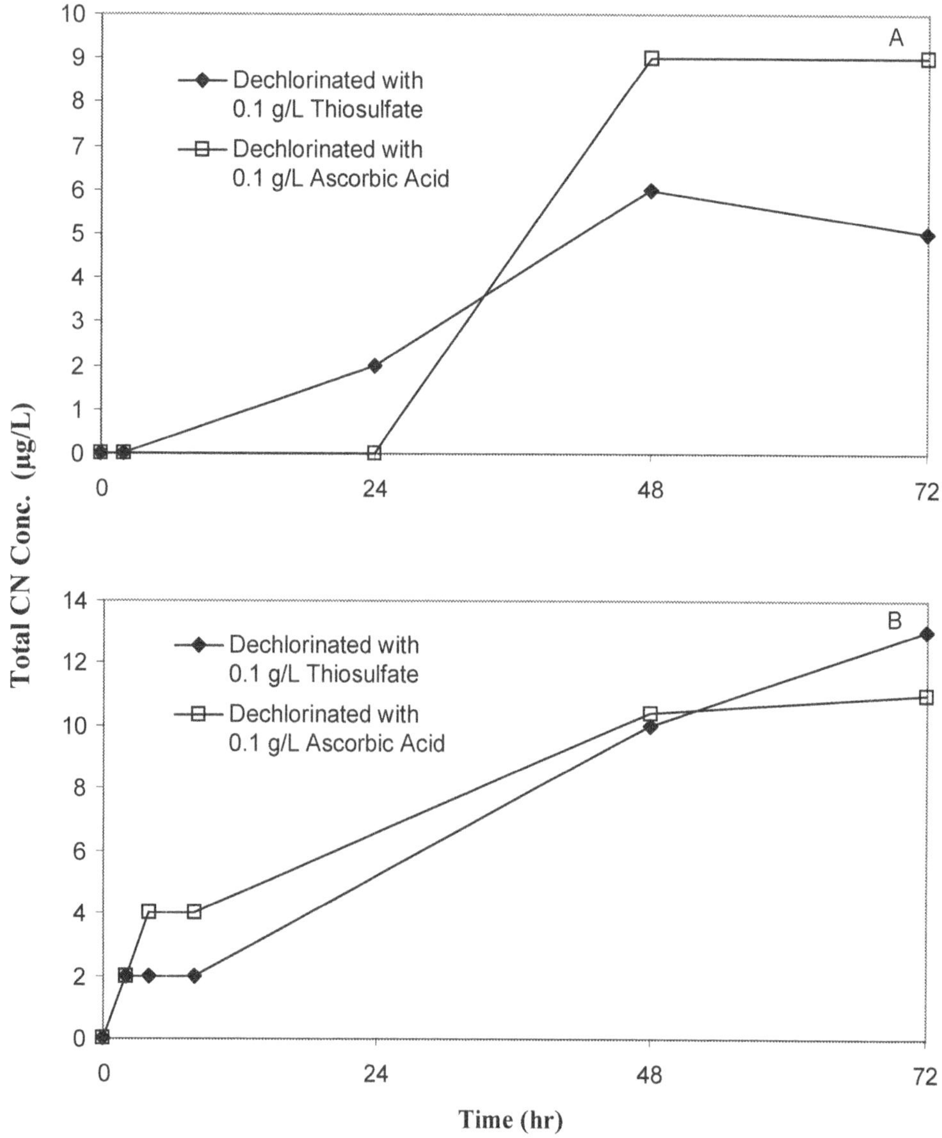

Figure 6-8. Detection of Total Cyanide from Chloramination (3 mM NaOCl + 2 mM NH_4Cl) of WWTP Secondary Effluents Followed by Dechlorination with Sodium Thiosulfate or Ascorbic Acid (A = Effluent A; B = Effluent B).

13 µg/L at a reaction time of 72 hours. The two different analytical dechlorinating agents gave the same range of concentrations with regards to formation of cyanide.

6.3.3.3 WWTP Effluents Spiked with Model Organics

Total cyanide was also detected from the chloramination study performed in several WWTP wastewater effluents spiked with 0.1 mM of L-serine (see Figure 6-9). The concentrations of total cyanide detected in these organics-spiked WWTP effluent samples were higher than those detected in unspiked WWTP effluents, demonstrating the effects of the precursor organic compounds in the chloramination process. Generally, samples dechlorinated with thiosulfate yielded less total cyanide than those dechlorinated with ascorbic acid.

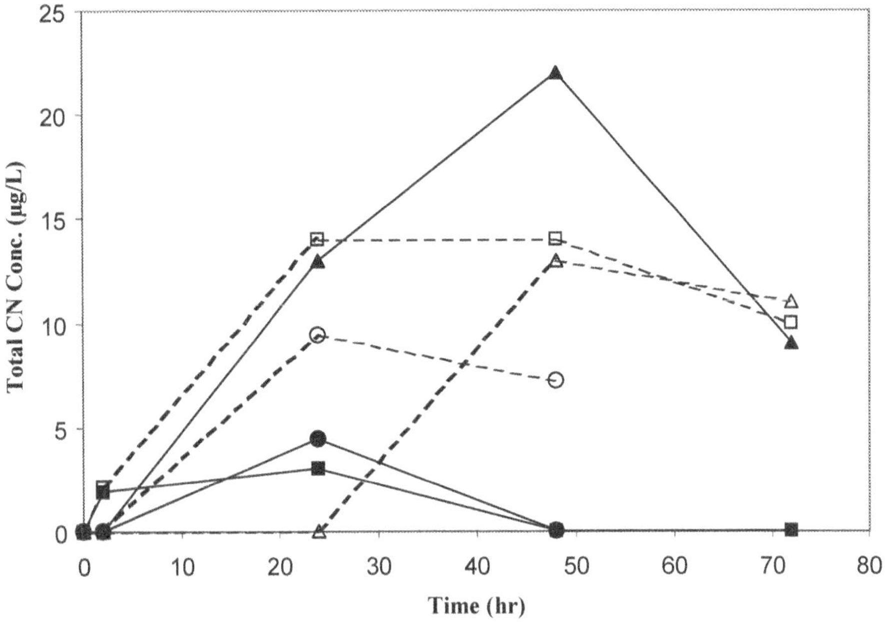

Figure 6-9. Detection of Total Cyanide from Chloramination (3 mM NaOCl + 2 mM NH_4Cl) of WWTP Secondary Effluents Spiked with L-Serine (0.1 mM). Closed symbols (■, ▲, ●) represent systems (Effluents A, B, and C) dechlorinated with 0.1 g/L thiosulfate. Open symbols (□, △, ○) represent systems dechlorinated with 0.1 g/L ascorbic acid.

6.3.4 Mechanism for Cyanogen Chloride or Cyanide Formation

Study of the mechanism for the formation of CNCl from chloramination of organic compounds has been very limited and most of the studies have focused on CNCl formation during drinking water treatment. No mechanisms for cyanide formation upon chloramination have been examined. The most recent relevant work was conducted by Pedersen et al. (1999) who demonstrated the formation of CNCl from the reaction of monochloramine with formaldehyde, a common DBP produced during ozonation or chlorination of waters containing natural organic matter. Based on the work presented here and on previous work of Pedersen et al. (1999), a reaction scheme for chloramination of L-serine is proposed in Figure 6-10.

The step (A) in the first reaction is a typical chloramination step of α-amino acids and its mechanism has been well understood (Kovacic et al., 1970). The product of step *A,N*-chloro-α-amino acid, is not stable and decomposes in step B to form chloride ion, carbon dioxide, and an imine (Stanbro and Smith, 1979). The imine hydrolyzes to ammonia and carbonyl in the second reaction according to Stanbro and Smith (1979). The carbonyl, following a similar reaction pathway as formaldehyde, reacts with monochloramine and yields *N*-chloroaminomethanol (Pedersen et al., 1999) which then decomposes to form *N*-chloromethanimine (third reaction) (Pedersen et al., 1999). *N*-Chloromethanimine can decompose by fragmentation to yield hydrogen cyanide, HCN (Pedersen et al., 1999). HCN is well known to react with chloramine to produce cyanogen chloride as the final reaction indicates (APHA/AWWA/WEF, 1998).

The potential formation of cyanide and cyanogen chloride from the chloramination of organic compounds in WWTP secondary effluent was addressed in this chapter. Chloramines form when chlorine reacts with residual ammonia. Studies with synthetic solutions demonstrated that cyanogen chloride could be formed in significant concentrations by chloramination of various model organics including L-serine, benzene, catechin, and humic acid. The amino acid, L-serine, yielded the largest concentrations of CNCl upon chloramination. Studies with WWTP secondary effluents demonstrated that cyanogen chloride is formed in the chloramination of WWTP secondary effluent, which indicates that precursors of CNCl exist in biotreated WWTP wastewater. To our knowledge, this is the first reported observation of the formation of cyanogen chloride in WWTP effluents.

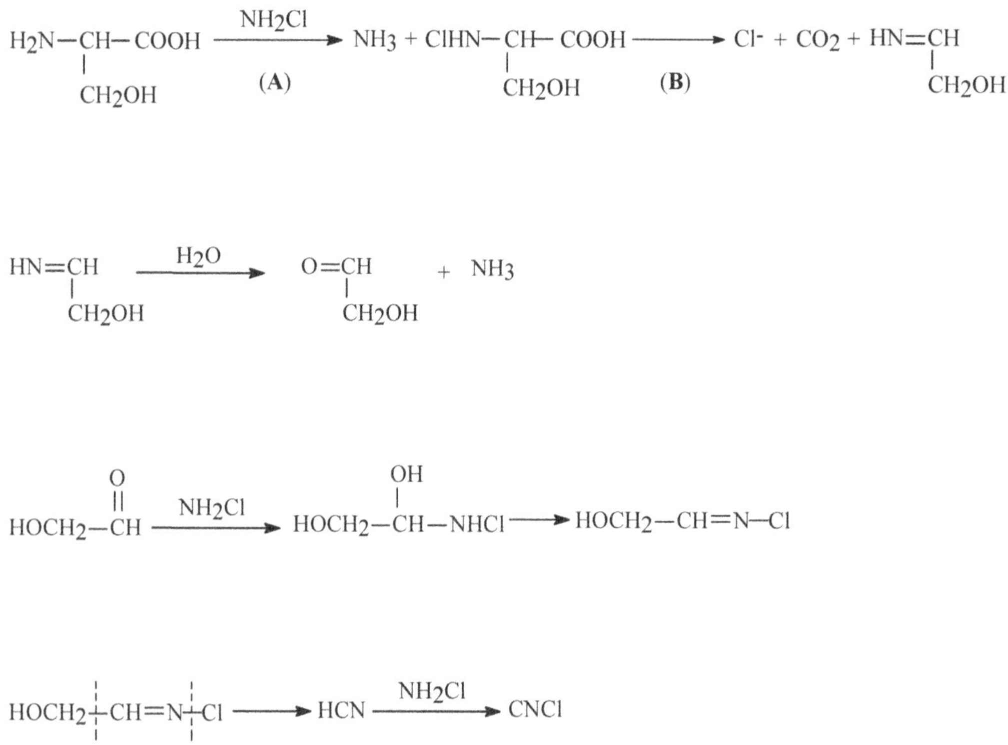

Figure 6-10. Proposed Reaction Scheme for Formation of Cyanide and Cyanogen Chloride from Chloramination of the Amino Acid L-Serine. Step A is Typical for Chloramination of α-Amino Acids. Step B is a Decomposition Step.

It was revealed that neither total cyanide nor free cyanide was detected in any of the chloraminated-but-not-dechlorinated synthetic solutions spiked with CNCl precursor organic compounds. However, detectable cyanide, always at low levels, was observed in chloraminated L-serine solutions that were also dechlorinated to prevent destruction of any free cyanide produced. In studies with chloraminated and dechlorinated WWTP secondary effluents, total cyanide was detected at concentrations as high as 13 μg/L.

Effects of analytical dechlorinating agents on the preservation of cyanogen chloride were also examined. Sodium thiosulfate can destroy CNCl effectively, and thus may not be suitable for use as a sample dechlorinating agent before analysis for CNCl. Ascorbic acid is an alternative dechlorinating agent for samples designated for CNCl analysis.

These findings indicate that chlorination of WWTP secondary effluent containing residual ammonia can lead to chloramination of organic compounds and the resulting production of cyanogen chloride and free cyanide at concentrations of 5-25 μg/L. These species may remain in WWTP final effluent if residual chlorine is low.

CHAPTER 7.0

EFFECTS OF NITROSATION ON THE FORMATION OF CYANIDE IN WWTPS SECONDARY EFFLUENT

7.1 Introduction

7.1.1 Nitrosation Process

Nitrite is a common constituent found in domestic wastewater secondary treatment effluents. White et al. (1983) reported nitrite nitrogen (NO_2-N) concentrations of 4-6 mg/L in secondary effluent and 0.02 mg/L in nitrification filter effluent of a municipal WWTP in California. Other researchers also reported several cases where NO_2-N of 2.0-3.3 mg/L in effluents from secondary treatment (Roberts and Casey, 1984; Ho and Roberts, 1984; Arber et al., 1984). Moreover, nitrite can accumulate in wastewaters exhibiting high nitrogen concentrations (Turk and Mavinic, 1989).

It has been reported that cyanide can be formed in aqueous solutions by nitrite (NO_2^-) reacting with organic compounds, although these organic compounds have not been clearly identified, and the reaction mechanisms have not been established. Koshimizu (1975) reported that aromatic amines and glycine could react with nitrite to form cyanide. Rapean et al. (1980) showed that nitrite formed HCN via nitrosation of a number of compounds, including methyl isobutyl ketone, phenol, catechol, ethylacetoacetate, pyruvic acid, oxalacetic acid, and 2,4,4-trimethyl pentanes. Kanno et al. (1978) reported that aliphatic and aromatic amines were reactive with nitrite and described a study of the formation of HCN from aniline showing that the nitrogen in the cyanide was derived from the nitrite ion. All these studies were conducted in synthetic solutions.

Nitrosation may play a role in the observed formation of cyanide in WWTPs. In nitrosation reactions, the nitroso group (-NO) is introduced into organic molecules. The nitrosation of amines and related compounds has been widely studied because it can result in the formation of nitrosamines, which are powerful carcinogens in animals and human beings. The nitrosation reaction is particularly important in the food processing industry because nitrite is generally present in foods, especially in cured meat (Izumi et al., 1989). However, very few studies of nitrosation reactions in water and wastewater treatment have been conducted. There is a significant opportunity for nitrosation to occur in WWTPs, especially in aerobic biotreatment where nitrification occurs, which involves the conversion of ammonia (NH_3) to nitrite (NO_2^-) and then to nitrate (NO_3^-) by nitrifying bacteria. Because nitrifying bacteria are sensitive organisms and are susceptible to a wide variety of inhibitors (Rittmann and McCarty, 2000), the nitrification process sometimes slows and nitrite build-up may occur thereby triggering the nitrosation process.

7.1.2 Artifacts of the Total Cyanide Analysis Technique

Even if nitrosation reactions are slow in the wastewater at ambient temperatures, the co-existence of nitrite and organic compounds in wastewater samples provides opportunity for nitrosation to occur in the cyanide analysis. The most commonly used total cyanide analysis technique uses a strong acid and a high temperature distillation procedure, which greatly favors the nitrosation process. For this reason nitrite has long been recognized as an interference in the total cyanide analytical procedure (Carr et al., 1997; APHA/AWWA/WEF, 1998). As suggested in *Standard Methods*, sulfamic acid is used to eliminate the nitrite interference by dosing up to 4 g/L of sample, following the reaction shown below:

$$HNO_2 + H_2NSO_3H \rightarrow N_2 + H_2O + H_2SO_4 \qquad (7\text{-}1)$$

The goal of the study was to address the possibility of formation of cyanide from nitrosation in WWTP secondary effluents as well as during total cyanide analysis if sample pretreatment is inadequate. The overall objective was to gain a better understanding of the potential for interactions between nitrite and organic compounds to lead to cyanide production in WWTPs.

7.2 Methods

7.2.1 WWTP Wastewater Used in the Study

The primary wastewater used in this study was collected at the postaeration tank (after sand filters) of a secondary treatment plant of the Metropolitan Water Reclamation District of Greater Chicago, Cicero, Illinois. Two batches of secondary effluent were collected at different times and designated as Effluent A and Effluent B. The wastewater had no detectable cyanide, thiocyanate or nitrite. The nitrate concentrations were 63.5 and 57.3 mg/L, respectively. Additional characteristics of both WWTP secondary effluents are described in Table 5-1.

7.2.2 Laboratory Studies

Laboratory studies were conducted to investigate the potential production of cyanide as a result of nitrosation reactions in secondary effluents or in the course of cyanide analysis. Experiments were performed first on aqueous solutions containing some model organic compounds to evaluate the potential for cyanide to be generated as a product of nitrosation. Experiments were also conducted with some WWTP secondary effluents to assess the effects of nitrosation in wastewater.

7.2.2.1 Nitrosation Study

Generally 10^{-3} mol/L of the pure compounds were used to increase the detection probability of any cyanide formed. However, in some cases, a much lower concentration of organic compounds (10^{-5} mol/L) was also used to test cyanide formation under conditions more representative of WWTP secondary effluent. All nitrosation experiments were done in batch, in pH-adjusted solutions, in amber glass bottles (catalog no. 02-911-6, Fisher Scientific, Pittsburgh, PA), with dosing of different levels of nitrite ions as $NaNO_2$ (Fisher Scientific, Pittsburgh, PA) ranging from 5 mg/L to 500 mg/L. In some cases, separate reaction batches were set up for samples corresponding to designated reaction times ranging up to 72 hours. The batches of nitrosated solutions were kept in a constant temperature bath (25 ± 1°C) before analysis.

7.2.2.2 Model Organic Compounds

Nitrosation was performed on aqueous solutions containing some model organic compounds including 1,2,4-trihydroxybenzene, phenol, *p*-toluidine, aniline, L-serine, benzoic acid, and glycine. These seven compounds were chosen as representative of potentially nitrite-reactive organics in WWTPs with serine and glycine as amino acids and the remaining compounds as aromatic hydrocarbons (Kanno et al., 1978; Carr et al., 1997). All chemicals were reagent grade or better and obtained from Fisher Scientific, Inc. (Pittsburgh, PA), Sigma Chemical Co. (St. Louis, MO), and Acros Organics (Fairlawn, NJ).

7.2.2.3 Analytical Methods

Total cyanide analysis was performed using the conventional strong-acid distillation method specified in *Standard Methods,* (APHA/AWWA/WEF, 1998 – Method 4500-CN-C). Hydrogen cyanide is liberated from the strongly acidified sample by a two-hour distillation process at a temperature of 125°C and purging with air. The HCN gas liberated is collected by passing the purge gas through an NaOH scrubbing solution. An aliquot of the NaOH solution is then used to determine the cyanide concentration by a colorimetric procedure. Analyses were performed using a MidiVap (BSI, Co., Oxford, MA) cyanide analysis apparatus which uses 50 mL samples.

WAD cyanide was analyzed according to the conventional weak-acid distillation method given in *Standard Methods*, (APHA/AWWA/WEF, 1998 – Method 4500-CN-I) using a similar procedure as total cyanide except that it is performed in a weak acid solution (acetate buffer, pH 4.5).

Free cyanide by microdiffusion was measured in accordance with an ASTM method (ASTM, 1998a). A 3-mL sample is buffered at pH 6 and placed in the outer ring of a two-compartment microdiffusion cell (Bel-Art Products, Inc., Pequannock, NJ) for four hours. The HCN diffuses through the common headspace above the two compartments and into sodium hydroxide solution in the center compartment. The cyanide in the NaOH solution is analyzed colorimetrically at 578 nm.

7.3 Results and Discussion

7.3.1 Formation of Cyanide in Total Cyanide Analysis Procedure

7.3.1.1 Synthetic Solutions

Total cyanide was detected in samples containing various amounts of nitrite ion and the model organic compounds. Under the analytical conditions (strong acid distillation, 125°C, without sulfamic acid addition to destroy nitrite), total cyanide was measured in significant concentrations in samples containing nitrite and the model organics as indicated in Figure 7-1. The total cyanide yield was in the range of 0.002 mg/L (MDL for total cyanide) to 5 mg/L with cyanide production varying for the different organics. Among the seven organics tested 1,2,4-trihydroxybenzene exhibited the largest cyanide yield (i.e., 0.12 mg/L for a nitrite dose of 50 mg/L, 4.7 mg/L for a nitrite dose of 250 mg/L, and 5.1 mg/L when nitrite was dosed at 500 mg/L). Cyanide formation from nitrosation of L-serine, glycine, and benzoic acid was not significantly greater than the method detection limit. It was found that almost all aromatic compounds tested, phenol, aniline, *p*-toluidinem, and 1,2,4-trihydroxybenzene—but not benzoic acid—have a significant capability to react with nitrite and to form cyanide under the testing conditions.

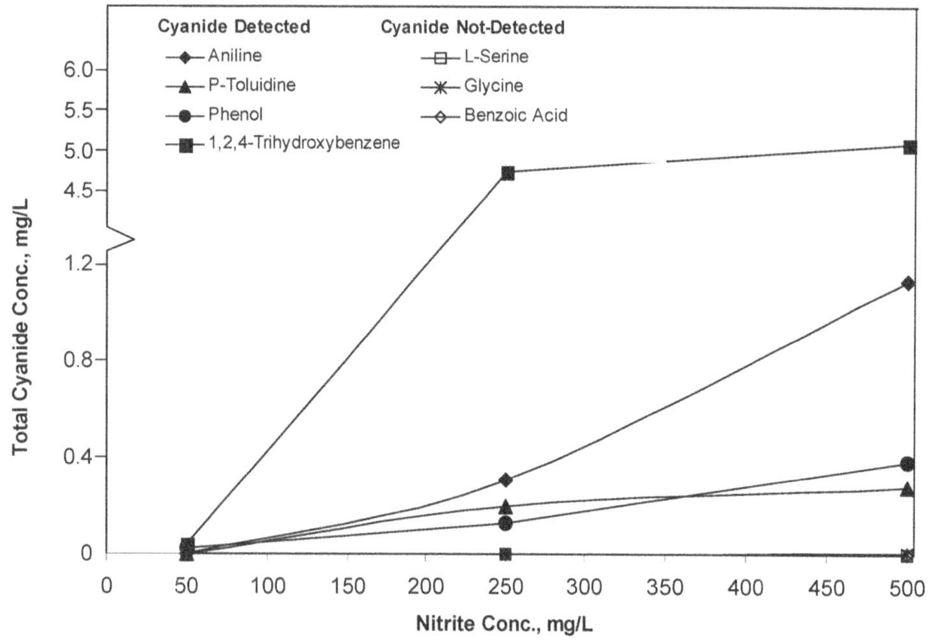

Figure 7-1. Total Cyanide Detected Versus Nitrite Dose for Nitrosation of Synthetic Solutions of 1 mM Organic Compounds under Strongly Acidic Distillation Conditions as Employed in Total Cyanide Analysis Procedure (T = 125°C, pH = 0.5).

7.3.1.2 WWTP Effluent with Nitrite Spiking

The nitrosation study was performed with WWTP secondary effluents. The total cyanide and nitrite concentration in the wastewater effluent used were both nondetectable. Therefore, nitrite was spiked in the effluent at concentrations from 10 mg/L to 50 mg/L. Two sets of experiments were performed, five days apart. A summary of results is presented in Table 7-1. In the absence of sample pretreatment with sulfamic acid, total cyanide was detected at 0.027 mg/L and 0.031 mg/L when WWTP Effluent A was spiked with 10 and 50 mg/L of nitrite, respectively. This indicates that there were nitrite-active substances in the wastewater secondary effluent. The cyanide increase with added nitrite was rather small, only 0.004 mg/L for an increase of 40 mg/L of spiked nitrite, probably caused by exhaustion of the nitrite-reactive compounds. When the same experiment was repeated five days later, it was found that cyanide formation decreased with the same amount of nitrite spiked (i.e., 0.018 and 0.026 mg/L of cyanide formed for 10 and 50 mg/L of nitrite spiked, respectively). This suggests that nitrite-active organic compounds in the wastewater effluent may decay slowly with time, perhaps because of biodegradation.

Table 7-1. Total Cyanide Analyses for WWTP Effluent A Spiked with Nitrite.

Exp. no.[a]	NO_2^- spiked (mg/L)	Total CN conc.[b] (mg/L)
1	10.0	0.027
	50.0	0.031
2	10.0	0.018
	50.0	0.026

[a] Two sets of experiments were performed, 5 days apart.
[b] Without sample pretreatment by sulfamic acid addition.

7.3.1.3 WWTP Effluents With Organics and Nitrite Spiking

Total cyanide analyses were also performed with WWTP Effluent A spiked with 10^{-5} mol/L of 1,2,4-trihydroxybenzene and 5 mg/L of nitrite. It was found that total cyanide was measured at 0.047 mg/L at 24 hours of reaction time. At 144 hours, the total cyanide concentration increased to 0.052 mg/L. This increase may be caused by sufficient supply of nitrite-active compounds (i.e., the organics spiked in the solution).

7.3.2 Formation of Cyanide in WAD Cyanide Analysis Procedure

The potential for cyanide formation in WAD cyanide analyses by the reaction of nitrite with organics was also investigated. Under the analytical conditions (mildly acidic distillation, 125°C, without sulfamic acid addition to destroy nitrite), WAD cyanide was also measured in significant concentrations in samples containing nitrite and some of the model organics as indicted in Table 7-2. The WAD cyanide measured was less than the total cyanide measured in similar experiments clearly showing the mildly acidic condition does not favor cyanide formation to the same degree as in the strong acid distillation of the total cyanide test. The cyanide yields from the WAD tests were 18-85% of the yields from the total cyanide analyses.

Table 7-2. Comparison of Total and WAD Cyanide Detected in Reagent (Deionized) Water Samples Spiked with Nitrite and Organics.

Organic Compounds	Initial Conc.		Cyanide Detected	
	Organic Cmpd. (mM)	NO_2^- (mg/L)	Total CN (mg/L)	WAD-CN (mg/L)
Aniline	1.0	50	ND [a]	ND
	1.0	250	0.310	0.197
	1.0	500	1.13	0.208
P-Toluidine	1.0	50	ND	ND
	1.0	250	0.197	0.168
	1.0	500	0.280	0.225
1,2,4-Trihydroxy-benzene	1.0	50	0.120	ND
	1.0	250	4.74	2.06
	1.0	500	5.10	4.18

[a] ND = Not detected. Detection limits were 0.002 mg/L for total and WAD cyanide.

These studies demonstrated that nitrite can indeed be a significant interference in total cyanide and WAD cyanide analyses employing acidic conditions and high temperature distillation. Generation of cyanide in the course of the analysis is an analytical artifact of concern, especially when low levels (μg/L) of cyanide are the focus. The potential importance of this analytical artifact was also noted by Carr et al. (1997). The use of an adequate dose of sulfamic acid is critical to eliminate the interference in total and WAD cyanide analyses (Carr et al., 1997; APHA/AWWA/WEF, 1998).

7.3.3 Effects of Sulfamic Acid on the Elimination of Nitrite Interference

Cyanide formation in the total and WAD cyanide analyses was effectively eliminated with sulfamic acid addition (4 g/L) for samples spiked with nitrite and most of the organic compounds tested. This indicates that sulfamic acid reduction of nitrite was significantly more rapid than whatever may be the rate-limiting step in cyanide formation. Only in the case of 1,2,4-trihydroxybenzene was total cyanide detected after sample pretreatment with sulfamic acid. Total cyanide concentrations of 0.011, 0.007, and 0.035 mg/L were measured in 1 mM trihydroxybenzene samples dosed with 50, 250, and 500 mg/L nitrite and 4 g/L sulfamic acid. Without sulfamic acid addition, the total cyanide yields for the same samples were 0.12, 4.74, and 5.10 mg/L, respectively. Thus, sulfamic acid addition is effective but will not necessarily eliminate nitrite interference at the dosing level recommend in *Standard Methods* (APHA/AWWA/WEF, 1998) for waters containing reactive organics.

7.3.4 Formation of Cyanide from Nitrosation in Synthetic Solutions

Nitrosation of organics may occur and produce cyanide in WWTPs before sample collection and analysis. Thus, the detection of total cyanide in the sample may not be solely attributed to the analytical artifact. To investigate the potential for in-plant formation of cyanide through reactions of nitrite, the nitrosation of selected organic compounds in reagent (deionized) water was evaluated under normal WWTP water temperature and nitrite/organic concentration conditions.

This study investigated nitrosation in aqueous solutions containing 1,2,4-trihydroxybenzene and phenol as two model aromatic compounds. Nitrite was added to batch solutions of these two compounds in reagent (deionized) water. The batches of nitrosated solutions were kept in a constant temperature bath (25 ± 1°C) before analysis at predesignated times for free cyanide by microdiffusion. It is emphasized that free cyanide is defined here as the cyanide that diffuses as HCN, at room temperature, light free, from a solution at pH 6. The testing condition is much more mild than those employed in total and WAD cyanide tests. Significant free cyanide formation was found in acidic solutions containing 0.01 mM of 1,2,4-trihydroxybenzene and 5 mg/L of nitrite as indicated in Figure 7-2. After 24 hours of reaction time, the cyanide yield was about 120 μg/L (for both initial pH of 2 and 4), which increased slowly to 135 μg/L (for initial pH of 4) and 150 μg/L (for initial pH of 2). The yield of free cyanide from nitrosation of 0.01 mM 1,2,4-trihydroxybenzene at the acidic condition was about 0.15 mg per 1.26 mg of parent compound or 57.7% on a molar basis. No cyanide was formed in similar experiments conducted at pH 7.1 and 10. It is clear that nitrosation is a pH-dependent rate-limited process. Because the pKa of HNO_2 is 3.15, it appears that HNO_2 is more reactive than NO_2^- at 25°C.

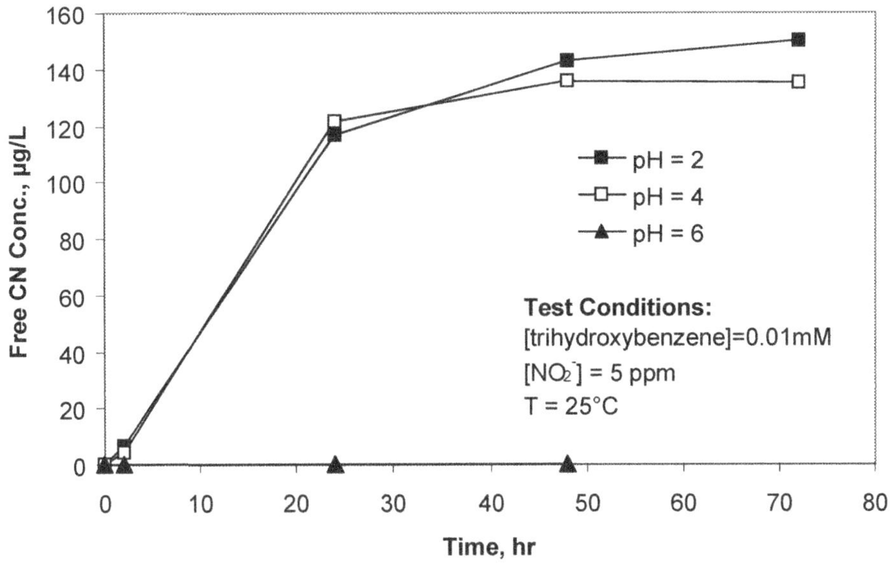

Figure 7-2. Free Cyanide Detected Versus Time for Nitrosation of Solutions of 1,2,4-Trihydroxybenzene in Reagent Water at Three pH Values and T = 25°C.

7.3.4.1 Effects of Temperature on Nitrosation Process

No free cyanide was formed by nitrosation of phenol at 25°C, another aromatic compound tested, even when dosing at higher concentrations (1 mM for phenol and 250 mg/L for nitrite) and in the acidic pH range. However, if the solution was initially subjected to heating at 50°C for two hours, cyanide formation was enhanced greatly. Figure 7-3 shows that no free cyanide was detected from nitrosation at 25°C at any pH ranges, whereas 15 µg/L of free cyanide was measured from nitrosation at 50°C and acidic pH ranges.

7.3.4.2 Effects of UV Irradiation on Nitrosation Process

The effect of UV irradiation (monochromatic light at 253.7 nm) on the formation of cyanide from nitrosation of phenol at 25°C was also tested. Details on the apparatus and methods for UV irradiation are described in Section 5.3.2.3 of this report. As shown in Figure 7-4, about 17 µg/L of free cyanide was generated by nitrosation of 1 mM phenol under UV irradiation compared to no production of cyanide from nitrosation in the absence of UV irradiation (see Figure 7-3). This demonstrated that UV irradiation substantially enhanced the formation of free cyanide from nitrosation of phenol at pH 6.

7.3.5 Formation of Cyanide from Nitrosation in WWTP Secondary Effluent

The nitrosation study was also performed with WWTP secondary effluents. The total cyanide and nitrite concentration in the wastewater Effluent A were both nondetectable. Therefore, nitrite was spiked in the effluent at doses of 10-50 mg/L. Two sets of experiments were performed five days apart. A fair amount of total cyanide, 27 µg/L and 31µg/L, was detected

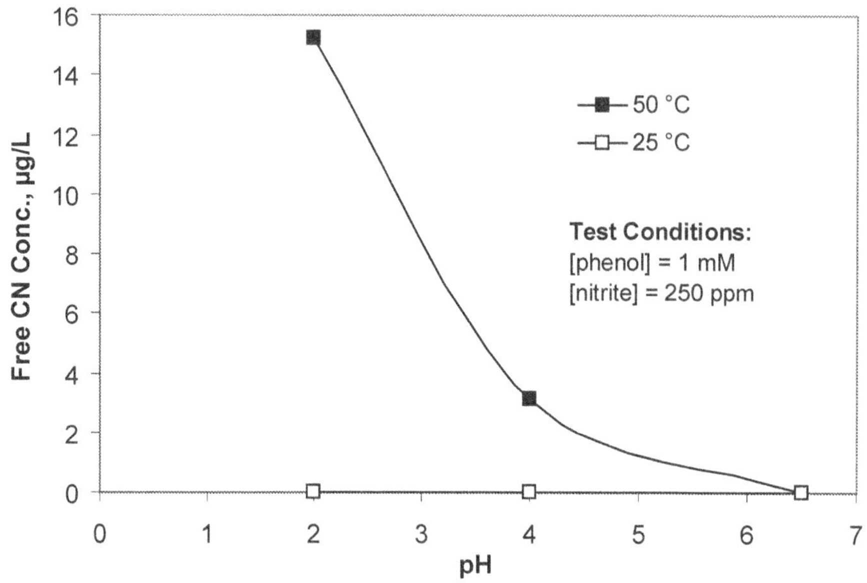

Figure 7-3. Cyanide Measured in 1 mM Phenol Solutions Subjected to Reaction with 250 mg/L Nitrite for Different pH Values and Two Temperatures.

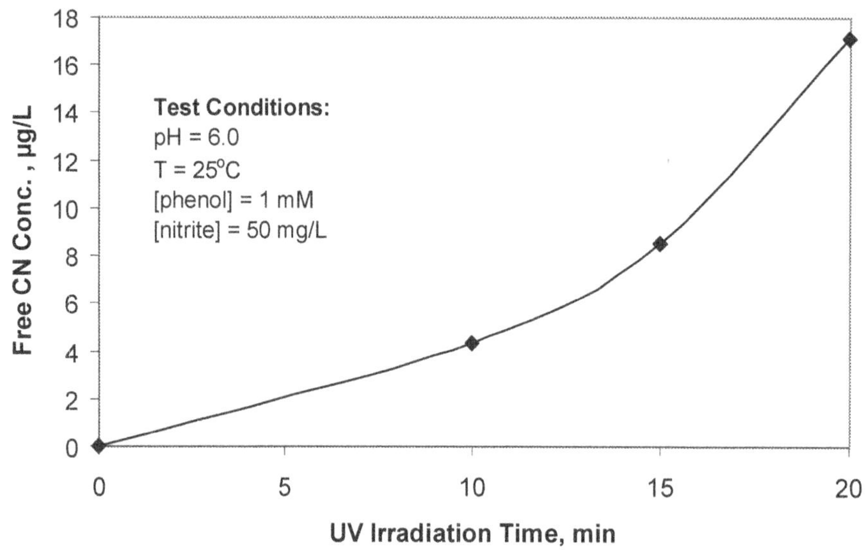

Figure 7-4. Cyanide Measured in 1 mM Phenol Solutions Subjected to Reaction with 50 mg/L Nitrite under UV Irradiation at pH 6.0 and 25°C.

via acidic total cyanide analysis (strong acid distillation) in WWTP effluents spiked with nitrite (10 mg/L and 50 mg/L, respectively) but without sulfamic acid addition revealing that nitrite-active organic compounds could be present in WWTP secondary effluents. However, no cyanide was detected in any samples that were pretreated with sulfamic acid indicating that no cyanide production would occur spontaneously in the plant via nitrosation even if sufficient nitrite was present. Thus, cyanide measured in the samples without sulfamic acid added was an analytical artifact.

To examine the effects of reactive organic compounds in WWTP effluent on the nitrosation process, experiments were conducted with Effluent A spiked with 10^{-5} mol/L of 1,2,4-trihydroxybenzene and 5 mg/L of nitrite. The effect of pH was also examined with two pH values used in the experiment, 7.1 (original pH without adjustment) and 3.0 (adjusted with HCl). Samples were treated with sulfamic acid in total cyanide analysis to remove NO_2^- interference. It was found that neither total cyanide nor free cyanide was detected in solutions at pH 7.1 indicating that nitrosation would not occur spontaneously in the plant at neutral pH.

Figure 7-5 shows that total cyanide and free cyanide were both detected in WWTP effluents spiked with nitrite and the model reactive organic compound 1,2,4-trihydroxybenzene at pH 3.0. This apparently was the result of nitrosation triggered by the protonation of nitrite [i.e., by nitrous acid (HNO_2)]. In these tests, the total cyanide concentrations were obtained after samples were pretreated with sulfamic acid, which indicates that cyanide was actually formed by the nitrosation reaction before nitrite was removed from the solution. Total cyanide (equivalent to free cyanide in these experiments) yield was about 0.01 mg/L for a reaction time of two hours and as high as 0.15–0.20 mg/L for a reaction time of 24-48 hours.

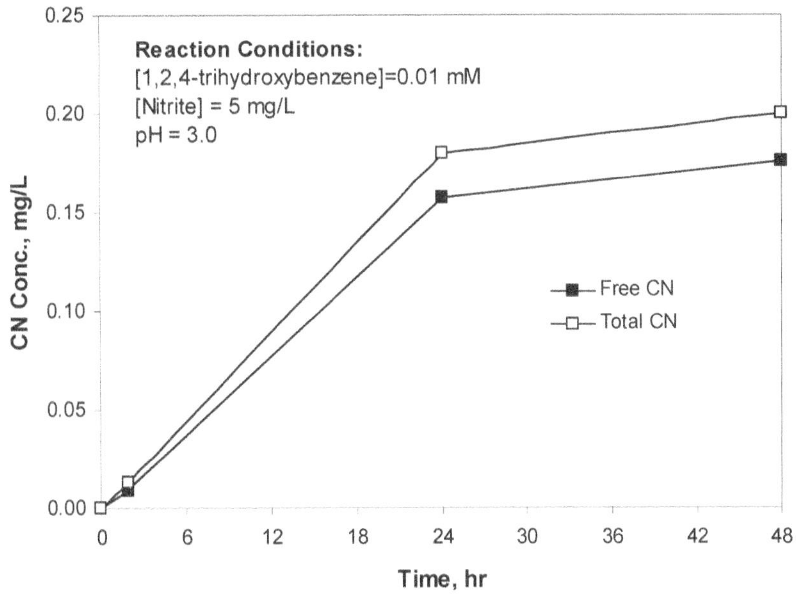

Figure 7-5. Total and Free Cyanide Measured in WWTP Secondary Effluent Spiked with 5 mg/L Nitrite and 0.01 mM 1,2,4-Trihydroxybenzene at pH 3.0.

7.3.6 Mechanism of Cyanide Formation from Nitrosation of Aromatic Compounds

The mechanism of aromatic nitrosation is much less known than any other nitrosation reactions and has been studied mechanistically in relatively few cases (Williams, 1988). Almost all studies performed only focused on the chemistry of major products (i.e., nitroso compounds). The mechanism for the production of cyanide in the nitrosation of some organic compounds (especially aromatic compounds) is still unclear.

Kanno et al. (1978) reported that aliphatic and aromatic amines were reactive with nitrite and proposed a reaction mechanism of aniline with nitrite as shown in Figure 7-6. However, the last step from nitrosophenol to the formation of free cyanide has not been validated and is likely not nearly as simple as implied.

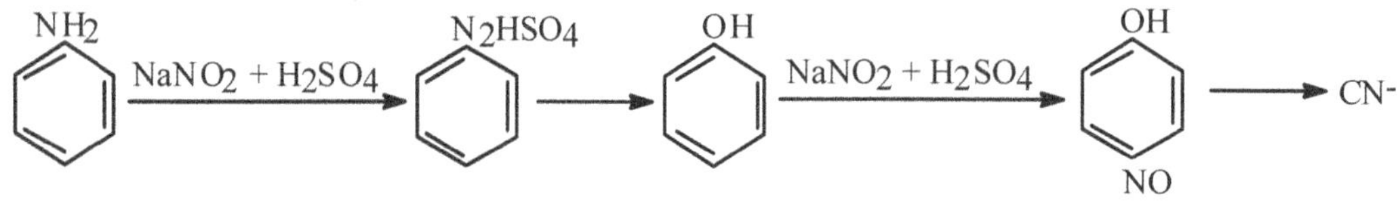

Figure 7-6. Proposed Reaction of Aniline with Nitrite (Proposed by Kanno et al., 1978).
The mechanism for formation of cyanide is not explained.

On the basis of a comprehensive literature review, we hypothesize that the general reaction scheme for the formation of cyanide from phenol is as shown in Figure 7-7. Phenol is readily nitrosated in dilute acid solution by nitrous acid at room temperature to give nitroso-phenol (I), which exists primarily as the tautomeric *p*-benzoquinone monooxime (II). The mechanism for this reaction and the chemistry of these products have been well studied (Williams, 1988; Thorn and Mikita, 2000). Under some circumstances (acidic condition and/or high temperature) continued nitrosation of nitroso-phenol can rupture the aromatic ring (Williams, 1988) and produce formaldehyde oxime (III), which can be dehydrated to hydrogen cyanide (March, 1992). The dehydration of aldoximes and similar compounds to form nitriles is well documented in the literature (March, 1992).

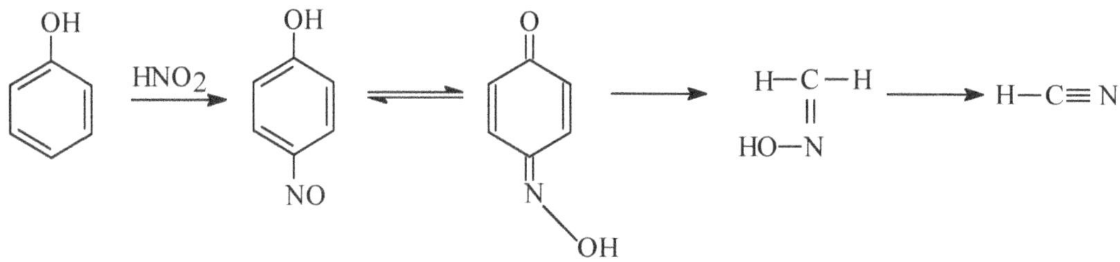

I II III

Figure 7-7. Proposed Mechanism for the Formation of Cyanide from Nitrosation of Phenol.

7.3.7 Mechanism of Cyanide Formation from Nitrosation of Aliphatic Compounds

There is very limited experimental evidence showing the formation of free cyanide from nitrosation of aliphatic compounds, and thus little information regarding reaction mechanisms. The formation of nitrile (R–CN) is very likely, however, and may potentially lead to the detection of cyanide in the samples under some circumstances (nitrile is generally not detected in conventional total cyanide analysis). In our experiments, no solid evidence of free cyanide formation was observed in the nitrosation process of aliphatic compounds. The mechanism in Figure 7-8 has been documented (March, 1992). Activated methyl and methylene carbons are

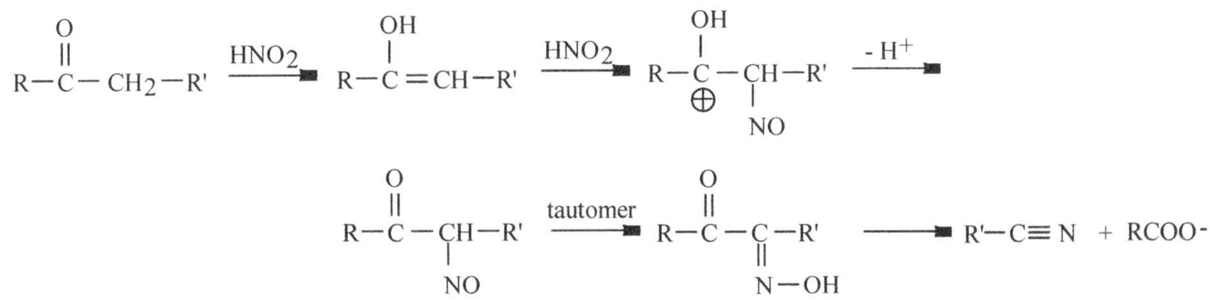

Figure 7-8. Mechanism for the Formation of Nitrile from Nitrosation of Aliphatic Compounds, Using Ketone as an Example (Proposed by March, 1992).

nitrosated to aldoximes and ketoximes, respectively. Aldoximes can be dehydrated to nitriles and ketoximes can also be converted to nitriles based on the second-order Beckmann rearrangements (March, 1992).

7.3.8 Summary

The presence of nitrite ion in WWTP secondary effluents may play an important role in the formation of low concentrations of cyanide through reactions with certain kinds of organic compounds. Laboratory studies were conducted with reagent water and WWTP secondary effluents spiked with nitrite and/or potentially reactive organic compounds to investigate the potential for nitrosation reactions to yield free cyanide.

Results from the laboratory studies with seven organic compounds (aniline, *p*-toluidine, phenol, 1,2,4-trihydroxybenzene, L-serine, glycine, and benzoic acid) revealed that cyanide can be formed by nitrosation, with the greatest production of cyanide observed for reaction with aromatic compounds. The most substantial free cyanide (CN^-) production observed in experiments by nitrosation at ambient condition was 0.15 mg/L after 72 hours of reaction of 0.01 mM 1,2,4-trihydroxybenzene with 5 mg/L nitrite at 25°C.

Experiments performed in WWTP effluents revealed that nitrite-reactive organic compounds could be present in secondary effluents. Experiments also demonstrated that no cyanide production would occur spontaneously in the WWTP secondary effluent (ambient condition, pH $\approx$7) via nitrosation even if sufficient nitrite was present. However, substantial free cyanide formation, from 0.15 to 0.20 mg/L, was observed at low pH in experiments with WWTP effluents when some reactive organics and nitrite were both spiked. Thus, the formation of cyanide through nitrosation was observed to be strongly pH dependent, being most significant at low pH and negligible at neutral to high pH. This result points to HNO_2 as being more reactive than the dissociated NO_2^- ion. Experiments also demonstrated that cyanide formation is favored at high temperature conditions and in the presence of UV irradiation.

It was also determined that free cyanide can be generated in the total and WAD cyanide analytical processes if nitrite and reactive organics are both present in the synthetic solution and if sample pretreatment with sulfamic acid to remove nitrite is insufficient. In our experiments, a large amount of nitrite ion spiked in the secondary effluent can easily lead to the detection of total cyanide if sufficient sample pretreatment is not performed. This analytical artifact can usually, but not always, be eliminated by adding the standard dose (4 g/L) of sulfamic acid recommended in *Standard Methods*, especially when strongly nitrite-reactive organics such as

1,2,4-trihydroxybenzene are present in the sample. In this case, sufficient sample pretreatment with sulfamic acid at the time of sampling, not at the time of analysis, is highly recommended. Additionally, the standard practice of preserving samples at pH 12 or above for cyanide analysis is also essential to prevent the nitrosation reaction from occurring spontaneously during the sample holding time because nitrosation is highly pH dependent.

A mechanism for the formation of cyanide ion from nitrosation of nitrite-reactive organic compounds is tentatively proposed. Available data and theory indicate that the conversion of organic compounds to either aldoximes or ketoximes is very likely involved in the formation of cyanide ion.

7.4 Conclusions

This study demonstrated that various cyanide species may be generated in municipal wastewater in the course of secondary treatment and/or disinfection with chlorine or UV irradiation. Four reaction groups/processes that may be involved in the formation of cyanide were investigated including chlorination, chloramination (from hypochlorite reaction with residual ammonia), nitrosation (reaction with nitrite), and UV irradiation.

Results from the laboratory study revealed several pathways through which thiocyanate (SCN^-) can be converted to free cyanide. Cyanide was detected in synthetic thiocyanate solutions undergoing incomplete oxidative destruction by chlorine because of relatively low chlorine dosing at all pH values tested (3.0, 6.5, and 11.0). Experiments performed in WWTP secondary effluents supported this observation. Free residual chlorine can easily destroy free cyanide in the solution, so if chlorine dosing is adequate, it will destroy all SCN^- and any free cyanide liberated by its reaction with SCN^-. Additionally, thiocyanate can be broken down to yield cyanide by UV irradiation, which is an important alternative to conventional chlorination for wastewater disinfection.

Chloramination can occur in WWTPs when chlorine reacts with residual ammonia present. It was revealed that cyanogen chloride could be detected in the chloramination of WWTP secondary effluents, which implies that precursors of CNCl can exist in the WWTP secondary effluent. Moreover, formation of cyanide from the chloramination of amino acid-related organic compounds was demonstrated. Detectable cyanide, always at low levels (ca. 10 μg/L), was observed in chloraminated synthetic solutions followed by dechlorination. Studies with WWTP effluents also demonstrated that total cyanide was detected in chloraminated and dechlorinated wastewater with observed total cyanide concentrations as high as 13 μg/L. Chloramination could be a potential pathway for cyanide formation observed in WWTPs operating with high ammonia content wastewater.

The presence of nitrite ion as a common constituent in domestic wastewater effluents may play an important role in the formation of cyanide through reaction with certain kinds of organic compounds (e.g., 1,2,4-trihydroxybenzene). Free cyanide (CN^-) was identified as a product of nitrosation of 1,2,4-trihydroxybenzene at 25°C. Substantial free cyanide formation was also observed at low pH in experiments with WWTP effluents when reactive organics and nitrite were both spiked. Formation of cyanide through nitrosation was strongly pH dependent, being most significant at low pH and negligible at neutral to high pH. This result points to HNO_2 as being more reactive than the dissociated NO_2^- ion. Sufficient sample pretreatment with sulfamic acid at the time of sampling, not at the time of analysis, is highly recommended.

CHAPTER 8.0

TRANSPORT AND FATE OF CYANIDE SPECIES IN SURFACE WATERS

8.1 Introduction

The transport, fate, and subsequent risk associated with cyanide released into the environment are generally linked to its interaction with trace elements, iron, and incident light. Moreover, given the low order of toxicity of complex cyanide [e.g., $Fe(CN)_6^{-4}$], regulators generally have taken a conservative approach, citing the potential for photodissociation into constituent components (i.e., HCN) upon reaching a receptor, thereby re-establishing toxic effects. This research has been aimed at establishing a more quantitative basis for assessing cyanide interactions in the environment including biological decay of HCN and air–water mass transfer of HCN under conditions representative of the environment. Heretofore, these areas have been characterized by an incomplete, contradictory, or at best qualitative knowledge base such that projecting realistic levels of environmental risk for cyanide and its compounds has not been possible under most conditions.

Accordingly, this task has focused on the fate and transport of common cyanide species in receiving waters. These efforts include an evaluation of biogeochemical properties, such as photolysis, volatilization, and biological utilization, and the development of a transport and fate model based on site data compiled at an industrial site in Tennessee. During the project, research goals under this task were as follows:

- To develop a quantitative assessment of the photolysis of complexed cyanide [i.e., $Fe(CN)_6^{-4}$] to form free cyanide (i.e., HCN), water-air mass transfer of free cyanide, and biotic transformation of free cyanide by a natural aquatic microfloral population using field-based and controlled laboratory observations;
- To develop a comprehensive numerical simulation framework (cyanide transport and fate model, or CTFM) to describe the longitudinal concentrations and reactions of complexed and free cyanide species in a receiving water affected by point and distributed inputs; and,
- To compile temporally and spatially variable field data for site-specific calibration of CTFM.

8.2 Methods and Materials

Field work was undertaken at an industrialized region near Knoxville, Tennessee, in the foothills of the Great Smoky Mountains. The specific site was located in a region underlain by a karst geology where it had been determined that some local springs were contaminated with low levels of hexacyanoferrate ($Fe(CN)_6^{-4}$), which has the potential to undergo photolysis to form free cyanide (HCN, CN^-) as illustrated below:

$$[Fe(CN)_6]^{-4} + H_2O + light \rightarrow [Fe(CN)_5H_2O]^{-3} + CN^-$$

(8-1)

$$[Fe(CN)_6]^{-4} + 6H_2O + light \rightarrow [Fe(OH)_2(H_2O)_4] + 2HCN + 4CN^-$$

This reaction has received considerable study *in vitro* (see Young, 1995, for a review) and is known to proceed to completion via photodecomposition. However, the reaction pathways are not well understood beyond the photolytic loss of the first cyanide anion.

During this study a surface water system receiving cyanide-contaminated water from several springs was monitored from November 1998 through August 2000. Sampling dates are shown in Table 8-1. The study site also contained two pilot-scale constructed wetlands, which were being used to assess the potential usefulness of such a treatment approach to reduce the concentration of cyanide and other contaminants in the effluent from one of the cyanide-contaminated springs.

Table 8-1. Field Sampling Dates.

Sampling trip	Site activities	Dates on site
1	Field data collection, water samples collected for return to laboratory	November 16-17, 1998
2	Field data collection, water samples collected for return to laboratory	March 15, 1999
3	Field data collection, tracer study, water samples collected for return to laboratory	July 11-14, 1999
4	Field data collection, volatilization study, water collected for return to laboratory	November 19-21, 1999
5	Field data collection, volatilization study, water, and sediment samples collected for return to laboratory	March 10-11, 2000
6	Volatilization study, sediment samples collected for return to laboratory	August 6-9, 2000

Samples were taken for analysis at 13 main sites over the course of the two years along Springfield Branch, Blue Pond Drainage, and Duck Spring (Figure 8-1) with samples occasionally taken at four additional sites. These locations were chosen for a variety of purposes including spring location, stream configuration, and ease of access. Site identifiers and descriptions are given in Table 8-2.

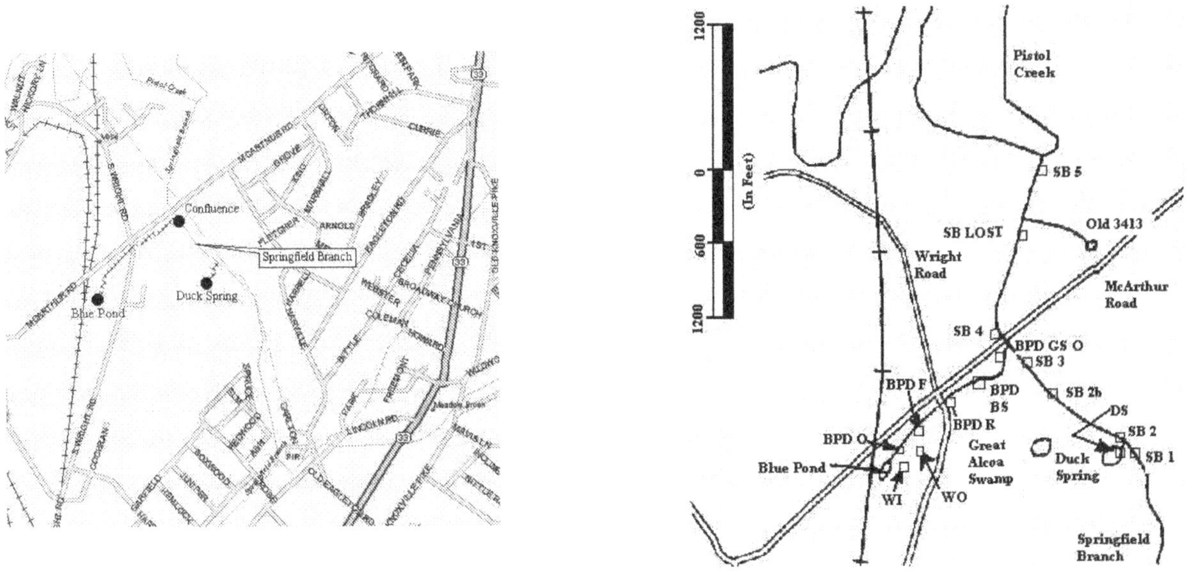

Figure 8-1. Two Views of the Study Area.

Table 8-2. Site Descriptions.

Site Identifier	Site Description	Distance from Confluence of Springfield Branch and Blue Pond Drainage (m)
BPD O	Blue Pond Drainage at dam with V-noch weir	368.5
BPD F	Blue Pond Drainage at four inch flume	292.9
BPD R	Blue Pond Drainage ~ 1.5 m downstream of Wright Road	213.7
BPD BS	Blue Pond Drainage before entering Great Alcoa Swamp	118.9
BPD GS O	Blue Pond Drainage at Great Alcoa Swamp Outflow before confluence with Springfield Branch	6.1
WI	Inflow pipe into constructed wetland 1 from Blue Pond Drainage ~ 12.2 m upstream from dam	383.1
BWC	End of constructed wetland 1 near pipe to constructed wetland 2	368.5
WO	Effluent pipe from constructed wetland before reentering Blue Pond Drainage	353.3
DS	Duck Spring as it enters the outflow channel to Springfield Branch	576.1
DS O	Duck Spring outflow channel ~1.5 m before confluence with Springfield Branch	564.2
SB 1	Springfield Branch ~ 38.1 m before confluence with Duck Spring outflow	600.8
SB 2	Springfield Branch at confluence with Duck Spring outflow	562.7
SB 2b	Springfield Branch ~ 137.2 m before confluence with Blue Pond Drainage	181.1
SB 3	Springfield Branch ~ 21.3 m before confluence with Blue Pond Drainage	21.3
SB 4	Springfield Branch ~ 9.1 m after confluence with Blue Pond Drainage	-9.1
SB LOST	Springfield Branch ~ 15.3 m before confluence with drainage from OLD 3413	-256.9
SB 5	Springfield Branch ~ 30.5 m before confluence with Pistol Creek	-428.2

Measurements were made at each sampling location (Figure 8-1 and Table 8-2) during each sampling period for depth, width, water velocity, pH, turbidity, temperature, and conductivity. These data provided fundamental information about the geometry and water quality of the system. Water samples from each location were returned to the laboratory to be analyzed for cyanide species, sodium, potassium, chloride, nitrate, nitrite, and phosphate.

8.2.1 Site Descriptions

8.2.1.1 Blue Pond Drainage

The Blue Pond Drainage extended from Blue Pond to Springfield Branch (Figure 8-1). Along this channel were several constructed features that made good location identifiers. The first was a dam across the channel with a V-notch weir located approximately 18 m downstream from the downstream boundary of the Blue Pond. Approximately 76 m further downstream was a 4" flume placed in the stream so that all water had to pass through it unless under flood conditions. From the flume, the drainage channel paralleled McArthur Road until it reached the Great Swamp. The flow became comparatively diffuse during passage through the Great Swamp, but upon exiting, flow concentrated and continued as a well-defined stream for a final 30.5 m to its confluence with Springfield Branch. Approximately 90% of the channel was protected from direct sunlight by vegetative cover.

Approximately 15 m before the weir, a pipe had been installed to divert a portion of the Blue Pond effluent into the first experimental wetland (see wetland description below). This water passed through both wetlands and reentered the drainage channel 15 m downstream from the dam.

Blue Pond is an abandoned quarry that is mainly spring fed but receives limited runoff from a nearby fallow field and surrounding woodland. Sediments with the visual appearance of flocculant masses of iron oxyhydroxides had accumulated to a depth of several meters in Blue Pond. These sediments had not compacted and remained flocculant.

8.2.1.2 Pilot-Scale Treatment Wetlands

Two experimental wetlands were constructed in series near the Blue Pond outflow. The first wetland was designed as a surface-flow wetland to expose influent water to the atmosphere and encourage oxygen transfer, both to permit oxidation of reduced iron and manganese species, and to provide an oxic environment for the support of natural wetland species. Water entering the first wetland passed through a small cascade to enhance aeration.

The second wetland was designed as a subsurface-flow wetland. In operation, however, metal oxide flocs formed in the first wetland remained in suspension as the water entered the subsurface cell. The influent flocs quickly coated the gravel medium of the wetland and plugged the pore spaces thereby reducing the hydraulic conductivity of the wetland bed to the point that the wetland became a surface-flow wetland (BHE Environmental, 2000). Subsequently, the second wetland was rebuilt to perform better in a surface flow mode.

Each wetland was planted with a different plant species to assess the influence of plant species in treatment effectiveness. The first wetland was planted with bulrushes; the second wetland was planted with cattails. Both wetlands were located in a comparatively open part of a small woodland area.

Experimental Wetland 1, as built, was a small shallow body of water, approximately 14.6-m long and 5.2-m wide with an average depth of 0.2 m. It was connected to the head end of the Blue Pond Drainage channel through an underground PVC pipe that diverted into the wetland approximately 3% of the Blue Pond outflow. Experimental Wetland 2 was similar in size and shape to Wetland 1. It was approximately 12.8-m long and 6.7-m wide with an average depth of 0.1 m, and was connected to Experimental Wetland 1 through an underground PVC pipe. Water leaving the wetland reentered the Blue Pond Drainage channel through a second PVC pipe just above the first access road across the drainage.

8.2.1.3 Duck Spring

Duck Spring was a small round spring-fed pool with a mean depth of approximately 0.4 m. The mouth of the spring was located in the rocky bottom opposite from the outflow channel, which led to Springfield Branch (see Figure 8-1). The outflow channel was approximately 28-m long and 2.4-3.4-m wide. The water depth in the outflow channel changes was 0.1-0.3 m at the center of the channel. The entirety of the outflow channel was protected from direct sunlight by the surrounding woodland canopy. Approximately 70% of the Duck Spring was similarly protected but a significant portion of the surface was relatively exposed.

8.2.1.4 Springfield Branch

Springfield Branch was a small stream of varying width, depth, and velocity. The reach of interest was 1067 m long, which covered the distance from a point a short distance above the Duck Spring outflow confluence to the confluence with Pistol Creek (see Figure 8-1). Through this area Springfield Branch received water from three known “point” sources including: the Duck Spring outflow, Blue Pond Drainage channel, and Old 3413, which was the drainage from a small, diffuse spring. Springfield Branch passed under one well-traveled road and was used as a water supply for cattle on a local farm.

8.2.2 Field Measurements

Discharges at each sampling site—except the V-notch weir (BPD O), 4-inch flume (BPD F) and wetlands sites (see Figure 8-1)—was estimated by numerically integrating measured flow velocities over measured cross-sectional areas. The cross-sectional area for a given site was computed by averaging depth and width measurements taken directly in at least three locations at the site, transversally, using a calibrated rod and weighted tape. The water depth at the flume was measured directly by tape against the side of the flume. The depth at the weir was measured at the notch apex.

Water flow velocity was not measured at the wetland sites because the flow rate could be measured with high accuracy as the measured volume collected from the flowing pipe over a known period of time. At every other site, the water velocity was measured at approximately the same transverse locations as the depth using a Marsh/McBirney model 201 portable water current meter (calibration data available on request; Snider, 2001).

A water sample was collected from near the middle of the stream at each site and tested for pH, conductivity, temperature, and turbidity. Two additional water samples were collected at each location. One of the latter samples was treated with sodium hydroxide to keep free cyanide in solution. This sample bottle was shielded from light by covering it with a minimum of three layers of aluminum foil to prevent photolysis of $Fe(CN)_6^{-4}$ that might be present in the sample. The other sample was not preserved with sodium hydroxide so accurate measures could be made of the absorbance spectrum and concentrations of chloride, nitrite, nitrate, phosphate, sodium, and potassium in solution. All samples where placed in a cooler with ice to inhibit biological activity and the loss of free cyanide from solution. The samples were then shipped by overnight courier to Rowley Laboratory at Clarkson University in Potsdam, New York. Upon receipt in Potsdam the samples were stored in the dark in a constant temperature room set at 3.7°C.

8.2.3 HCN Volatilization Rate

The rate of HCN volatilization from waters in the Blue Pond–Springfield Branch system was assessed in a straightforward way using a simple flux chamber and collection system. The device had to meet the following functional requirements:

- Be large enough to collect from a representative surface area (0.2 m^2 was selected);
- Be small enough to permit efficient displacement of chamber contents and avoid chamber accumulation of HCN (V ~ 20 L, depending on installation; HRT of 0.5-1.2 hour possible, depending on pump settings);
- Be capable of securing in a position to allow extended, unattended operation (device anchored by staking and tethering; sampling accomplished by portable vacuum pump capable of drawing 0.5 L/min for 8 hours by rechargeable battery was selected);
- Efficiently collect HCN from the effluent stream (NaOH solution was used for collection purposes).

The approach to the measurement of HCN flux involved the following assumptions:

- The rate of HCN loss was sufficiently small that the concentration of HCN in the water beneath the chamber would remain constant during the test;
- The rate of HCN mass transfer from water to air was not affected by the position and operation of the flux chamber;
- All HCN lost from water to air was captured by the apparatus.

A schematic diagram of the flux chamber, including gas collection system, is shown in Figure 8-2. The device consisted of an inverted 15-L translucent plastic storage box with a plastic foam flotation collar. The flux chamber had a projected area of 0.18 m^2 (0.48 m x 0.37 m) and a 0.11-m headspace. Influent and effluent gases were transported into and out of the chamber using plastic tubing to flush evolved volatile gases into the gas collection bottle. Air was drawn through the chamber at a known rate by pulling a mild vacuum on the chamber with a small air pump. Exit gases flowed into a small bubble diffuser that was immersed in a 0.25-molar solution of sodium hydroxide to capture the cyanide. The setup was run six hours and 18 hours at the point of highest concentration in the first constructed wetland and Duck Spring. The high concentrations were chosen to provide the best available conditions to observe volatilization. The airflow rate, sampling time, and sky conditions were noted during each test. Ambient air was sampled in a similar fashion at each site to permit correction for the cyanide in the air that was entering the flux chamber. The concentration of cyanide in solution was determined by using the colorimetric methods described below.

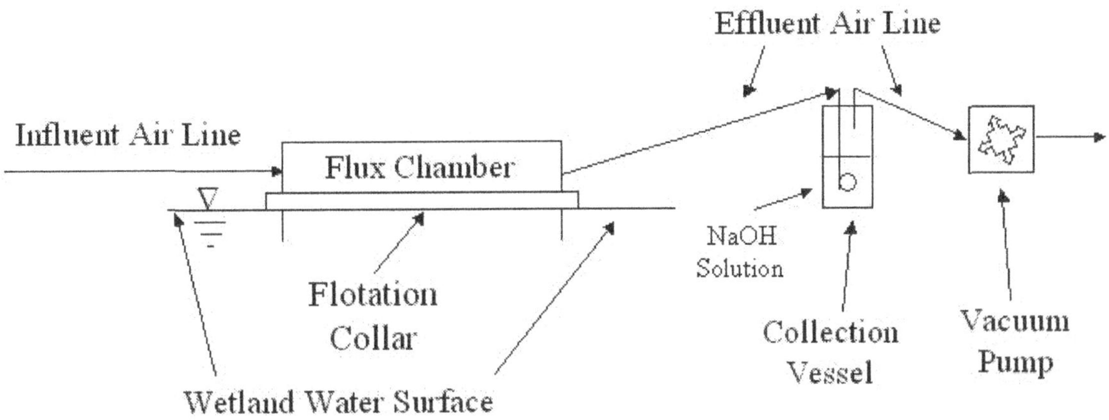

Figure 8-2. Flux Chamber for Measuring HCN Volatilization Rate.

The mass flux of free cyanide was computed by mass balance using elapsed time of the test, the air flow rate, the measured HCN concentration and volume of the collection solution, and similar data for ambient air to isolate the effects of HCN volatilization.

8.2.4 Chemical Analytical Methods

8.2.4.1 Total Cyanide

Total cyanide was measured by digesting 300 mL of sample according to APHA (1992) Method 4500-CN-C. By this method all cyanide in the original sample was converted into free cyanide, which volatilized under acidic conditions and was trapped in a strongly basic NaOH scrubbing solution. At the conclusion of the digestion step, the NaOH solution was diluted to 0.04 M, filtered through a 0.45-mm pore diameter filter, and transferred to a 250-mL high-density polyethylene container, which was stored at 3.7°C for the colorimetric finish by APHA (1992) analysis 4500-CN-D. Absorbance was measured against a reagent blank at 578 nm on a Pye Unicam SP6-550 UV/visible spectrophotometer. Calibration standards for this analysis were linear over the concentration was 0.02-0.2 mg CN/L.

8.2.4.2 Ion Chromatographic Methods

Ion chromatography (IC) was used to measure concentrations of NO_3^-, NO_2^-, Cl^-, and $Fe(CN)_6^{-4}$. IC analyses were performed on a Dionex ion chromatograph model 4000i/4270, which was equipped with either a conductivity or UV/visible detector depending on the chemical species to be quantified. Table 8-3 shows the components used for each analysis. Sample preparation for the IC required that the samples be vacuum filtered through a 0.45-mm pore diameter filter. The samples were then injected into the sample loop of the IC instrument and separated by differential adsorption to the resin in the column.

Table 8-3. Ion Chromatography Analysis Parameters.

Analysis	Type of ion chromatography	Separation column	Guard column	Suppressor
Anions	Isocratic	AS 4A	AG 4A	Yes
Iron cyanide	Gradient	AS 11	AG 11	No

Two different methods of IC were used. These are termed gradient and isocratic elution. In gradient IC, the eluent composition changes by either ratio or type to separate the ions and remove them from the column. Isocratic IC uses the same eluents in the same ratio for the entire run with a regenerant to help remove the anion buildup that would interfere with the analysis from the column. All columns used were Dionex IonPac® columns.

8.2.4.3 Iron Cyanide

Iron cyanide complexes are very stable, which allows them to be separated through anion exchange chromatography. The method used for the $Fe(CN)_6^{-4}$ analyses reported here were essentially those recommended in Dionex Application Note 55 (1988). Without modification, the procedure had a detection limit for $Fe(CN)_6^{-4}$ of 0.04 mg/L as CN. Unfortunately, nearly all samples collected from the field were below this detection limit. A higher level of accuracy was eventually possible by the acquisition of an apparatus to provide sample preconcentration. The latter took a large sample from a secondary concentration pump and put it through a concentrator column attached to a secondary injection valve, which bypassed the auto-sampler used for routine analyses. Once the desired amount of sample was concentrated on the column, the injection valve switched to inject the concentrated sample into the Dionex unit. Sample volumes of 5 and 20 mL could be concentrated in this fashion, which decreased the detection limit for $Fe(CN)_6^{-4}$ to 3 mg and 0.2 mg/L, respectively (Hopke, 2000).

8.2.4.4 Anions

Anions [other than $Fe(CN)_6^{-4}$] were analyzed by IC as described in Analysis 4110 B (APHA,1992). To prevent column fouling, the sample solution was passed through a 0.45-mm pore diameter filter before being injected into the ion chromatograph. The sample passed through an anion suppressor before entering the conductivity detector. In the anion suppressor, the separated anions were converted to their highly conductive acid forms and the carbonate-bicarbonate eluent was converted to weakly conductive carbonic acid (APHA, 1992).

8.2.4.5 Diffusible ("Free") Cyanide

Free cyanide (HCN, CN^-) is the most toxic of cyanide, the form most susceptible to volatilization and biodegradation. Because of the low concentration of total cyanide in solution, the method ASTM Determination of Free Cyanide by Microdiffusion (1995) was used to estimate the free cyanide concentration in solution. The procedure involved a colorimetric finish per ASTM Method D 4282-95 (ASTM, 1998a) with absorbance measured at 582 nm on a Pye Unicam SP6-550 UV/visible spectrophotometer. The relationship between cyanide concentration and absorbance was linear at this wavelength for the concentration of 0.0023-0.1 mg CN/L.

8.2.4.6 Atomic Absorption Spectrometry

The cations (K and Na) were analyzed directly (without pretreatment) using a Perkin-Elmer model 3110 atomic absorption spectrometer according to the procedures recommended by Perkin-Elmer (1994). The wavelength and slit width used for each analysis are shown in Table 8-4.

Table 8-4. Analysis Variables Used for Each Cation in the Atomic Absorption Spectrometry.

Element	Wavelength (nm)	Slit width
K	766.5	0.7
Na	330.2	0.7

8.2.5 Assessing Biological Utilization of HCN

Observations to be discussed in the results indicated significant loss of total cyanide as Blue Pond water passed through the wetland system. Several species of bacteria are known to be capable of degrading cyanide compounds (Goncalves et al., 1998) so it was assumed that observed losses of total cyanide during wetland treatment may have been caused by native microfloral utilization of the HCN as a carbon source, nitrogen source, or electron donor. To provide material to test for possible biological utilization of HCN in the Blue Pond Drainage and wetland system, a composite sample of sediments and included water was obtained by randomly collecting bottom material from several locations within the first constructed wetland. The material was homogenized on-site and transferred for shipment and storage into 1-L containers. After collection, the material was maintained in the dark on ice and shipped by overnight courier to Clarkson University, where it was stored in a constant temperature room at 3.7°C. Sediment material was collected during the March and August 2000 sampling trips. Preliminary comparisons showed that the biological utilization activity was similar between the March and August sediment samples.

Various experimental designs were used to assess the occurrence of biological utilization of HCN, test enrichment effects from nutrient additions, and study aspects of utilization kinetics. The samples from March 2000 were used for initial determinations of the existence of the potential for biological utilization and the effect of nutrient additions on the utilization rate. These experiments were performed as small batch replicates, which were sacrificed for observation over time. Table 8-5 outlines the treatment combinations used during experiments performed as replicate sacrifice experiments.

Table 8-5. Replicate Sacrifice Experiments.

Factor	Sediment slurry batch		
	March 2000*	August 2000*	Control (no sediment)*
HCN concentration range (mg/L)	0-0.5	0-0.1	0-0.5
Tryptic soy broth	+	-	+
Wastewater	+	-	+
No amendment	+	+	+

* inclusion (+) or exclusion (-) of factor in treatment combinations for experimental series

The samples from August 2000 were used to determine the dependence of the utilization rate on the temperature and concentration. The experiments using the August material were performed in large batches with resampling of the solution over time. These experiments were performed in a constant temperature room set at 10, 20, or 30°C. Table 8-6 presents the factorial design for the large batch experiments on temperature and concentration.

Table 8-6. Treatment Combinations for Batch/Repeat Sample Study of HCN Utilization Kinetics.*

↓ HCN Concentration	Temperature →		
	10°C 0.014 mg/L	20°C 0.014 mg/L	30°C 0.014 mg/L
	10°C 0.04 mg/L	20°C 0.04 mg/L	30°C 0.04 mg/L
	10°C 0.1 mg/L	20°C 0.1 mg/L	30°C 0.1 mg/L

* All treatment combinations included wetland sediments at a solid:solution ratio of 11:1000 (w/v)

A typical experiment to determine the biological utilization rate of cyanide involved mixing 200 mL of the sediment-water slurry (11 mg/mL) with 5 mL of 0.025-M phosphate buffer (pH = 7). Any nutrient amendments were also added at this point. An experimental control consisting of deionized water and no sediment slurry was treated in the same manner to evaluate whether some of the reduction in cyanide could be attributed to volatilization. The solution was stirred on a magnetic stir plate while it equilibrated to the room temperature. Once equilibration had occurred, 1 mL of KCN solution in 0.2% sodium hydroxide solution was added to the buffered slurry. The KCN addition varied in concentration to create a cyanide concentration between 0.005 and 0.5 mg CN/L. The pH was circumneutral.

At this point, the replicate sacrifice experiments were mixed rapidly for one minute, separated into 25-mL lots, placed in separate containers, and attached to a wrist action shaker. At known points in time, containers of the water–sediment slurry and control solution were removed from the shaker. When each sample was removed, one drop of 50% NaOH solution was added to the solution and mixed quickly to increase the pH and prevent any future biological utilization from occurring. The solution was then filtered through a 0.45-mm pore diameter filter and analyzed for the CN concentration in solution. Replicate samples were sacrificed until the cyanide concentration in solution was no longer detectable or until the last replicate was sacrificed.

The large batch solution was rapidly mixed for 1 minute before a 25-mL sample was taken to measure the initial cyanide concentration. One drop of 50% sodium hydroxide solution was added to the sample to quickly increase the pH and prevent any future biological utilization from occurring. The remainder of the solution was covered with parafilm and stirred continuously at a rate sufficient to suspend the sediment but without causing air entrainment through the air-water interface. Periodically and at known times, 25-mL samples were removed for immediate analysis of cyanide concentration. Additional samples were removed until the cyanide concentration in solution had reached the detection limit.

Samples acquired during these experiments were immediately preserved with one drop of 50% NaOH solution, vacuum filtered through a 0.45-mm pore diameter filter and the cyanide concentration analyzed. Two methods of colorimetric analysis were used depending on the cyanide concentration. If the solution had a concentration of 0.2-0.5 mg CN/L, then the solution was analyzed using the colorimetric method for total cyanide. When the solution had a concentration <0.2 mg CN/L, the sample was analyzed using the colorimetric method for free (diffusible) cyanide. The detection limit was ~0.003 mg CN/L.

The rate of cyanide removal at each concentration was visualized by plotting the natural log of the concentration over time. When the data points were plotted in this manner, it was apparent that a lag phase occurred before cyanide began to decline. To calculate a meaningful rate of cyanide removal, the duration of the lag phase required estimation. To do this, linear regressions were performed on the log-transformed concentration time series for the date from each experiment using a minimum of three data points to define the linear portion for each. The regression with the least type I error rate was selected as most appropriate to characterize the utilization rate. The slope from the selected regression amounted to an estimate of the first-order rate coefficient for biological utilization. The intercept of the regression line with the natural log of the initial cyanide concentration provided an estimate of the duration of the lag phase. Figure 8-3 provides a graphical representation of this process. Subsequently, all data from an experimental treatment combination were adjusted for the lag period to estimate the utilization rate. Graphically, this amounted to a translation of the time axis to reset time to zero coincident with the end of the lag period.

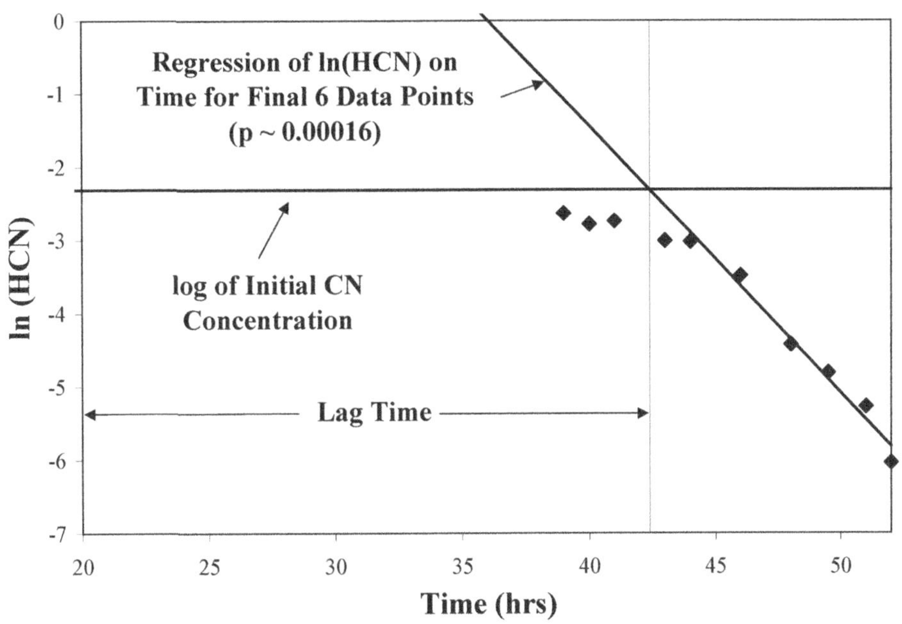

Figure 8-3. Estimation of Time Lag During Bioutilization Experiments.

Finally, analysis of the lag-adjusted rate data involved a minimum squared error fit of an integrated first-order rate law expression to all observations from each treatment combination simultaneously by allowing the rate coefficient to be a simple Arrhenius function of temperature as shown in the following equation:

$$C_t = C_o \; e^{-[(k_{20}\theta^{(T-20)})\,t]} \qquad (8\text{-}2)$$

where:

C_t	=	HCN concentration at time t
C_o	=	initial HCN concentration at the end of the lag period
k_{20}	=	utilization rate coefficient, fit by calibration to 20°C
θ	=	temperature correction coefficient, fit by calibration
T	=	temperature of interest, °C
t	=	elapsed time, hour

The use of this relationship assumes that the utilization rate has a first-order dependence on HCN concentration. This assumption seems justified from the post-lag phase data, which gave compelling and consistent evidence of a semi-log relationship between concentration and time (see Figure 8-3, for example).

8.2.6 Estimation from Censored Data

Quantification of the influent and effluent concentration of cyanide species was crucial to calibration of work to estimate the transformation rates for cyanides in the system. As described previously, water samples were collected from the inlet and outlet of the pilot wetland for the measurement of total cyanide, iron cyanide, and free cyanide. Other data on cyanide concentrations were available in the BHE Environmental report (2000), and these also were used for calibration purposes. In both sets of observations on cyanide concentrations, many values were below MDL. Such data sets are said to be left-censored because data values below the MDL are not available.

From the literature, many methods exist for estimating summary statistics from censored data. One of them is simple substitution—replacing less than MDL values by zero, by the MDL value, or by some value between zero and MDL, such as one half the MDL. After substitution, the mean and variance of the whole data set may be computed. Another method involves ignoring the values below MDL, and computing mean and variance using only the remaining detected values. Either of these two methods is determined to perform poorly for estimating the mean and variance for censored data in comparison with other procedures (Helsel, 1990). The other two distributional methods—maximum likelihood estimation (MLE) and probability plotting—are considered better than the first two methods, and the MLE approach can be much more precise than others. Thus, MLE was the method of choice for estimation from censored cyanide data.

Before using MLE to estimate the mean and variance of censored data, it was necessary to determine whether the data followed a normal or lognormal distribution. In the wetland data set, all cyanide species measurements were left-censored except total cyanide concentration in the wetland influent. Hence, MLE was not needed for these data. The distribution type of the data can be evaluated by checking how closely the data points match a straight line when plotted on normal probability axes or log-probability axes. This assessment was made by determining and comparing data plotted on normal and lognormal probability axes. From the comparison, it seemed clear that HCN influent and effluent concentrations derived from normal distributions as the latter always had higher coefficients of determination ($R^2 \geq 0.933$, $N \geq 9$). Thus, at an MDL of 0.000742 mg/L, the MLE approach determined the mean HCN influent concentration was 0.0043 mg/L, and the HCN effluent mean was 0.000289 mg/L.

When conducting data analysis on total cyanide effluent concentrations by using MLE, it is found that a required parameter (λ) exceeded the range of the MLE calculation table. Thus, probability plotting method was used to estimate the mean of total cyanide effluent concentration with MDL of 0.002 mg/L. Using normal probability plot of the total cyanide effluent data, the 0.5 quintile was estimated to be less than zero. Thus a log-probability plot of the data, which also fit the data well ($R^2 \geq 0.920$, $N \geq 9$), was used to estimate the mean of total cyanide effluent concentration. The concentration of complex cyanide (assumed to be hexacyanoferrate) was estimated by the difference between total and free cyanide by assuming that the main forms of cyanides in the system are metal complex and free cyanide.

The estimation results are summarized in Table 8-7.

Table 8-7. MLE Estimates of Cyanide Species in Pilot-Scale Treatment Wetland.

	Total cyanide	Free cyanide	Iron cyanide
Influent (mg/L)	0.0365	0.0043	0.032
Effluent (mg/L)	0.0021	0.00029	0.0018
Removal efficiency (%)	94.2	93.3	94.3

8.3 Development of the Cyanide Transport and Fate Model

8.3.1 Observations of Cyanide Loss in Blue Pond Pilot Wetlands

During the first two years of the project, considerable observational evidence had accumulated to suggest cyanide species were significantly nonconservative in surface waters. The observed reactivity of cyanide in a natural receiving water is exemplified in Figure 8-4. Depicted therein are average concentrations (with standard error bars) of total cyanide for all samples collected at discrete locations in the pilot-scale constructed wetland. It was necessary to use total cyanide for this assessment because the concentrations of HCN (diffusible cyanide) were generally at or below the MDL for samples from the wetland. The wetland samples were collected as part of other routine project sample activity over five trips to the field site to support the modeling effort. The overlaid solid line is the concentration trend predicted by simulating the total cyanide loss reaction as a simple first-order decay.

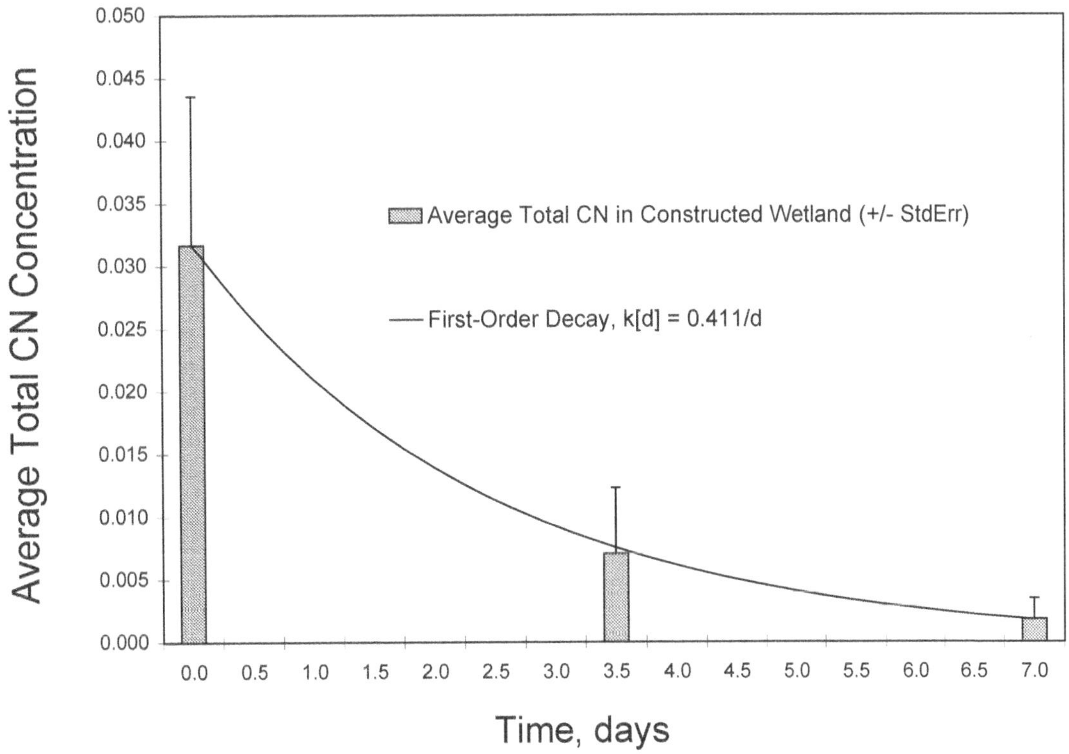

Figure 8-4. Observed Loss of Total Cyanide in Constructed Wetland as a Function of Nominal Hydraulic Residence Time.

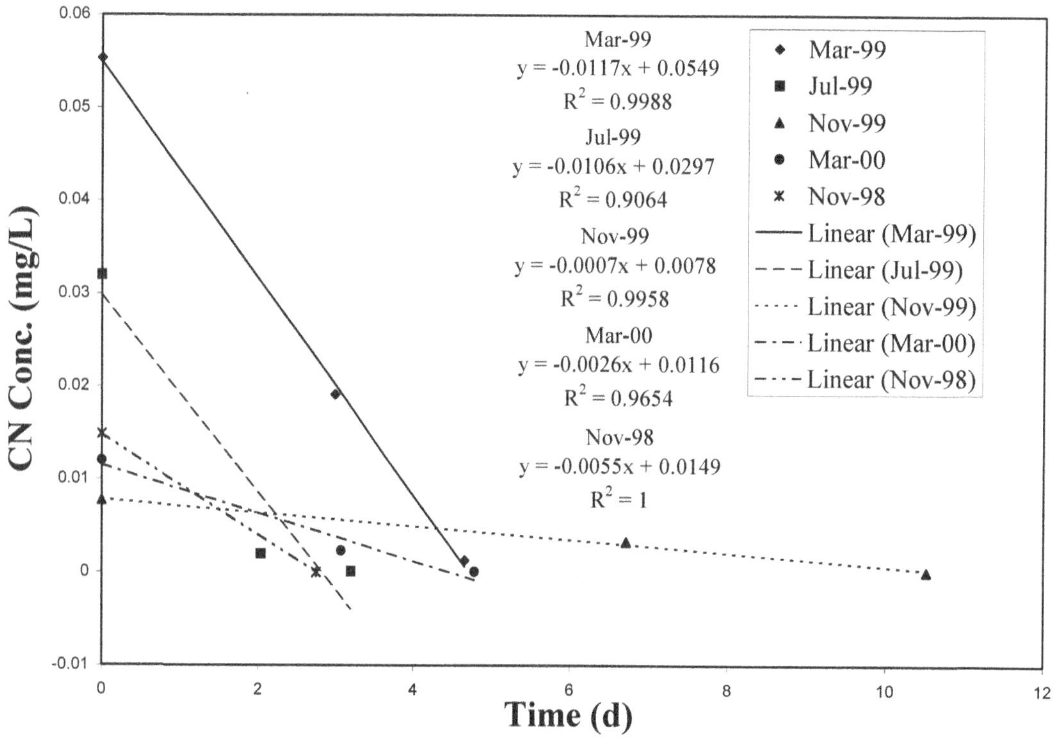

Figure 8-5. Total Cyanide Concentration in System Influent, Midpoint, and Effluent as a Function Hydraulic Residence Time as Observed on Several Sampling Dates.

A somewhat different picture of total cyanide loss through the pilot treatment wetland when the data are plotted from individual sampling trips during which the influent concentration and the hydraulic residence time varied considerably because of seasonally varying climatological and meteorological conditions. These data are given in Figure 8-5.

The linear trendlines through these data suggest a zero-order removal rate for total cyanide in the wetlands, which varied between 0.0007 and 0.0117 mg/(L·d) with an average loss rate of 0.0064 mg cyanide/(L·d) over the five monitoring trips. Although the total cyanide loss appeared to be zero-order, the rate displayed some concentration dependence. This concentration dependence suggested a first-order character as well. That is, influent with a high concentration of cyanide experienced a relatively rapid cyanide loss whereas low influent cyanide concentrations exhibited a lower loss rate. The flow differed for each data set so the hydraulic retention time varied as well. Shown in Figure 8-6 is a plot of the overall loss rate of total cyanide as a function of the median concentration of total cyanide for data collected during each sampling period and available directly from the preceding figure.

These data suggest the existence of some form of rate saturation at high concentrations and also a possible threshold concentration above which the cyanide loss rate exhibits concentration-dependence. This type of behavior is consistent with the kinetics of an allosteric enzyme, and a least squares fit of an equation that describes such a kinetic model was used to generate the trendline in Figure 8-6. This equation has the following form:

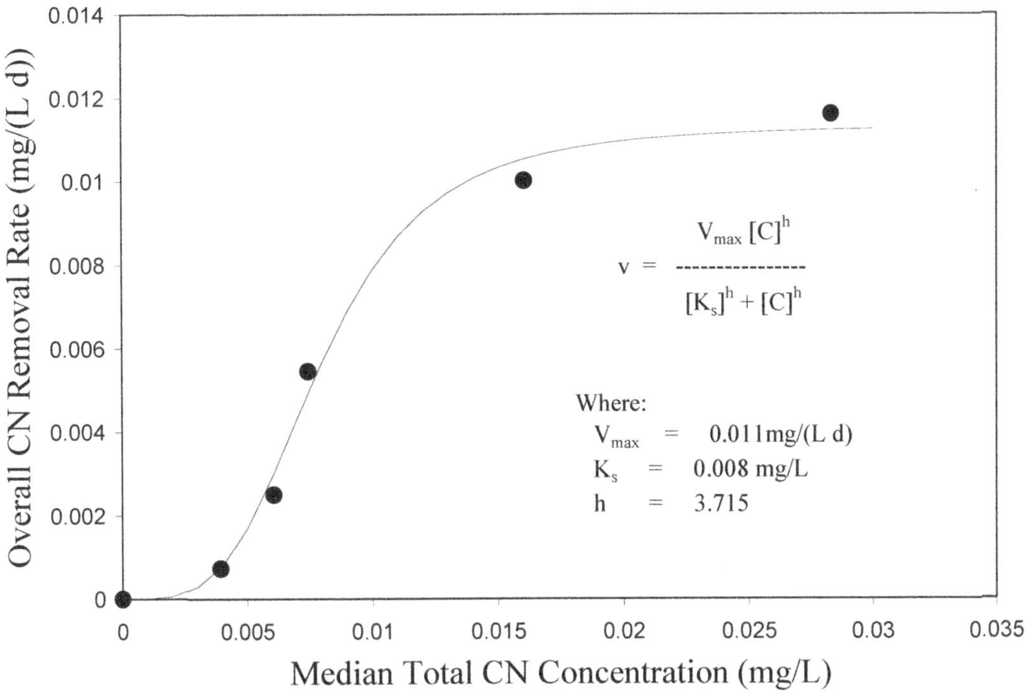

Figure 8-6. Overall Rate of Total Cyanide Loss in Pilot Wetland System as a Function of Median Total Cyanide Concentration—Plotted Function Follows Allosteric Enzyme Kinetic Model.

$$v = \frac{V_{max} C^h}{K_S^h + C^h} \tag{8-3}$$

where:

v	=	overall total cyanide loss rate, mg/(L·d)
V_{max}	=	maximum overall total cyanide loss rate, mg/(L·d)
C	=	total cyanide concentration, mg/L
K_s	=	half saturation constant, mg/L
H	=	empirical exponent

Although the limited data from the pilot wetland fit the allosteric enzyme model quite well ($r^2 = 0.9893$, $P < 0.001$), there is no obvious reason how such a model could provide a fundamental explanation for the behavior of cyanide in the pilot wetland.

The obvious question that arises from viewing such results is what happened to all the cyanide? On the basis of the known biogeochemistry of cyanide species, it seems most likely that the cyanide was lost from the system through volatilization or biotic utilization. But to decide which of these and other processes most influenced the fate of cyanide, and under what circumstances, required more knowledge and understanding. Under such conditions, a simulation tool (a "model") can be very useful to simplify the problem, improve understanding, and lead to better management decisions. Development and application of such a tool is the objective of the modeling work reported herein.

8.3.2 Initial Model Development Activities

Initial development work on the model of cyanide transport and fate advanced along a commonly taken path often called "the finite segment approach" (Thomann and Mueller, 1987; Chapra, 1997). This approach is termed more properly as a finite difference approach because it involves spatial and temporal discretization and any of several possible finite difference approximations to partial differential equations that represent material balances in continuous domains of space and time. In any case, application of the developed model strives to simulate changes in concentrations of complex cyanide and hydrogen cyanide spatially and temporally in the system of interest. This approach has been widely used for both fluvial and standing waters. However, the approach becomes less efficient, computationally, under certain conditions. Unfortunately, as will be described below, application of a generic finite difference model to the specific conditions at the field site revealed the existence of the problem, and it became clear that it would be necessary to pursue an alternative modeling approach.

8.3.3 Generic Finite Difference Model

In the present model development, it is assumed that mixing processes keeps the stream vertically and laterally (cross stream) well mixed. Gradients are expected, therefore, only along the longitudinal axis of the stream. Also, for the present, hydrodynamic dispersion is assumed to be a negligible process, whereas advection and physical/chemical reactions to be important processes contributing to the fate and transport of cyanide in the surface waters. The chemical reactions, including photolysis of complex cyanide, volatilization of free cyanide, and biodecay of free cyanide all are assumed to follow first-order kinetics.

Considering an arbitrary segment from a sequence of individually uniform segments that comprise the system or reach of interest, the overall mass balance equation around segment *i* may be stated as follows:

Rate of mass change in segment *i* = Net advective mass transport ± Mass loss or gain due to chemical reactions ± sources or sinks

The rate of the change of the mass in segment *i*, therefore, may be written as:

$$\frac{dm_i}{dt} = V_i \frac{dC_i}{dt} \tag{8-4}$$

where:

- m_i = mass of chemical in segment *i*, M
- V_i = volume of segment *i*, L^3
- C_i = chemical concentration in segment *i*, L^3/T
- t = elapsed time, T

This assumes that the volume of segment *i* does not change with time.

The net advective transport of mass is the difference between the mass flux upstream and the mass flux downstream (M/T), which may be expressed as follows:

$$Q_{i-1,i}C_{i-1} - Q_{i,i+1}C_i \tag{8-5}$$

where:

- $Q_{i-1,i}$ = upstream flow into segment *i*

$Q_{i,i+1}$	=	ownstream flow out of segment *i*
C_{i-1}	=	concentration in the center of segment *i-1,* approximated to represent the concentration at the interface between segment *i-1* and segment *i*
C_i	=	concentration in the center of segment *I,* using the same approximation of the interfacial concentration

The concentration approximation constitutes a backward differencing scheme for the numerical approximation of the basic differential equations.

The loss or gain of mass due to first-order reactions (M/T) is as follows:

$$\pm K_i C_i V_i \tag{8-6}$$

where

K_i = the overall first-order reaction rate, 1/T

Finally, the source or sink (M/T) of chemical in segment *i* is given as follows:

$$\pm W_i \tag{8-7}$$

The complete mass balance around segment *i* is then as follows:

$$V_i \frac{dC_i}{dt} = Q_{i-1,i} C_{i-1} - Q_{i,i+1} C_i \pm K_i C_i V_i \pm W_i \tag{8-8}$$

There are two cyanide species involved in modeling the transport and fate of cyanide in surface waters, hexcyanoferrate (II) ($Fe(CN)_6^{-4}$), and hydrogen cyanide(HCN). Let C_{FeCN} be the molar concentration of $Fe(CN)_6^{-4}$, and C_{HCN} be the molar concentration of HCN. Also, assume the molar stoichiometry of photolysis to be as follows:

$$Fe(CN)_6^{-4} \rightarrow 6\ HCN \tag{8-9}$$

Mass balances regarding the cyanides around segment *i* may thus be written as follows:

$$V_i \frac{dC_{FeCN,i}}{dt} = Q_{i-1,i} C_{FeCN,i-1} - Q_{i,i+1} C_{FeCN,i} - Kph_i C_{FeCN,i} V_i \pm W_{FeCN,i} \tag{8-10}$$

$$V_i \frac{dC_{HCN,i}}{dt} = Q_{i-1,i} C_{HCN,i-1} - Q_{i,i+1} C_{HCN,i} + 6Kph_i C_{FeCN,i} V_i - (Kv_i + Kd_i) C_{HCNi} V_i \pm W_{HCN,i} \tag{8-11}$$

where:

Kph_i	=	photolysis rate of hexcyanoferrate, 1/T
Kv_i	=	volatilization rate of hydrogen cyanide, 1/T
Kd_i	=	biodecay rate of hydrogen cyanide, 1/T
$W_{FeCN,i}$	=	sources or sinks of hexacyanoferrate, mole/T
$W_{HCN,i}$	=	sources or sinks of hydrogen cyanide, mole/T

Let the first segment is a boundary segment. The mass balance equations for the first segment can be written as follows:

$$V_1 \frac{dC_{FeCN,1}}{dt} = Q_{u1} C_{FeCN,u} - Q_{1,2} C_{FeCN,1} - Kph_1 C_{FeCN,1} V_1 \pm W_{FeCN,1} \tag{8-12}$$

$$V_1 \frac{dC_{HCN,1}}{dt} = Q_{u1} C_{HCN,u} - Q_{1,2} C_{HCN,1} + 6Kph_1 C_{FeCN,1} V_1 - (Kv_1 + Kd_1) C_{HCN,1} V_1 \pm W_{HCN,1} \qquad (8\text{-}13)$$

where:

Q_{ul} = upstream flow discharging into the first segment, L^3/T
$C_{FeCN,u}$ = upstream input concentration of hexacyanoferrate, M/L^3
$C_{HCN,u}$ = upstream input concentration of hydrogen cyanide, M/L^3

Now back to the mass balance equations around segment *i* – divided by V_i and rearranged, the following equations are obtained as the general equations for the *i*th segment:

$$\frac{dC_{FeCN,i}}{dt} = \pm \frac{W_{FeCN,i}}{V_i} + \frac{Q_{i-1,i}}{V_i} C_{FeCN,i-1} - \left(\frac{Q_{i,i+1}}{V_i} + Kph_1 \right) C_{FeCN,i} \qquad (8\text{-}14)$$

$$\frac{dC_{HCN,i}}{dt} = \pm \frac{W_{HCN,i}}{V_i} + \frac{Q_{i-1,i}}{V_i} C_{HCN,i-1} + 6Kph_i C_{FeCN,i} - \left(\frac{Q_{i,i+1}}{V_i} + Kv_i + Kd_i \right) C_{HCN,i} \qquad (8\text{-}15)$$

For the first segment, following the similar procedure, the equations are as follows:

$$\frac{dC_{FeCN,1}}{dt} = \left(\pm \frac{W_{FeCN,1}}{V_1} + \frac{Q_{u1}}{V_1} C_{FeCN,u} \right) - \left(\frac{Q_{1,2}}{V_1} + Kph_1 \right) C_{FeCN,1} \qquad (8\text{-}16)$$

$$\frac{dC_{HCN,1}}{dt} = \left(\pm \frac{W_{HCN,1}}{V_1} + \frac{Q_{u1}}{V_1} C_{HCN,u} \right) + 6Kph_1 C_{FeCCN,1} - \left(\frac{Q_{1,2}}{V_1} + Kv_1 + Kd_1 \right) C_{HCN,1} \qquad (8\text{-}17)$$

The complete set of equations (for n segments) to be solved simultaneously is thus given by the following set of reactions designated 8-18:

$$\frac{dC_{FeCN,1}}{dt} = \left(\pm \frac{W_{FeCN,1}}{V_1} + \frac{Q_{u1}}{V_1} C_{FeCN,u} \right) - \left(\frac{Q_{1,2}}{V_1} + Kph_1 \right) C_{FeCN,1} + 0 + \Lambda \qquad + 0$$

$$\frac{dC_{HCN,1}}{dt} = \left(\pm \frac{W_{HCN,1}}{V_1} + \frac{Q_{u1}}{V_1} C_{HCN,u} \right) + 6Kph_1 C_{FeCN,1} - \left(\frac{Q_{1,2}}{V_1} + Kv_1 + Kd_1 \right) C_{HCN,1} + 0 + \Lambda \qquad + 0$$

$$\frac{dC_{FeCN,2}}{dt} = \pm \frac{W_{FeCN,2}}{V_2} \qquad + \frac{Q_{1,2}}{V_2} C_{FeCN,1} \quad + \quad 0 \quad - \left(\frac{Q_{2,3}}{V_2} + Kph_2 \right) C_{FeCN,2} + 0 + \Lambda \quad + 0$$

$$\frac{dC_{HCN,2}}{dt} = \pm \frac{W_{HCN,2}}{V_2} \quad + \quad 0 + \frac{Q_{1,2}}{V_2} C_{HCN,1} + 6Kph_2 C_{FeCN,2} - \left(\frac{Q_{2,3}}{V_2} + Kv_2 + Kd_2 \right) C_{HCN,2} + 0 + \Lambda + 0$$

$$\frac{dC_{FeCN,n}}{dt} = \pm\frac{W_{FeCN,n}}{V_n} + 0 + \Lambda \quad + 0 \quad + \frac{Q_{n-1,n}}{V_n}C_{FeCN,n-1} + 0 - \left(\frac{Q_{n,d}}{V_n} + Kph_n\right)C_{FeCN,n} \quad + \; 0$$

$$\frac{dC_{HCN,n}}{dt} = \pm\frac{W_{HCN,n}}{V_n} + 0 + \Lambda \quad + 0 + \frac{Q_{n-1,n}}{V_n}C_{HCN,n-1} + 6Kph_nC_{FeCN,n} - \left(\frac{Q_{n,d}}{V_n} + Kv_n + Kd_n\right)C_{HCN,n}$$

where:

$Q_{n,d}$ = flow discharging from *n*th segment to downstream, L^3/T

These equations constitute a set of coupled first-order, time-dependent differential equations. To more conveniently find the numerical solutions to these equations, the set of equations may be re-expressed in matrix form, or as follows:

$$(D) = [A](C) \tag{8-19}$$

where:

(D) = 2n vector of time derivatives
(C) = (2n+1)×1 vector of concentrations
[A] = (2n+1)×2n matrix of coefficients, loads, flows, and volumes

This notation facilitates adapting the model for a site-specific application when solved using numerical techniques.

A preliminary simulation for six segments over a 10-day period was conducted by choosing the following values for parameters:

Q = flow rate, constant at 30 m^3/day
V = volume, varying from 50, 20, 20, 10, 40, 30 m^3 for each segment
Kph_i = photolysis rate of hexcyanoferrate, constant at 10/day for each segment
Kv_i+Kd_i = decay rate of hydrogen cyanide, volatilization rate + biodecay rate, constant at 0.1/day
$C_{FeCN,1}$ = initial concentration of FeCN in segment 1, constant at 8.3 mmol/m^3
$C_{FeCN,u}$ = upstream concentration of FeCN in segment 1, constant at 8.3 mmol/m^3
$C_{FeCN,i}$ = initial concentration of FeCN in segment i≠1, constant at 0 mmol/m^3
$C_{FeCN,u}$ = upstream concentration of FeCN in segment i≠1, constant at 0 mmol/m^3

No loading of hexcyanoferrate into the stream was assumed. The results are shown in Figure 8-7.

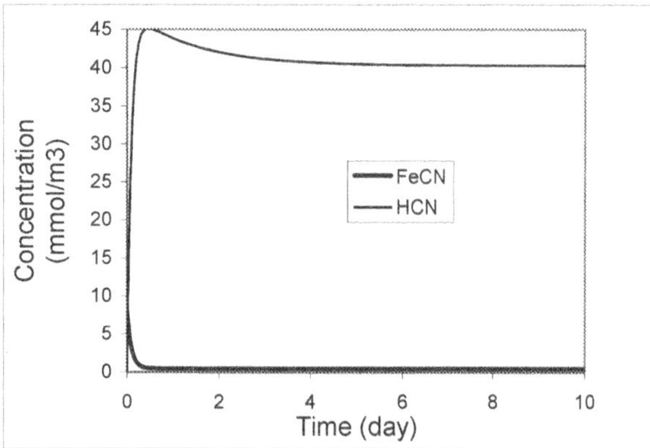

Variation of concentrations of cyanides in Segment 1

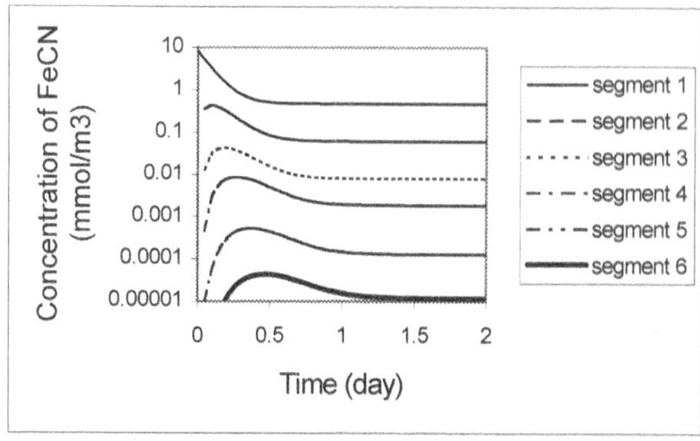

Variation of concentration of FeCN in each segment during two days of simulation; (Note: vertical scale is logarithmic)

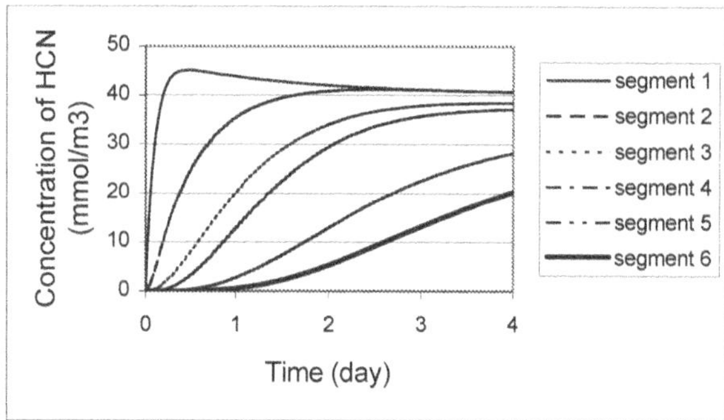

Variation of concentration of HCN in each segment during 4 days of simulation.

Figure 8-7. Generic Simulation of Finite Difference Model of Cyanide Transport and Fate in a Surface Water.

As shown by these data, the concentration of $Fe(CN)_6^{-4}$ decreases rapidly in first half day and then levels off to a constant value, whereas the concentration of HCN increases in the beginning and reaches a peak value at one-half day, then decreases with time. After about four days, the concentration of HCN in segments 1 and 2 approach steady state. The second and third figures show that the concentration of $Fe(CN)_6^{-4}$ and concentration of HCN all decrease with distance. The preliminary simulation results are consistent with the hypothesis that HCN is gained from the photolysis of $Fe(CN)_6^{-4}$ and also lost because of volatilization and biodecay.

8.3.4 Alternative Model Framework

The geometry of the stream at the field site (narrow, shallow, short hydraulic residence time) led to difficulties in applying the planned finite difference scheme (described and tested above) to solve the mass balance equations. The specific problem was numerical dispersion of a greater magnitude than the actual dispersion observed during tracer trials. Modifying the computational approach to the geometric constraints could have been achieved by greatly increasing the discretization of the spatial domain, but this was considered both resource-intensive and a somewhat arbitrary trade-off. Ultimately, it was not selected as the approach to resolve the problem. Rather, the problem was resolved by recasting the conceptual modeling approach from a fixed spatial domain (Eulerian) to a movable one (Lagrangian).

This change was preceded by considerable work on obviating the problem of numerical dispersion as a computational issue for application to the selected field site. This was done by adapting the approach termed Lagrangian parcels (Yapa et al., 1996). The Lagrangian approach is described in this section with results that demonstrate clear performance improvements when applied to conditions at the selected field site.

As described above, the governing equations for cyanide fate and transport include advection, dispersion, and reaction. The reactions include photolytic decay of iron cyanide [FeCN, ~hexacyanoferrate, $Fe(CN)_6^{-4}$] to produce free cyanide (HCN), whereas HCN undergoes loss by volatilization and biodecay. All fate-determining reactions are assumed to be first order, an assumption that is generally very well supported by existing process experimentation, as will be seen.

The governing equation for chemical transport in a channel can be written as follows:

$$\frac{\partial C}{\partial t} = -u\frac{\partial C}{\partial x} + D_x\frac{\partial^2 C}{\partial x^2} \tag{8-20}$$

where:

C	=	concentration, M/L^3
u	=	advection velocity of the channel, L/T
D_x	=	longitudinal dispersion coefficient, L^2/T
X	=	distance, L
t	=	time, T

This equation describes concentration varying with time because of advection and dispersion, where the first term in the right-hand side represents advection and second term represents dispersion. A first-order reaction can be added to yield the advection-dispersion-decay model as follows:

$$\frac{\partial C}{\partial t} = -u\frac{\partial C}{\partial x} + D_x\frac{\partial^2 C}{\partial x^2} - kt \tag{8-21}$$

where:

k	=	first-order reaction rate coefficient, 1/T

When applied to free (HCN) and complex cyanide (FeCN) fate and transport, the governing equations obtain the following:

$$\frac{\partial C_{FeCN}}{\partial t} = -u\frac{\partial C_{FeCN}}{\partial x} + D_x\frac{\partial^2 C_{FeCN}}{\partial x^2} - k_{ph}C_{FeCN} \tag{8-22}$$

$$\frac{\partial C_{HCN}}{\partial t} = -u\frac{\partial C_{HCN}}{\partial x} + D_x\frac{\partial^2 C_{HCN}}{\partial x^2} + k_{ph}C_{FeCN} - (k_b + k_v)C_{HCN} \tag{8-23}$$

where:

C_{FeCN} = concentration of FeCN, M/L^3
C_{HCN} = concentration of HCN, $M/^{L3}$
k_{ph} = photolysis rate of FeCN, 1/T
k_b, k_v = biodecay rate and volatilization rate of HCN, 1/T

As described in the previous section, a common method to solve these equations involves discretization of the spatial domain based on appropriate uniformity conditions, development of a finite difference representation for the governing equations over each segment of the spatial grid, and application of a suitable solution scheme to integrate the resulting set of coupled equations. But for conditions specific to the application of interest, as noted previously, the solution can degrade because of numerical dispersion. The latter is a computational artifact that results from using finite difference approximations to represent the continuous spatial and temporal derivatives in the governing equations (Thomann and Mueller, 1987). The approximation errors may be limited by using a sufficiently fine spatial grid and small time steps to satisfy solution stability criteria. However, this demands excessive computational time.

For example, using a forward-time, backward-space integration provides one finite difference solution scheme for which there is no spatial grid size constraint to ensure a positive solution. Because of this, this approach is widely used in water quality models to solve the appropriate mass balance differential equations. This approach, however, suffers from numerical dispersion caused by backward spatial differencing and explicit time differencing. The use of central spatial differences would eliminate numerical dispersion because of spatial truncation. It is also possible to adjust the approach to minimize numerical dispersion caused by time differences. However, the grid size and time step must be small enough to meet the scheme stability and solution positivity conditions. For the field site of interest for the current work, these requirements lead to an excessive computational requirement.

The Lagrangian parcel method (LPM) is an alternative that yields much less numerical dispersion. Conceptually and computationally, however, it is quite different from the conventional approach. The LPM came into use about 20 years ago and has been increasingly used in environmental modeling such as oil spill model in deep ocean (Yapa and Zheng, 1997), contaminant transport model in groundwater and soil (Abulaban, 2000) and other applications (Clark, 1996; Martin and McCutcheon, 1999).

In this method, a pollutant is represented by an ensemble of a large number of small parcels, and each parcel has a mass and a set of time-dependent spatial coordinates associated with it. During each time step, all parcels are displaced according to a drift velocity and a fluctuation velocity component at their respective locations. The fluctuation component of velocity is considered to be caused by turbulent diffusion and can be represented by the characteristics of a random walk, given by the following:

$$u_{ran} = \sqrt{\frac{2D_x}{\Delta t}} R_{nx} \tag{8-24}$$

where:

u_{ran} = random walk velocity in the x direction, L/T

D_x = contaminant diffusivity in the x direction (in one dimensional domain, D_x is the longitudinal dispersion coefficient), L^2/T

R_{nx} = normally distributed random number, generated using a Gaussian distribution with mean of zero and variance of 1

Δt = the time step, T

The displacement caused by random walk may be computed as the product of the time step and the fluctuation velocity. After all particles have been displaced during a given time step, the concentration in each grid cell of the discretized spatial domain can be computed based on the mass of pollutant particles in the cell volume.

There are several advantages using this method over finite difference method. First, this method is inherently stable because the time steps are selected based on the grid size and local flow velocity. Second, using this method introduces significantly less numerical dispersion compared to the finite difference method because the former does not depend on difference approximations to the governing equations. Third, the simulations are more efficient than those of the finite difference method, because the computations follow pollutant parcels and do not require solving a system of coupled equations. And fourth, it is comparatively easy to handle the spatial and temporal variation of parameters using the LPM, and easy to extend the model from one- to two- and three-dimensional problems. Because of the advantages with this method, it was used in modeling the fate and transport of cyanide in surface water. Figure 8-8 is a flow chart illustrating the structure of the model and the sequence of the computations.

Because the model involved two species, FeCN and HCN, a linked-list method was used in the program to handle insertion and deletion of particles for two species to achieve more efficiency (Yapa et al., 1996).

8.3.5 Suitability of Lagrangian Parcel Method for Modeling

Initial testing of the approach involved a comparison results obtained for simulation of a pulse input between the LPM, a finite difference scheme and the analytical solution to the governing equation:

$$C(x,t) = \frac{M}{A\sqrt{4\pi D_x t}} \exp\left(-\frac{(x-ut)^2}{4D_x t}\right) \qquad (8\text{-}25)$$

where:

M = mass of tracer input

U = drift velocity

D_x = longitudinal dispersion coefficient

A = cross-sectional area of the channel

8.3.5.1 Response to Pulse Input

The example involved an instantaneous release of tracer—1350 mg of neutrally buoyant inert material spilled instantaneously in a channel with uniform velocity of 0.5 m/s. The longitudinal dispersion coefficient was set at 0.2 m/s^2. Figure 8-9 shows the concentration distribution with respect to distance in the channel after 900 s of simulation using LPM and a forward time and backward space first-order explicit finite difference scheme. These figures also show the concentration distribution with respect to time at certain distance in the channel (x = 300 m) using the three methods. The results for the analytical solution coincide nearly exactly with those from applying the LPM. It is reasonably clear from these data that the results from LPM match very closely with the analytical solution, and significant numerical dispersion is introduced by the finite difference scheme.

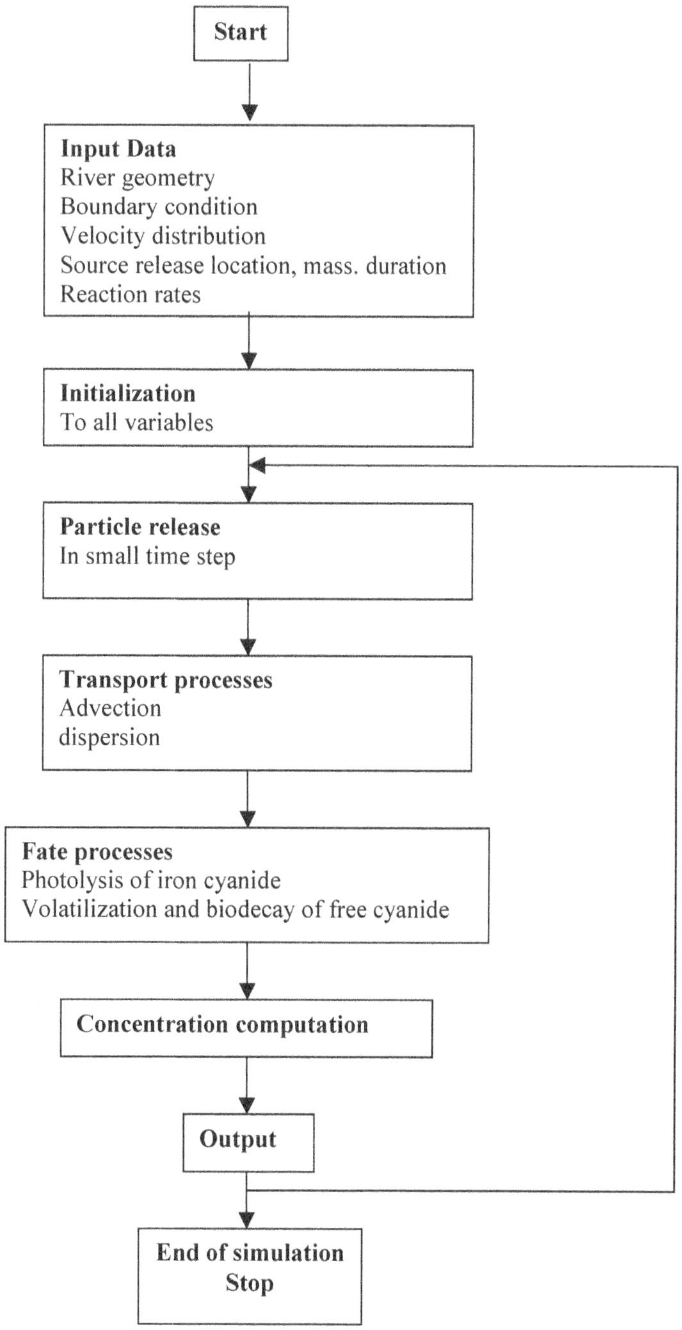

Figure 8-8. A Flow Chart of the Program for Modeling Fate and Transport of Cyanide Using LPM.

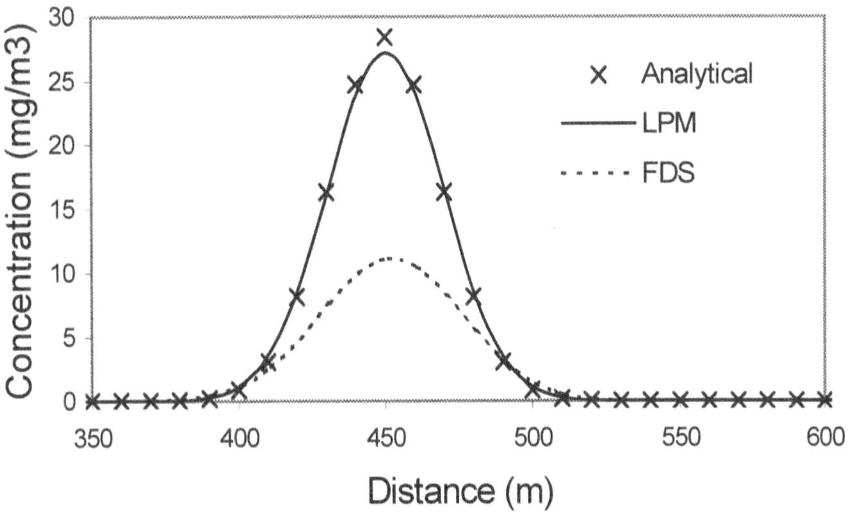

Concentration versus distance at t=900s

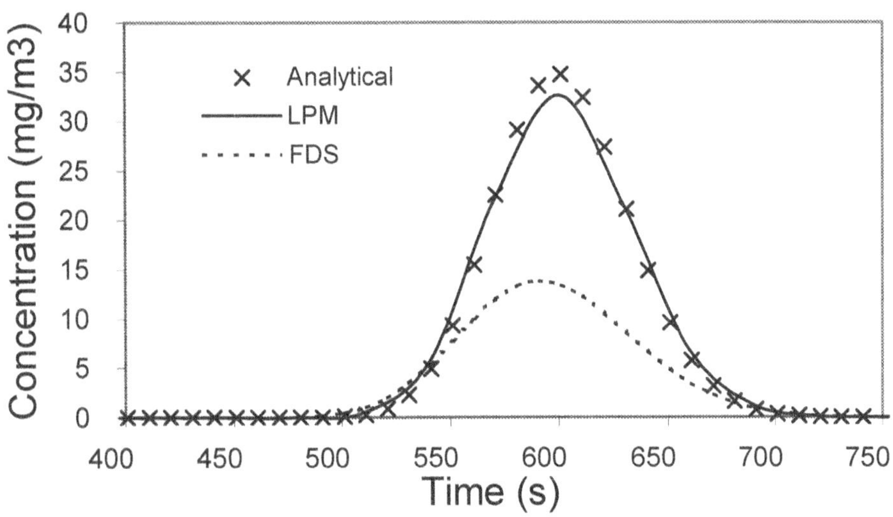

Concentration versus time at x=300m

Figure 8-9. Comparison of Modeling Approaches for Simulating Tracer Transport .

When reaction is incorporated into the model, the analytical solution is given by the following equation:

$$C(x,t) = \frac{M}{\sqrt{4\pi D_x t}} \exp\left[-\frac{(x-ut)^2}{4D_x t} - kt \right] \tag{8-26}$$

Figure 8-10 displays the concentration distribution with respect to distance with a reaction rate of 0.001 s^{-1} at time of 900 s, and the concentration distribution versus time at x = 300 m with the same reaction rate. It again is reasonably clear from the comparison that the result from LPM is nearly identical to the analytical solution.

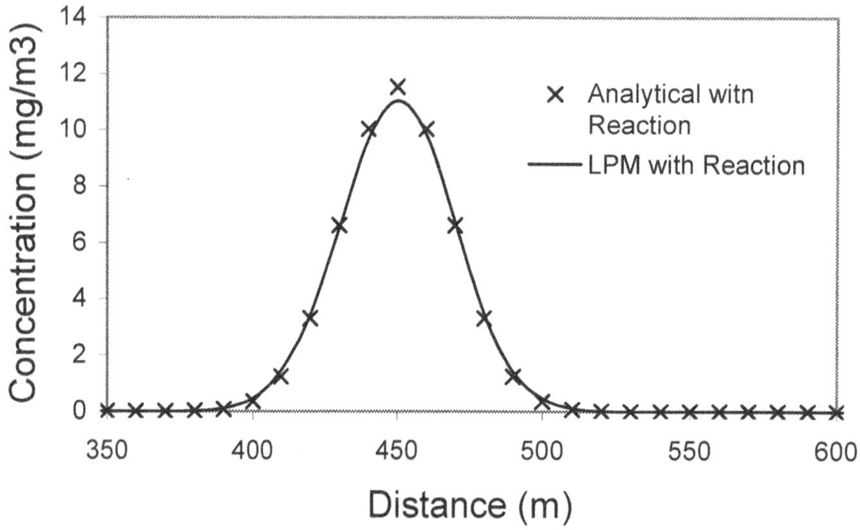

Concentration versus distance with reaction rate of 0.001 s^{-1} at t = 900 s.

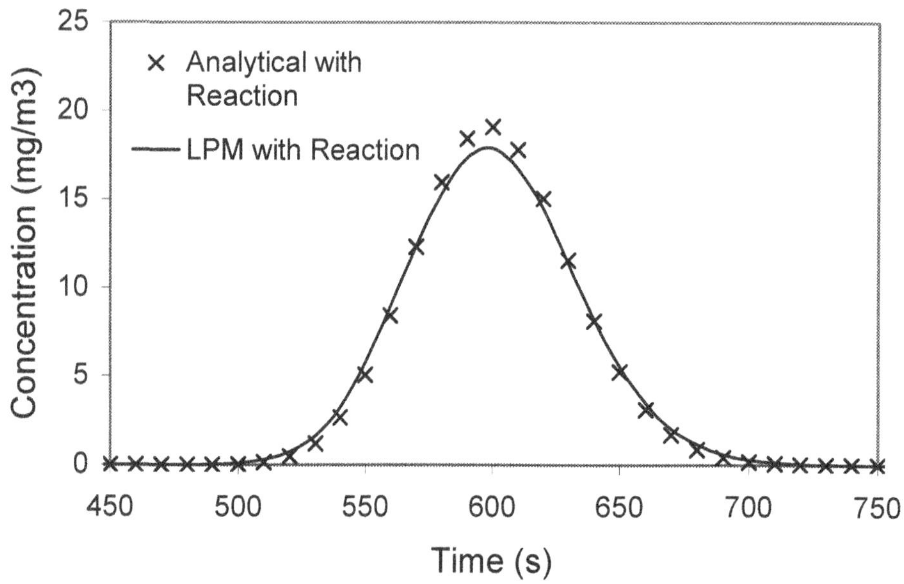

Concentration versus time with reaction rate of 0.001 s^{-1} at x = 300 m.

Figure 8-10. Simulation of Cyanide Reaction by LPM and Analytical Solution.

8.3.5.2 Response to Continuous Input

The analytical solution to the governing equation for a continuous source of pollutant can be expressed:

$$C(x,t) = \frac{C_0}{2}\left[e^{\frac{ux}{2D_x}(1-\gamma)} erfc(\frac{x-ut\gamma}{\sqrt{4D_x t}}) + e^{\frac{ux}{2D_x}(1+\gamma)} erfc(\frac{x+ut\gamma}{\sqrt{4D_x t}}) \right] \tag{8-27}$$

where:

$$\gamma = \sqrt{1 + 4\eta} \tag{8-28}$$

$$\eta = \frac{kD_x}{u^2} \tag{8-29}$$

C_0 the input concentration immediately following a step change.

Figure 8-11 shows the spatial concentration distribution for a step release at time = 900 s for k = 0 and k = 0.001 s^{-1}, respectively, whereas Figure 8-12 shows the change in concentration for a step source at a sampling point 300 m downstream for no reaction and reaction rate of 0.001 s^{-1}, respectively. A value for the initial concentration of 3 mg m^{-3} was used in the simulations. It can be seen from the simulation results that the comparison of the LPM with the analytical solution is very good.

8.4 Experimental, Observational, and Computational Assessment of Cyanide Fate-Determining Processes

To be able to adequately describe the fate of cyanide using a modeling approach requires a substantial amount of groundwork—preliminary data gathering. Especially needed is the best possible information on the fundamental processes depicted in the governing equations of the model.

8.4.1 Hydraulics and Transport

8.4.1.1 Blue Pond Drainage Channel

In addition to the routine calibration data collection, additional fieldwork was conducted to measure the residence time distribution (RTD, "time of travel") for the watercourse that drains from the Blue Pond to Springfield Branch and on to Pistol Creek. This was done to permit resampling of the same water parcel during downstream transport, in effect, a Lagrangian sampling scheme. By doing the latter, it was intended to better observe the effects of photolysis and volatilization on the aqueous cyanide species than could be done with a scheme that required the system to be at steady state. Although the latter is commonly assumed, it may not actually exist.

The approach taken for the tracer tests was a classical assessment of the residence time distribution for a tracer monitored at a fixed location downstream from a pulse input. In this case the tracer was KCl, and specific conductance was monitored to observe downstream movement of the tracer.

It may be observed in the results from the LPM that there can be some small oscillations of concentration around the constant value. That is because of the nature of the random walk process that forms the basis for the method.

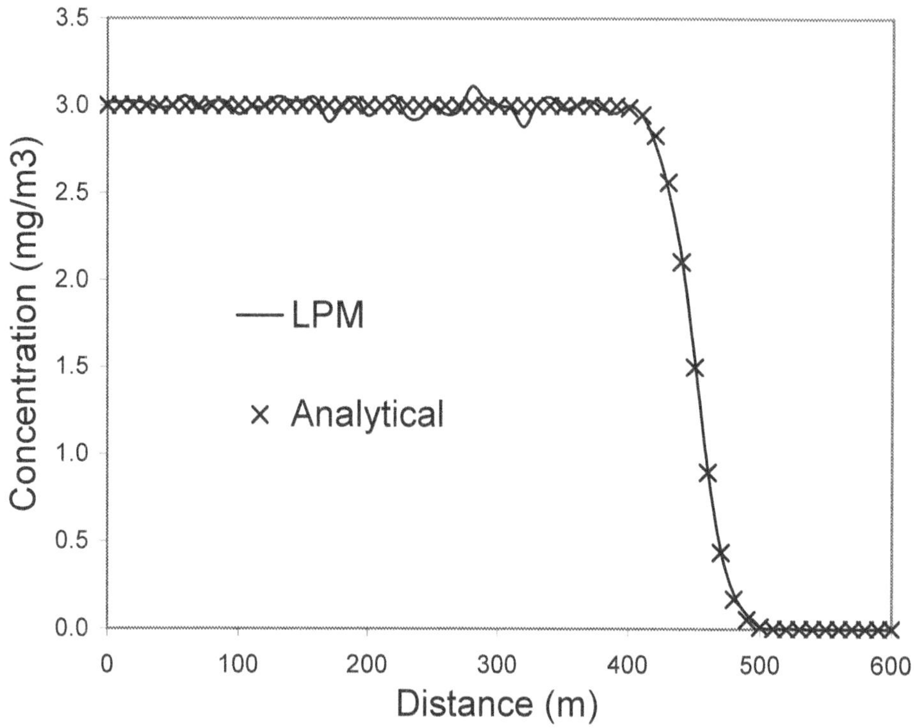

Spatial concentration distribution for continuous release of tracer at t = 900 s.

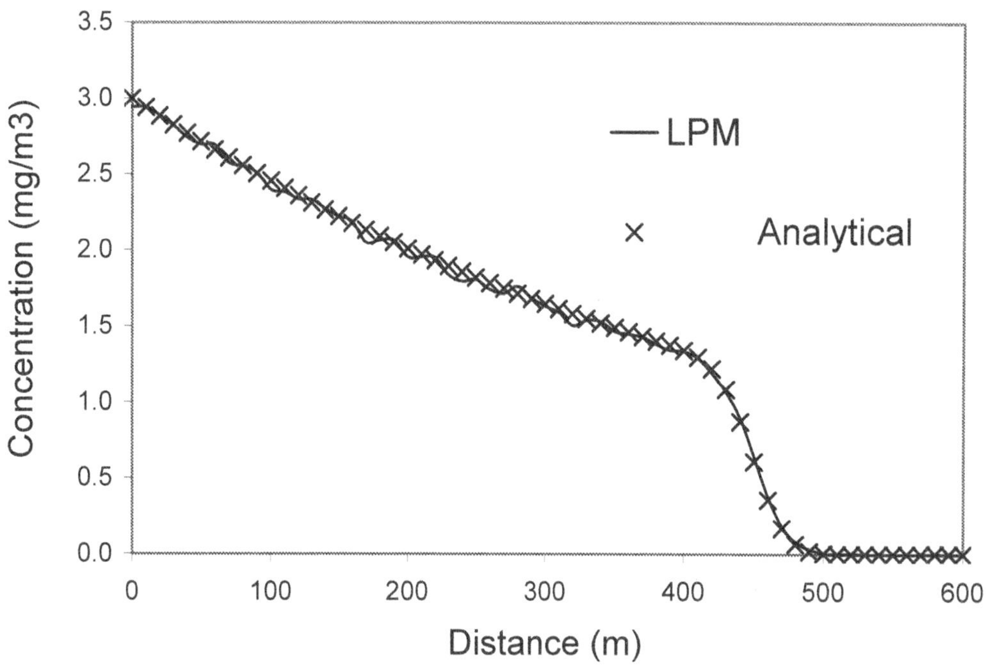

Spatial concentration distribution for continuous release with reaction rate of 0.001 s^{-1} at t = 900 s

Figure 8-11. Simulation of Continuous Input Using LPM and Analytical Solution to Governing Equation.

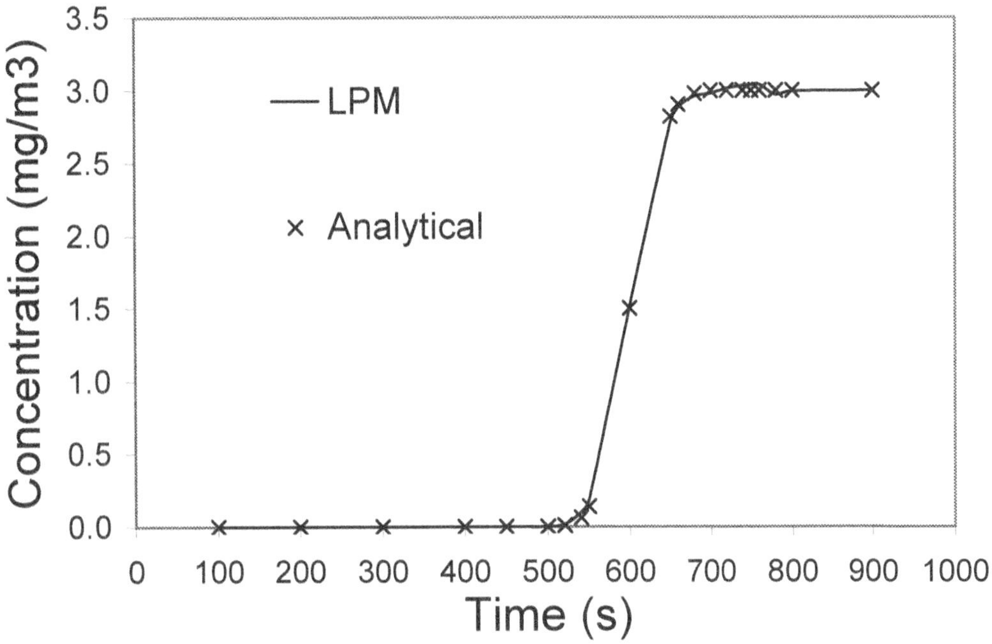

Temporal concentration distribution for continuous release of tracer at x = 300 m

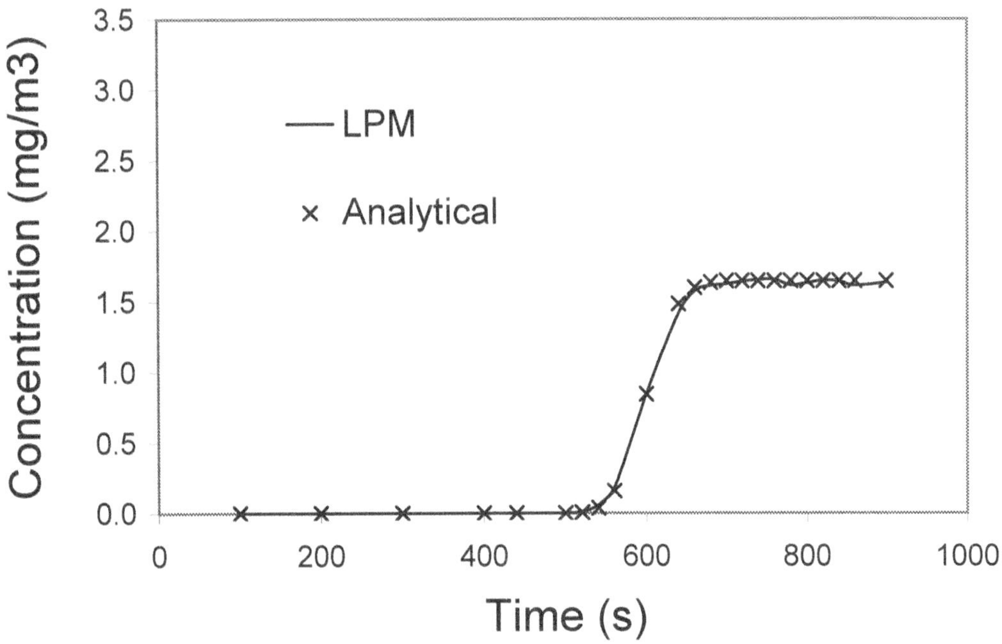

Temporal concentration distribution for continuous release with reaction rate of 0.001 s^{-1} at x = 300 m

Figure 8-12. Simulation of Continuous Input Using LPM and Analytical Solution to Governing Equation.

Shown in Figure 8-13 are data from two tracer tests. One pair of curves presents a view of the time course of specific conductance, whereas the second pair shows the time series for total cumulative mass of tracer passing the monitoring point. Both trials gave similar results and indicate that approximately 50% of the tracer had passed the monitoring location after 40-42 minutes, and that the peak concentration of the tracer pulse would require on the order of 35-37 minutes to reach the monitoring point. These data, combined with measured distances of travel, were used to estimate peak concentration travel times to other locations along the drainage. In turn, these times were used to establish sampling times for the other locations. In addition, as will be discussed subsequently, these data were used to provide estimates of the dispersion coefficient for hydraulic transport in the Blue Pond Drainage.

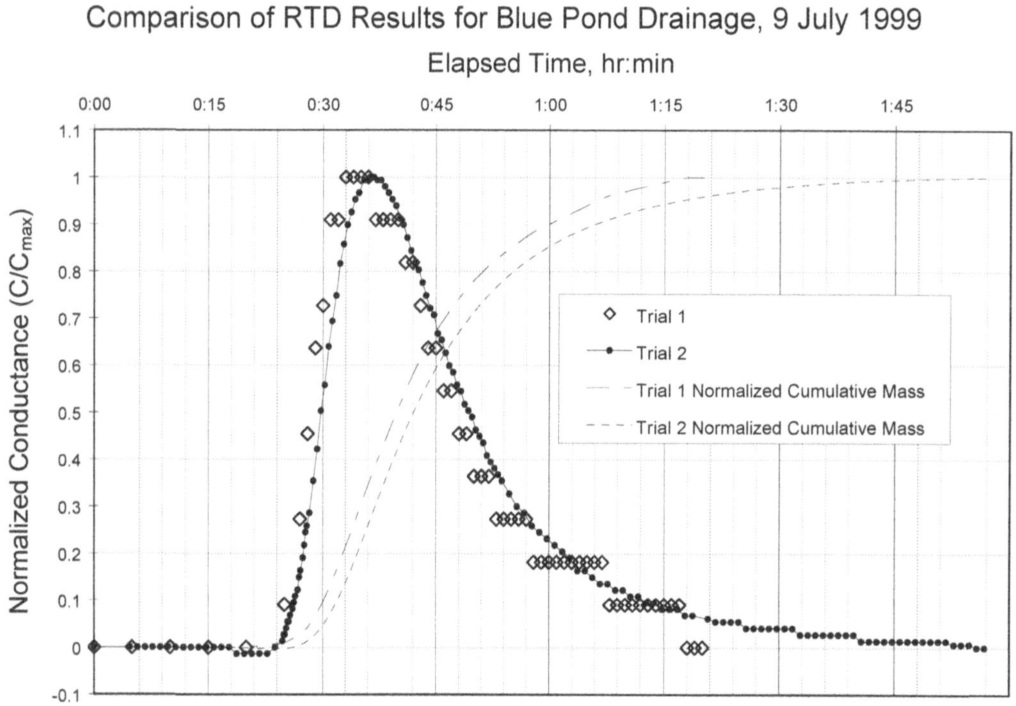

Figure 8-13. Results from Two Tracer Tests of Residence Time Distribution in a Reach in the Blue Pond Drainage.

8.4.1.2 Treatment Wetland Hydraulics

Because of the clear nonconservative behavior exhibited by cyanide in the pilot-scale treatment wetland (two cells in series), it was considered crucial to understand as clearly as possible the hydraulic behavior of the wetland. With that information and data on cyanide species concentrations through the wetland, it would be possible to expect best accuracy for estimates of cyanide fate process parameters based on model calibrations.

Development of a predictive hydraulic model for the wetland began with the analysis of hydraulic residence time distribution E(t) [=RTD(t) for pulsed input]. The latter was estimated from a dye tracer study, the data for which were obtained by recasting results reported by BHE Environmental, Inc. (see Figure 8-14; BHE Environmental, 2000). For the tracer study, a pulse input of rhodamine, a nonreactive fluorescent water tracing dye, was applied to the wetland inflow and rhodamine concentration monitored at the outflow until the concentration approached baseline levels of fluorescence.

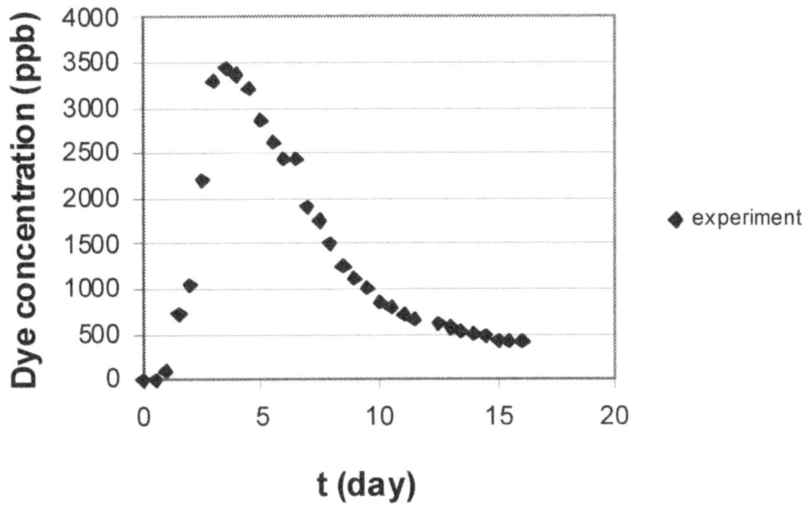

Figure 8-14. Rhodamine Concentration Plotted Against Time for Blue Pond Constructed Wetland (Data from BHE Environmental, 2000).

The "Experimental" curve shown in Figure 8-15 is the RTD(t) as a function of time. The RTD(t) is a normalization of the measured concentration [C(t), data in Figure 8-14], achieved by applying the following equations to the concentration data:

$$\mathrm{RTD(t)} = \frac{\mathrm{C(t)}}{\mathrm{C_0}} \quad (8\text{-}30)$$

$$\mathrm{C_0} = \int_0^\infty \mathrm{C(t)dt} \quad (8\text{-}31)$$

It should be noted that is was necessary to extend the monitoring period to allow the final effluent concentration to approach zero. This was done by linear extrapolation and the period was extended from 16 days to 22 days (Figure 8-15).

The RTD(t) (Figure 8-15) clearly indicated the treatment wetland behaved as a non-ideal reactor. The following assessment contains results for four hydraulic models that were examined for adequacy to characterize the net hydraulic behavior of this non-ideal reactor.

Model 1 – CMFRs in Series. The first model to describe non-ideal flow through the wetland represents the reactor as a cascade or series of N equal-sized completely mixed flow reactor (CMFR in series). To use this model, the parameter N must be determined. The statistical parameters, mean and variance of tracer data are related to N and may be used to estimate a value for N as shown by the relationships depicted below.

$$\mathrm{N} = \frac{\mathrm{t_m}^2}{\sigma^2} \quad , \quad \mathrm{t_m} = \frac{\int_0^\infty \mathrm{tRTD(t)dt}}{\int_0^\infty \mathrm{RTD(t)dt}} \quad (8\text{-}32)$$

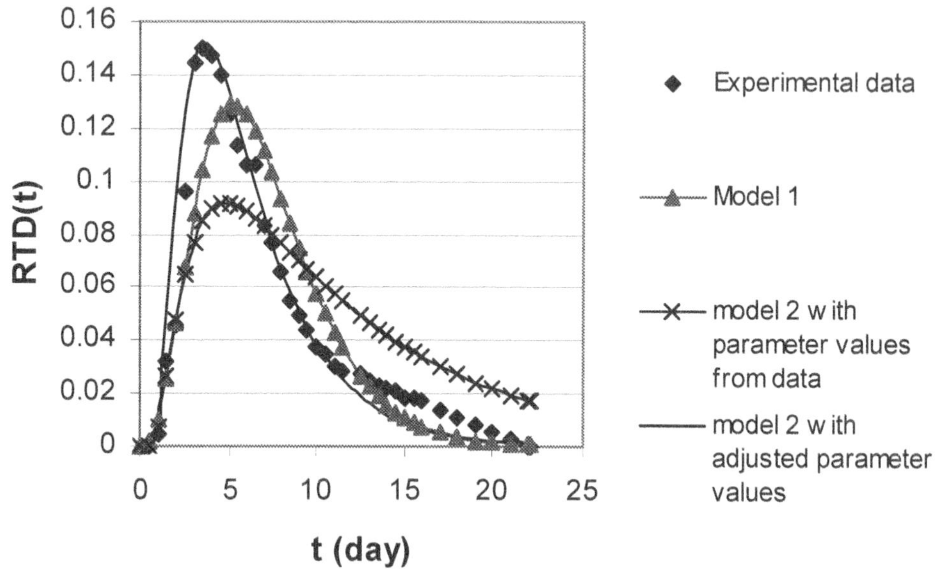

Figure 8-15. Hydraulic Residence Time Distribution for Experimental Values and Model Curves Based on Model 1 (CMFRS in Series) and Model 2 (PFRD) with Parameter Values from Experimental Data and Adjusted Parameter Values for Least Squares Fit.

$$\sigma^2 = \frac{\int_0^\infty (t - t_m)^2 RTD(t)dt}{\int_0^\infty RTD(t)dt} \qquad (8\text{-}33)$$

In these equations, t_m is the mean of the system residence time distribution, which is also the theoretical hydraulic residence time of the complete reactor, and σ^2 is the variance of the residence time distribution. Finally, the expression for the hydraulic residence time distribution function, RTD(t), is as given by the following equation:

$$RTD(t) = \frac{1}{t_m}\frac{N^N}{(N-1)!}\left(\frac{t}{t_m}\right)^{n-1}\exp\left(-\frac{Nt}{t_m}\right) \qquad (8\text{-}34)$$

On the basis of the tracer data analysis, the values of mean, t_m, and variance, σ^2, are obtained with 6.95 day and 17.91 day^2, respectively, and N, the conceptual number of smaller reactors into which the complete reactor is divided, is 3. Thus, the E(t) curve for 3 CMFRs in series can be obtained by substituting the parameter values into the preceding equation. The predicted curve using this approach is that for Model 1 shown in Figure 8-15.

Model 2 – Dispersed Plug Flow Reactor (PFRD). The second model investigated was a plug flow reactor with dispersion (PFRD), which assumed non-ideal behavior was caused by some dispersion in a flow pattern that otherwise would be considered ideal for a plug flow reactor. The RTD(t) function for the PFRD can be expressed as given in the equation below, where D/V_XL is a dimensionless group that represents the single parameter of the model, and is often referred to as the dispersion number or the inverse of Peclet number.

$$RTD(t) = \frac{1}{2t_m\sqrt{\pi\left(\frac{D}{V_X L}\right)\frac{t}{t_m}}}\exp\left(-\frac{\left(1-\frac{t}{t_m}\right)^2}{4\left(\frac{D}{V_X L}\right)\frac{t}{t_m}}\right) \tag{8-35}$$

$$\frac{\sigma^2}{t_m^{\ 2}} = 2\frac{D}{V_X L} + 8\left(\frac{D}{V_X L}\right)^2 \tag{8-36}$$

The value of the dispersion number can be estimated from the mean and variance of the tracer data by the following relationship, which applies to the open boundary.

Using the variance and mean of the tracer data, estimated in the first model, the above equation could be solved for the dispersion number, which equals 0.237. The model RTD(t) curve for a reactor with dispersion number of 0.237 can be developed from the preceding equations, and has been plotted as Model 2 in Figure 8-15. When both models are compared with the tracer data, one may reasonably conclude that neither of the models fit the experimental data particularly well.

For a better fit to the data, both the dispersion number and the mean hydraulic residence time were adjusted to minimize the squared error [i.e., (model estimated concentration – measured concentration)^2]. The two best-fit parameter values were determined to be a dispersion number of 0.197 and mean hydraulic residence time of 4.4 days. The predicted output curve using these parameters is also shown in Figure 8-15. This curve appears to provide a better fit to the data than the others. However, at long residence times (over 11 days), the model still does not fit the data well.

Model 3 – Two PFRDs in Parallel. The third model was selected to further reduce error incurred in describing the hydraulics of the constructed wetland. That model involved multiple PFRDs, the most simple of which would be two PFRDs in parallel. The total hydraulic residence time distribution function RTD(t) was assumed to be the sum of the two individual PFRD RTD(t) functions by giving each PFRD its individual flow weighting. This may be expressed as shown in the equation below, where w_1 and w_2 are flow weights of PFRD #1 and PFRD #2, respectively, and $RTD_1(t)$ and $RTD_2(t)$ are the hydraulic residence time distribution functions of the two reactors, which follow the form of the equation used for Model 2, with individual dispersion numbers and individual mean hydraulic residence times.

$$RTD(t) = w_1 RTD_1(t) + w_2 RTD_2(t) \tag{8-37}$$

Using sum of the squares minimization, the best-fit parameter values for the PFRDs in parallel are given in Table 8-8.

The predicted outflow curves using these parameters are shown in Figure 8-16. The prediction results show that small portion of flow in the system has very long residence time, whereas large portion takes a relatively short time in the system, behaving like 'short-circuiting'. Comparing the three model results, it is evident that two PFRDs in parallel matched the experimental data significantly better than models for individual reactors could.

Table 8-8. Model 3 Best-Fit Parameter Values for the PFRDs in Parallel.

	Weight	Dispersion number	Mean hydraulic residence time
Reactor 1	29%	0.100	10.3
Reactor 2	71%	0.112	4.2

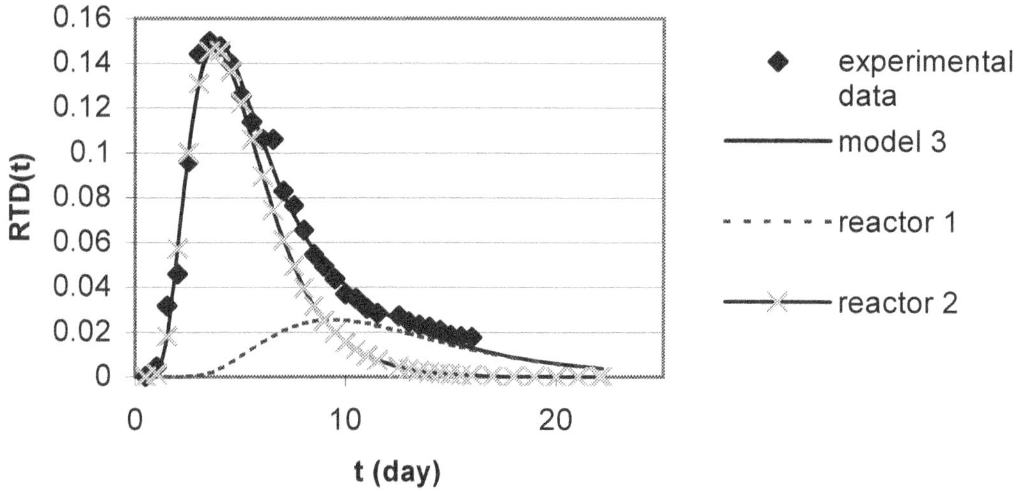

Figure 8-16. Hydraulic RTD for Model 3, Two PFRDs in Parallel.

Model 4 – Two PFRDs in Series. The fourth hydraulic model developed treated the system as two cells in series, which is closer to the actual physical layout than the other hydraulic models examined. That model, accordingly, was expressed as two PFRDs (dispersed plug-flow reactors) in series with individual dispersion coefficients. The PFRD#1 and PFRD#2 were considered to have the same dimensions as the cell 1 and cell 2, respectively, as determined for Model 3.

The cross-sectional averaged dye concentration can be obtained by simultaneously solving a system of coupled partial differential equations, shown in the following two equations. These relationships may be derived from Fick's law and include longitudinal dispersion and advection. In these two equations, C is the tracer concentration in the reactor, D is longitudinal dispersion coefficients, U is advective velocity and subscripts 1 and 2 represent PFRD#1 and PFRD#2, respectively:

$$\frac{\partial C_1}{\partial t} = D_1 \frac{\partial^2 C_1}{\partial x^2} - U_1 \frac{\partial C_1}{\partial x} \tag{8-38}$$

$$\frac{\partial C_2}{\partial t} = D_2 \frac{\partial^2 C_2}{\partial x^2} - U_2 \frac{\partial C_2}{\partial x} \tag{8-39}$$

The above two equations can be solved to obtain tracer concentration in the wetland effluent as a function of time. This solution may be expressed as:

$$C(t) = \int_0^t \frac{M}{4\pi\sqrt{D_1 D_2 t(t-\tau)}} \exp\left(-\frac{(L_1 - U_1\tau)^2}{4D_1\tau} - \frac{(L_2 - U_2(t-\tau))^2}{4D_2(t-\tau)}\right) d\tau \tag{8-40}$$

where:

M	=	input mass of tracer per unit cross section area, M/L^2
L_1, L_2	=	physical lengths of PFRD#1 and PFRD#2, respectively, L
U_1, U_2	=	flow velocity through PFRD#1 and PFRD#2, respectively, L/T
τ	=	time for the tracer to reach PFRD#2, T.

M was estimated to be about 85.8 g m^{-2} by analyzing the RTD(t) data. U_1 and U_2 were assumed to be the equal to 4.3 m/d by using the flow and cross-section area from field data. To fit the data, the dispersion coefficients in each reactor were assumed to be 0.2 and 20 m^2 d^{-1}. The predicted output curve using the parameters as well as the observations is shown in Figure 8-17.

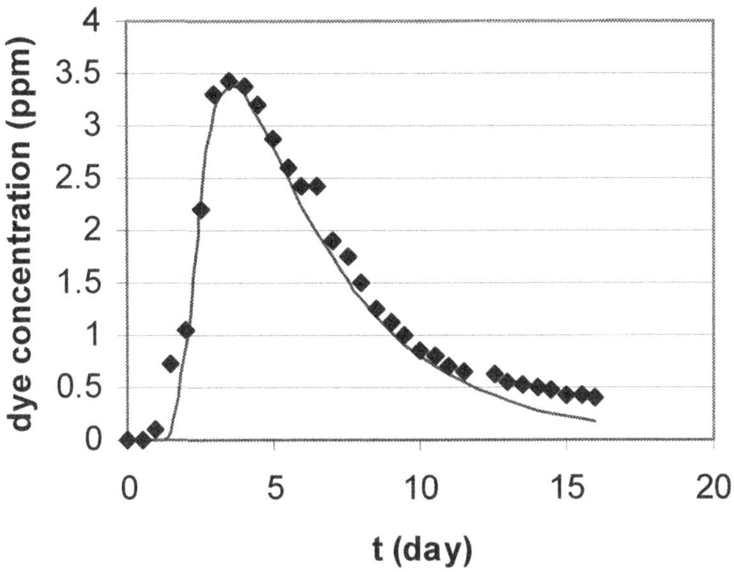

Figure 8-17. Dye Concentration Versus Time for Model 4, Two PFRDs in Series. ♦, Experimental Data; —, Model 4.

A preliminary comparison between the results for Model 3 (two PFRDS in parallel) and Model 4 (two PFRDS in series), indicates that the Model 3 approach is superior. This leads to a seeming contradiction. That is, the analysis indicates that two PFRDs in parallel represents hydraulic conditions in the wetland more accurately than two PFRDs in series, and does so even though the actual system has a physical appearance like a system in series. The fact that two PFRDs in parallel more accurately reflect the observations indicates that indeed influent water "short-circuits" as it flows through the wetland, and the dual flows represent water masses with different hydrodynamic characteristics. The data do not permit determining the extent to which the first or the second cell or both are involved in the flow segregation.

8.4.2 Volatilization

Assessment of the magnitude and significance of volatilization on water phase concentrations of HCN required information on rates of mass transfer, and modeling of this process required knowledge of the mass transfer coefficient. Considerable effort was directed at improving knowledge of these issues and involved both paper-based (literature searching) and experimentally based work.

8.4.2.1 Mass Transfer Coefficient of HCN from in Situ *Measurements*

As described previously, direct measurements of HCN flux were undertaken during the March 2000 and August 2000 trips to the field site. In March 2000, two trials in first cell of the treatment wetland were available to quantify the HCN flux. Volatile HCN was trapped in NaOH solution and analyzed directly. The results showed that the gross mass flux of HCN for the two trials was 2.26 μg m^{-2} hr^{-1} and 4.15 μg m^{-2} hr^{-1}, respectively. Corrected for the ambient air, the net flux of HCN for the two trips was 0.4 μg m^{-2} hr^{-1} and 2.29 μg m^{-2} hr^{-1}. During August 2000, HCN flux measurement was conducted in wetland system and Duck Spring. The analysis results showed that the gross mass flux of HCN for the two trials in wetland was 2.36 μg m^{-2} hr^{-1} and 2.05 μg m^{-2} hr^{-1}, respectively, and the mass flux in Duck Spring was 18.81 μg m^{-2} hr^{-1} and 14.70 μg m^{-2} hr^{-1}, respectively. Corrected for HCN in ambient air, the net flux of HCN for two trials in wetland was 2.31 μg m^{-2} hr^{-1} and 2.00 μg m^{-2} hr^{-1}, respectively, and that in Duck Spring was 16.92 μg m^{-2} hr^{-1} and 12.81 μg m^{-2} hr^{-1}.

Accordingly, the overall mass transfer coefficient in liquid phase for HCN can be estimated based on the measured flux, concentration of HCN in water and air, and Henry's law constant of HCN. The concentration of HCN in water was estimated to be 9.36 μg/L in the wetland in March, 3.58 μg/L in the wetland in August and 23.8 μg/L in Duck Spring in August based on our measurements. As a first approximation, it may be assumed that the HCN may disappear immediately after volatilizing from water and thus the HCN concentration at the water-air interface approaches zero. Then, based on the gross flux and aqueous concentration, the overall mass transfer coefficient for HCN was obtained to be 0.24 x 10^{-3} m hr^{-1} and 0.44 x 10^{-3} m hr^{-1}, respectively for the two trials in wetland in March, 0.66 x 10^{-3} m hr^{-1} and 0.57 x 10^{-3} m hr^{-1}, respectively for the two trials in wetland in August, and 1.33 x 10^{-3} m hr^{-1} and 1.04 x 10^{-3} m hr^{-1}, respectively in Duck Spring. As a second approximation, the HCN in the ambient air was analyzed to be about 0.0014 μg/L in wetland in March, 0.00054 μg/L in the wetland in August and 0.021 μg/L in Duck Spring in August. Using the literature-based value of Henry's law constant for HCN, 2.2 x 10^{-3} (unitless) at 12°C (the monthly average temperature in March 2000), and 4.7 x 10^{-3} (unitless) at 25°C (the monthly average in August 2000), the HCN concentration in water equilibrating with ambient air was estimated to be 0.61 μg/L in wetland in march, 0.11 μg/L in the wetland in August and 4.47 μg/L in Duck Spring in August. From the net flux, therefore, the overall mass transfer coefficient can be estimated to be 0.05 x 10^{-3} m hr^{-1} and 0.26 x 10^{-3} m hr^{-1} for the two trials in wetland in March, 0.67 x 10^{-3} m hr^{-1} and 0.58 x 10^{-3} m hr^{-1} in the wetland in August, and 1.76 x 10^{-3} m hr^{-1} and 1.33 x 10^{-3} m hr^{-1} in Duck Spring in August. The results indicate that the overall liquid phase mass transfer coefficient for HCN is temperature dependent (i.e., the mass transfer coefficient increases with the increase of temperature).

8.4.2.2 Theoretical Estimation of Mass Transfer Coefficient from Two-Film Theory

Volatilization of HCN from water to air can be described by the widely used two-film theory. The overall liquid-phase mass transfer coefficient could be estimated from both liquid-phase resistance and gas-phase resistance as follows:

$$\frac{1}{K_{OL}} = \frac{1}{K_L} + \frac{1}{H * K_g} \qquad (8\text{-}41)$$

where:

K_{OL} = overall liquid-phase mass transfer coefficient
K_L = liquid-film mass transfer coefficient
K_g = gas-film mass transfer coefficient
H = Henry's law constant

To use this model, it is necessary to measure or estimate the mass transfer in each phase and the Henry's law constant. The overall mass transfer coefficient of a chemical may depend on liquid-phase or gas-phase resistance or both. It has been suggested that the mass transfer in liquid phase mainly controls the volatilization when Henry's law constant is greater than 4.6 atm L mol^{-1}, and that mass transfer in the gas phase is rate-controlling when it is less than 0.013 atm L mol^{-1}. If Henry's law constant is between these two values, resistance from both phases is significant. The values of Henry's law constant for HCN were reported as 0.073–0.122 atm L mol^{-1} (see Table 8-9).

Table 8-9. Literature Values of Henry's Law Constant (H) for HCN.

Temperature (°C)	H (atm L mol^{-1})	H (unitless)	Reference
25	0.122	0.005	Bodek et al.,1988. *Environmental Inorganic Chemistry.*
Not given	0.073	0.003	Doudoroff, 1976. EPA-600/3-76-038. USEPA/ORD/ERL-Duluth, Duluth MN.
Not given	0.104–0.114	0.0043–0.0047	Smith and Mudder, 1991. *The Chemistry and Treatment of Cyanidation Wastes.* Mining Journal Books Limited, London
25	0.115	0.0047	Avedesian, 1983. *Canadian Journal of Chemical Engineering.* 61 (6). P801-806

Thus, HCN mass transfer resistance in both phases may be significant, so an accurate value for the Henry's law constant for HCN as a function of temperature is important to accurately quantify the volatilization rate and for application in a model of cyanide transport and fate. Bodek et al. (1988) indicated the relationship between Henry's law constant for HCN and temperature may be described as follows:

$$\log H_{HCN} = -\frac{1272.9}{T} + 6.238 \qquad (8\text{-}42)$$

where:

H_{HCN} = Henry's law constant of HCN, mmHg/M
T = temperature, K

This equation was determined for HCN concentrations of 0.01–0.5 M and temperatures of 20–95°C (Bodek et al., 1988). The relationship for HCN was also reported by Yoo et al. (1996) used saturation pressure, ionization constants, and experimental isothermal data to develop the following relationship:

$$\ln H(T) = 9.5850 - 0.03147T + 3.1704 \ln T - 6302.0/T \qquad (8\text{-}43)$$

where:

H(T) = Henry's Law constant for HCN, atm L/mol
T = temperature, K

Using this relationship to estimate Henry's law constant for HCN at 25°C yields a value of 0.056 atm L mol^{-1}, which is half that of other literature values at 25°C.

Previous research indicated no matter how much mass transfer is limited by gas or liquid film, transfer rates vary with environmental conditions. The overall mass transfer coefficient (K_{OL}), liquid-film transfer coefficient (K_L) and gas-film transfer coefficient (K_g) are not constant. For example, K_L is related to the degree of turbulence in water. And K_g depends on the wind speed, which determines the thickness of liquid film. Because it is not practical to measure the K_L and K_g for all chemicals in different ambient conditions, reference compounds, whose mass transfer coefficients have been measured in a wide range of environmental conditions may be used. Oxygen has been used as a reference compound to estimate liquid-film mass transfer coefficient for many other chemicals K_L (Chapra, 1997; Mills et al., 1982) using the following relationship:

$$K_L = K_{L,O}\left(\frac{32}{M}\right)^{0.25} \tag{8-44}$$

where:

M	=	the molecular weight of HCN
$K_{L,O}$	=	oxygen- transfer coefficient in the water phase

The gas- film mass transfer coefficient can be estimated from water vapor transfer rate in air by:

$$K_g = K_{g,water}\left(\frac{18}{M}\right)^{0.25} \tag{8-45}$$

where:

$K_{g,water}$ = water vapor transfer rate in air (Chapra, 1997; Mills et al., 1982)

Typical environmental values of K_L and K_g as suggested by Liss and Slater (1974) are 0.2 m hr^{-1} for oxygen transfer and 30 m hr^{-1} for water transfer, respectively. Because the coefficients vary with environmental conditions, the relationship between the coefficient and perturbing factors is important. Numerous formulae have been developed for predicting oxygen reaeration in streams and rivers. Three among these are most commonly used including: O'Connor-Dobbins, Churchill, and Owens-Gibbs (Chapra, 1997). All were developed based on surface renewal model for different types of streams using experimental or empirical approaches. Among these three, the Owen-Gibbs relationship is best applied to shallower systems, such as those of interest at the field site. This relationship can be described as follows:

$$K_a = 5.32\frac{U^{0.67}}{H^{1.85}} \tag{8-46}$$

where:

K_a	=	oxygen reaeration rate, 1/d
U	=	water velocity, m/s
H	=	water depth, m

Furthermore, $K_{L,O}$ can be estimated by the product of K_a and H. Similarly, the gas-film coefficient for water ($K_{g,water}$) is related to wind speed and can be approximated as follows (Chapra, 1997; Mills et al., 1982):

$$K_{g,water} = 168 U_w \tag{8-47}$$

The preceding discussion indicates that the overall mass transfer coefficient for HCN in the liquid phase can be estimated as a function of velocity, depth, wind speed, and temperature. In August 2000, the average monthly temperature was 25°C. Accordingly, the Henry's law constant was chosen from the literature to be 4.7×10^{-3}. The water depth in the wetland and Duck Spring was 0.195 m and 0.234 m, respectively. The water velocity at the two sites was 4.33×10^{-5} m s^{-1} and 0.062 m s^{-1}, respectively. The wind speed in the flux chamber was estimated to be about 0.049 m s^{-1}, which is much lower than the outside wind speed (3.4 m s^{-1} in August 2000). Using these parameters, the theoretical overall mass transfer coefficient for HCN was estimated to be 0.0149 m d^{-1} in the wetland and 0.033 m d^{-1} in Duck Spring in August. When these values are compared with the actual values estimated from flux measurements in August 2000 (see above and Figure 8-18), it is clear that the two-film theory can be applied to estimate mass transfer coefficient under specific environmental conditions for HCN from water to air.

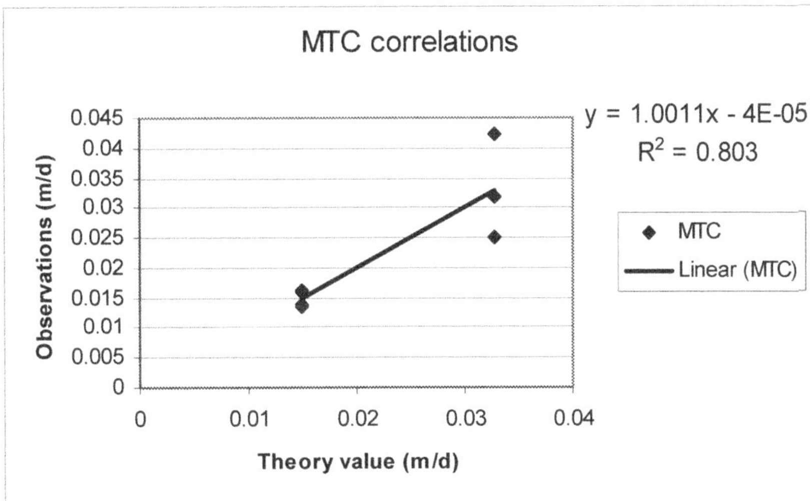

Figure 8-18. Mass Transfers Coefficient Correlations.

8.4.3 Biological Utilization of HCN

8.4.3.1 Effects of HCN Concentration and Temperature

The initial biodegradation experiments were performed at free cyanide concentrations ranging from 0.005 to 0.5 mg CN/L. As suggested by the plot of HCN concentration as a function of time in one group of experiments, shown in Figure 8-19, the rate of cyanide utilization generally decreased as the concentration increased. Using a simple first-order decay model, estimates of the rate coefficient decline from 0.123 to 0.002 hr^{-1} as the initial cyanide was varied from 0.005 to 0.5 mg/L.

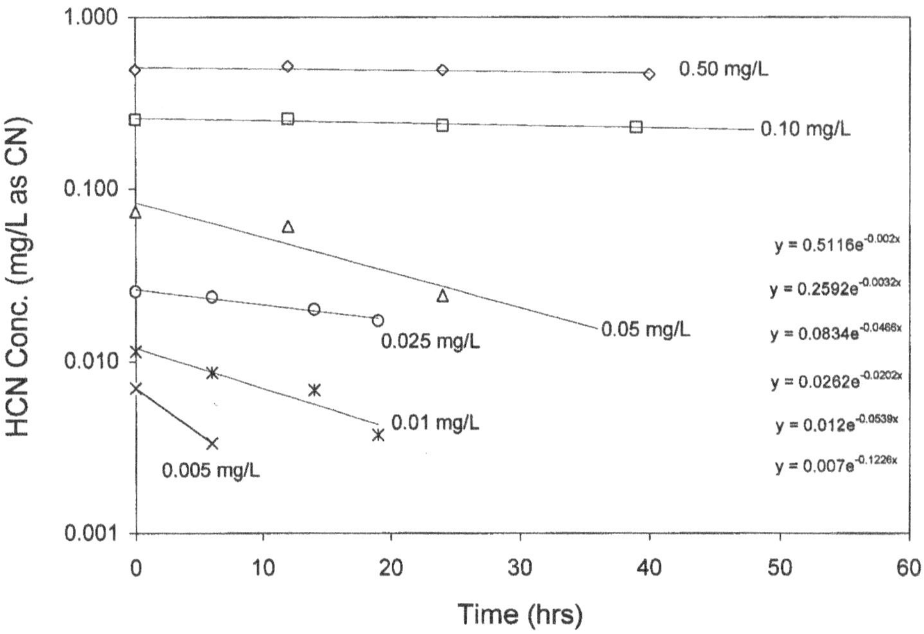

Figure 8-19. Preliminary Experiments on HCN Utilization by Wetland Microflora Showing Effects of Initial HCN Concentration.

Although these data are not plotted to demonstrate the phenomenon, the rate at which cyanide was used changed over the course of an exposure such that the rate accelerated. Such a lag phase was observed before cyanide biological decay began in every experiment. The length of the lag phase changed with temperature and cyanide concentration. Figure 8-20 shows the time lag in the appearance of biological decay for three different cyanide concentrations and Figure 8-21 shows a similar dependence of the lag period on incubation temperature.

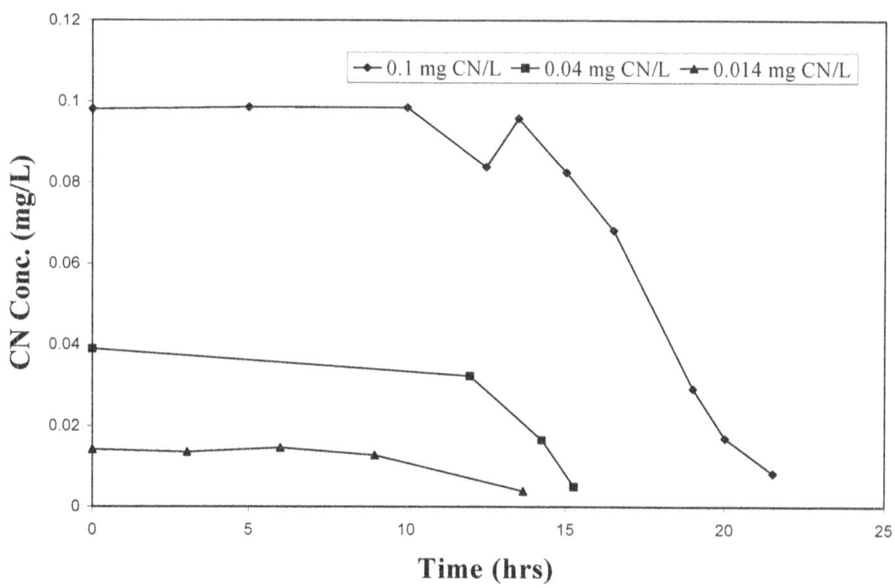

Figure 8-20. Effect of Initial Cyanide Concentration on Onset of Biological Utilization of HCN.

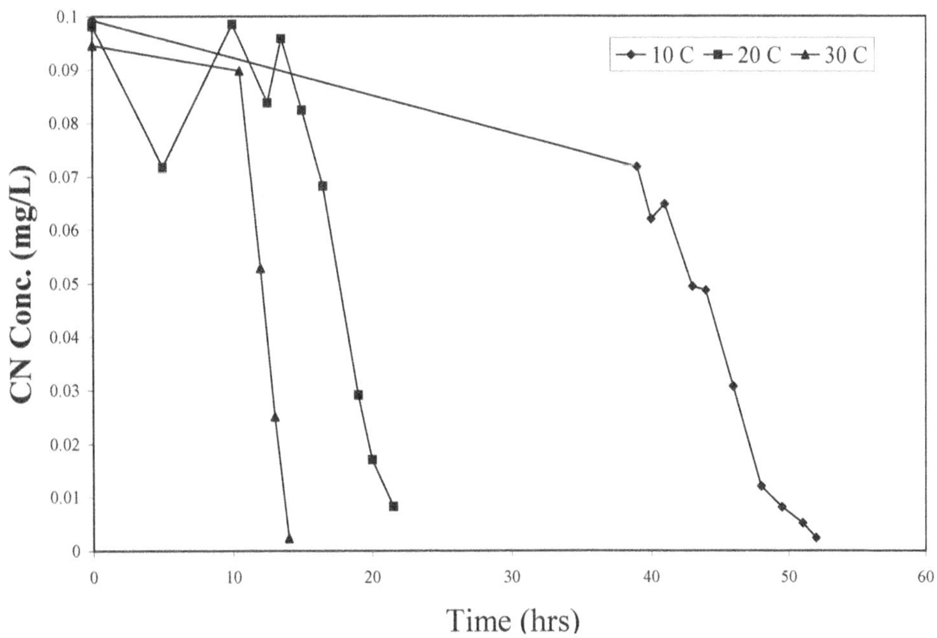

Figure 8-21. Effect of Incubation Temperature on Onset of Biological Utilization of HCN.

As the cyanide concentration increased, the lag phase before the onset of degradation also increased. At 20°C, an initial concentration of 0.014 mg CN/L had a lag phase of 7.9 hours, whereas an initial concentration of 0.1 mg CN/L had a lag phase of 16.5 hr. This was an increase of 8.6 hours for a cyanide concentration difference of 0.086 mg/L.

Figure 8-21 shows how the lag time changed with temperature. At 10°C a lag phase of 42.4 hours was measured while at 30°C the lag phase was reduced to 11.8 hours. This indicated that the lag phase increased when the cyanide concentration increased and decreased as the temperature increased.

These data show the time lag to the onset of bioutilization depended simultaneously on HCN concentration and temperature. Figure 8-22 and the least-squares fitted empirical equation given below summarize the effects of concentration and temperature on the lag phase. This relationship has an R^2 value of 0.9779 ($P < 0.0001$).

$$\text{Lag Period, hr} = (4842.7 * C + 34.043) * T^{-(3.4477*C + 0.7998)} \qquad (8\text{-}48)$$

8.4.3.2 Temperature Correction of Utilization Rate

The generalization of the HCN utilization rate was achieved by assuming first-order kinetics to account for the concentration dependence observed *in vitro* and by assuming an Arrhenius-type temperature dependence. With this approach, a squared error minimization analysis was performed on time lag adjusted data to estimate a value for the utilization rate coefficient at the standard temperature of 20°C (k_{20}) and for the temperature adjustment coefficient (θ). On the basis of this approach, the value found for k_{20} was 0.342 hr^{-1}, the value for θ was determined to be 1.033, and the standard error of concentration estimates during the fitting exercise was 3% of estimated concentrations. The accuracy of the estimated k_{20} and θ values for summarizing the results of the *in vitro* analyses are illustrated in Figure 8-23, which shows the measured and predicted concentration values for each postlag phase data point.

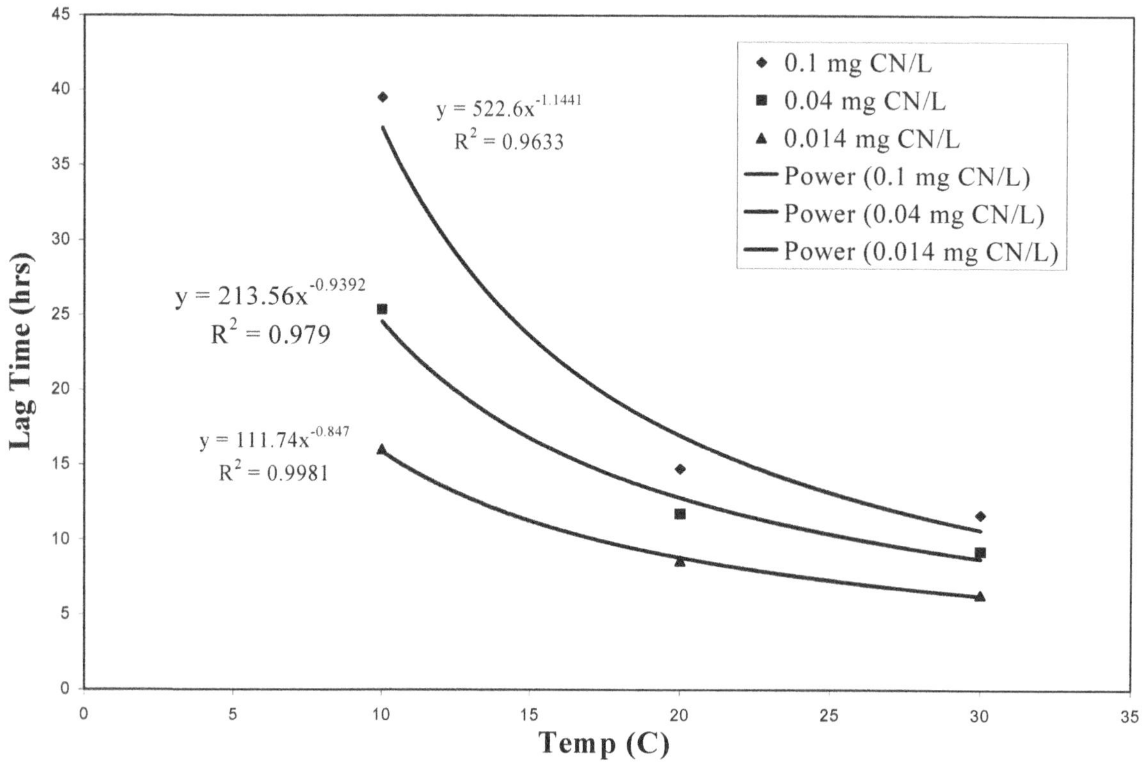

Figure 8-22. Effects of Temperature and Concentration on Length of Lag Period to the Onset of HCN Bioutilization.

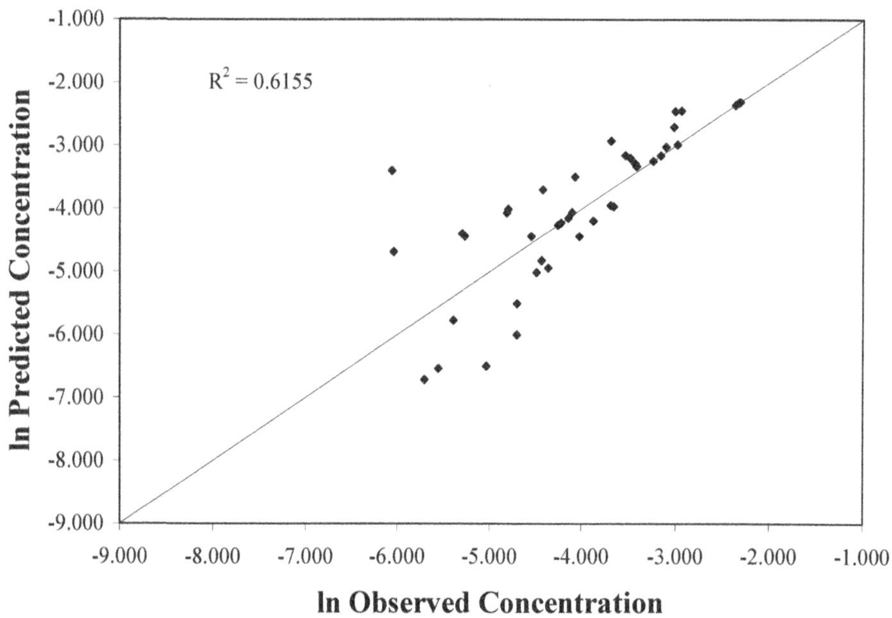

Figure 8-23. Measured and Predicted Concentrations for Postlag Phase Data Pairs.

8.5 Model Calibration

As described previously, the Lagrangian parcel method was adopted as the method of choice for modeling the transport and fate of cyanide in surface water. After developing and testing the model formulations, calibration was necessary and aimed at assessing the closeness of fit between model predictions and the data on cyanide and related observables from the field site.

8.5.1 Calibration to Average Field Conditions

To execute the model, several parameters were required. Discharge, velocity, and water depth were obtained from the field data collections. Longitudinal dispersion was estimated by analyzing the tracer test conducted during July 1999. The rate of "FeCN" photolysis [i.e., $Fe(CN)_6^{-4}$ and the lumped rate of HCN decay [the sum of volatilization and biodecay rates] were estimated by calibration. This involved adjusting parameters in the model to find values that allowed model predictions to match the observed FeCN and HCN concentration data. For the calibration step, this was done using trial-and-error with a combination of qualitative and quantitative assessments of estimation error. The volatilization rate of HCN was estimated theoretically based on the Henry's law constant for given conditions of temperature, wind speed, water velocity, and water depth as described previously. The biodecay rate of HCN was estimated as the residual between the lumped decay rate (quantified by calibration) and the volatilization rate of HCN (calculated directly). Also, the model estimated the loadings of complex and free cyanide to the stream.

The cyanide concentrations in two streams observed in the field, the Blue Pond Drainage and Springfield Branch, were used for the calibration work. In the Blue Pond Drainage, the Blue Pond itself was assumed to be the only cyanide source to the drainage, whereas two cyanide sources—Duck Spring and the Blue Pond Drainage effluent—were assumed to comprise the totality of sources to Springfield Branch. During model implementations, the simulation time was sufficiently long to allow the cyanide concentration distribution along the stream to reach steady state.

There were five trips to the field to collect samples and field data. Because the discharge, velocity, and water depth varied spatially and temporally in each stream, time-, and space-averaged conditions in each stream were used as a baseline condition for the calibration. Average discharge, velocity, depth, temperature, and wind speed values in the two streams are tabulated below. The average temperature and wind speed were obtained from five monthly average values coincident with the field data collection activity. The discharge, velocity, and depth were averaged from four observation sets rather than five because of flooding during the July 1999 trip. Values for these parameters are reported in Table 8-10.

Table 8-10. Average Hydrologic and Weather Data for the Blue Pond Drainage (BPD) and Springfield Branch (SB).

	Temperature (°C)	Discharge (m^3/s)	Depth (m)	Velocity (m/s)	Wind speed (m/s)
BPD	13	0.017	0.087	0.192	2.45
SB	13	0.272	0.273	0.204	2.45

Figure 8-24 shows the simulation results of fitting the observed FeCN and HCN concentration distribution in the Blue Pond Drainage under the average conditions. The distance was measured from the confluence with Springfield Branch. The cyanide source was about 470 m away from the confluence. Some of the model parameters selected during the calibration to the averaged data for the Blue Pond Drainage shown in Table 8-11.

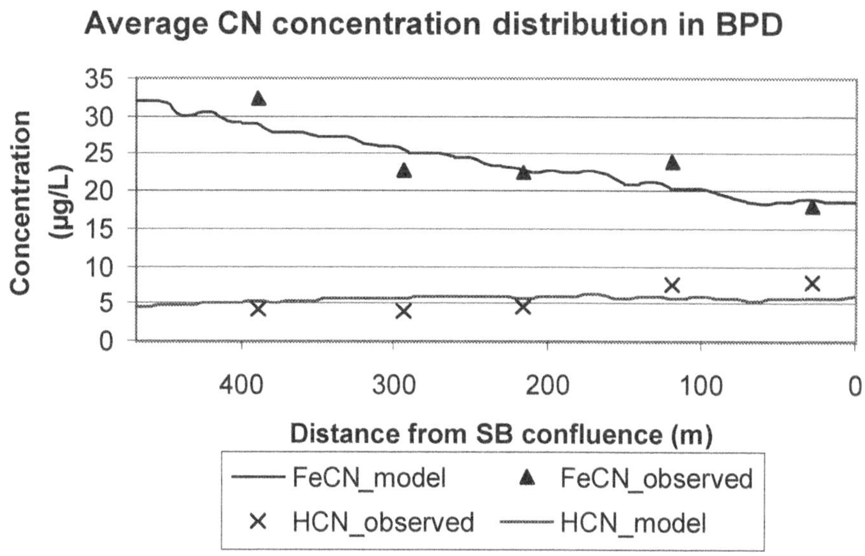

Figure 8-24. Calibration of Model to Average Cyanide Concentration Distribution in the Blue Pond Drainage.

Table 8-11. Calibration Rate and Cyanide Load in the Blue Pond Drainage (BPD) and Springfield Branch (SB).

	K_{ph} (/d)	K_v (/d)	K_b (/d)	load_FeCN (mg/s)	load_HCN (mg/s)
BPD	20	9.3	46.7	0.55	0.08
SB	10	2.7	34.7	16.5[a] 0.81[b]	1.60[a] 0.32[b]

a Cyanide source 1 to SB, Duck Spring.

b Cyanide source 2 to SB, BPD.

The photolysis rate used in this simulation was 20/d, and the lumped decay rate for HCN was 56/d. The volatilization rate of HCN was computed to be 9.3/d, so the biodecay rate could be estimated to be 46.7/d. In the volatilization rate calculation, the Henry's law constant was taken to be 0.0023 (unitless), which is the average of two values from the two formulae describing the relationship between Henry's law constant and temperature. This was necessary because there was no exact literature value for the Henry's Law constant at specific temperatures other than 25°C, and generally the temperature was around 10°C during field sampling events. Also the two references contained no information on uncertainty in the constants. The average values from the two sources were therefore used to estimate the mass transfer coefficient of HCN theoretically for all the model calibrations.

Figure 8-25 shows the simulation results for calibrating the model to observed FeCN and HCN concentrations in Springfield Branch under average conditions. In Figure 8-25, distance is given with respect to the confluence with the Blue Pond Drainage with upstream positive and downstream negative. By calibration, it was determined that the photolysis rate for FeCN was 10/d, and the lumped decay rate of HCN was 37/d. The volatilization rate of HCN was estimated from theory to be 2.7/d, which meant that the biodecay rate was about 34.7/d.

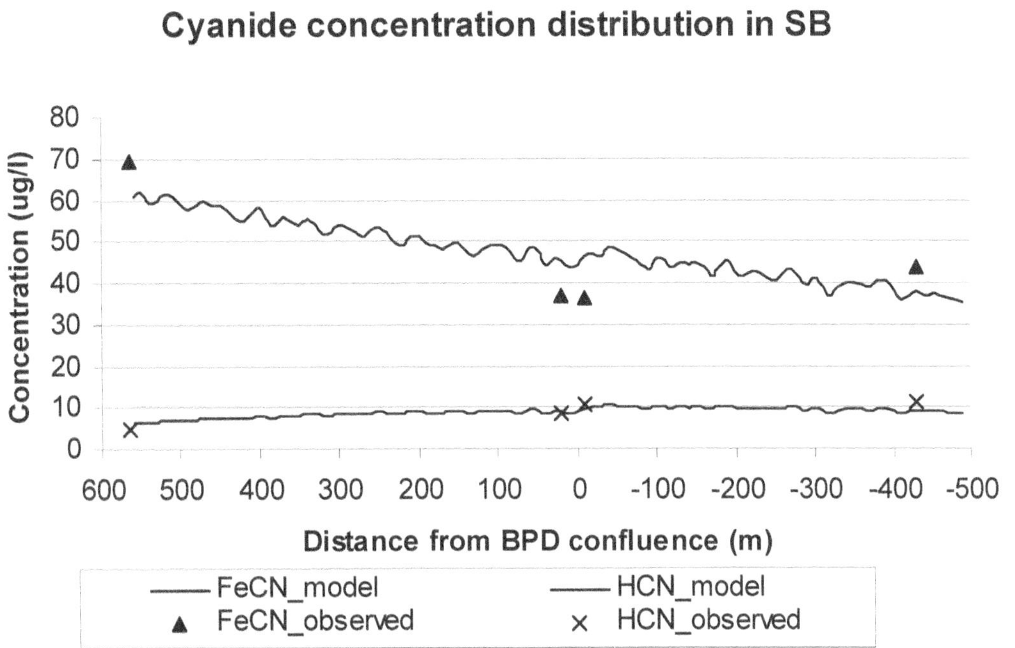

Figure 8-25. Cyanide Concentration Distribution in Springfield Branch under Average Conditions.

In addition, the load of complex cyanide and free cyanide to each stream can be estimated using the model. Under the average conditions, the FeCN and HCN load from the Blue Pond to the Blue Pond Drainage was determined to be about 0.55 mg s^{-1} and 0.08 mg s^{-1}, respectively. The load of FeCN and HCN from Duck Spring to the Springfield Branch was about 16.54 mg s^{-1} and 1.60 mg s^{-1}, respectively. And the load of FeCN and HCN from the Blue Pond Drainage to Springfield Branch in the confluence was about 0.81 mg s^{-1} and 0.32 mg s^{-1}.

The reaction rates and cyanide loads in each stream were summarized in the previous tabulation. It can be seen from the calibration results that the photolysis rate (K_{ph}) in Blue Pond Drainage is lower than that in Springfield Branch. This is consistent with the hypothesis that photolysis decreases with increases of water depths. This decrease occurs because the sunlight intensity decreases in deeper water, which due to depth averaging, leads to a lower net rate. It is clear from these data that Springfield Branch is much deeper than Blue Pond Drainage. In addition, water depth and water velocity affect volatilization of HCN significantly. The shallower and more turbulent nature of the water of the Blue Pond Drainage is more favorable to volatilization of HCN than in Springfield Branch. Though the water velocity in Springfield Branch is slightly greater than that in the Blue Pond Drainage, it can be seen from the results that the turbulence effect does not compensate for the effect of water depth on the volatilization rate. The results also show that the biodecay rate exceeds the volatilization rate of HCN, and exceeds the photolysis rate of FeCN in both the Blue Pond Drainage and Springfield Branch. However, the mechanism for biodecay is still not very clear. Further research, like model sensitivity analysis and uncertainty analysis, is very necessary to test which process is the most significant contributing fate of free cyanide in the stream.

8.5.2 Calibration to Sampling Events

Other simulations were conducted to examine the behavior of the model for individual sets of observed data and to determine the ways in which the model parameters might be adjusted to achieve satisfactory simulation accuracy. Shown in Figure 8-26 are simulation results and observed cyanide concentrations longitudinally through the Blue Pond Drainage during the July 1999 and March 2000 sampling events, respectively.

The parameters used to produce the model calibration for the Blue Pond Drainage observations during July 1999 and March 2000 included the photolysis and biodecay rates. These and other data required by the model have been summarized in Table 8-12.

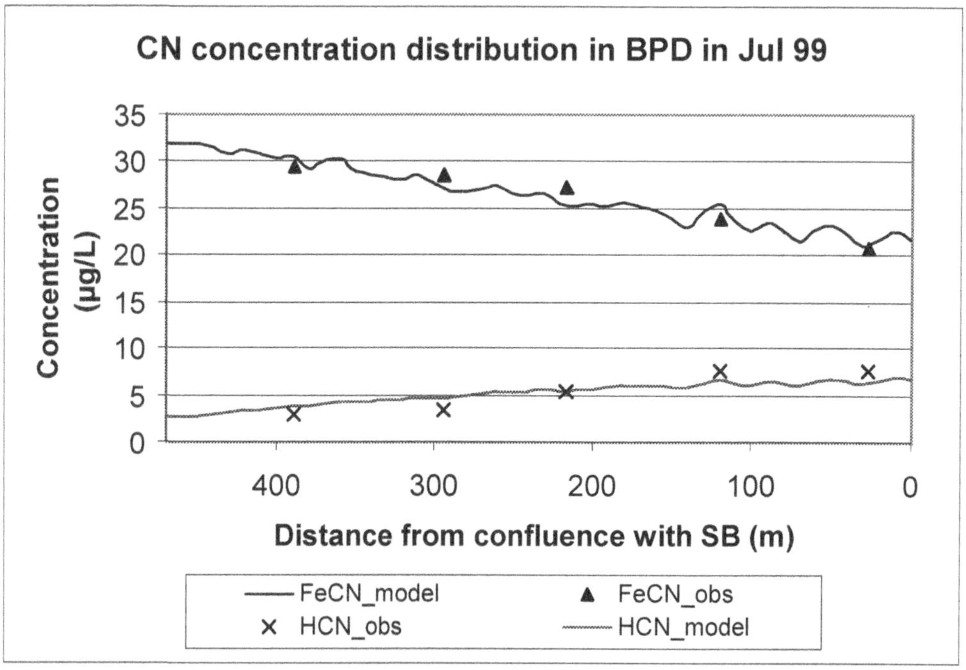

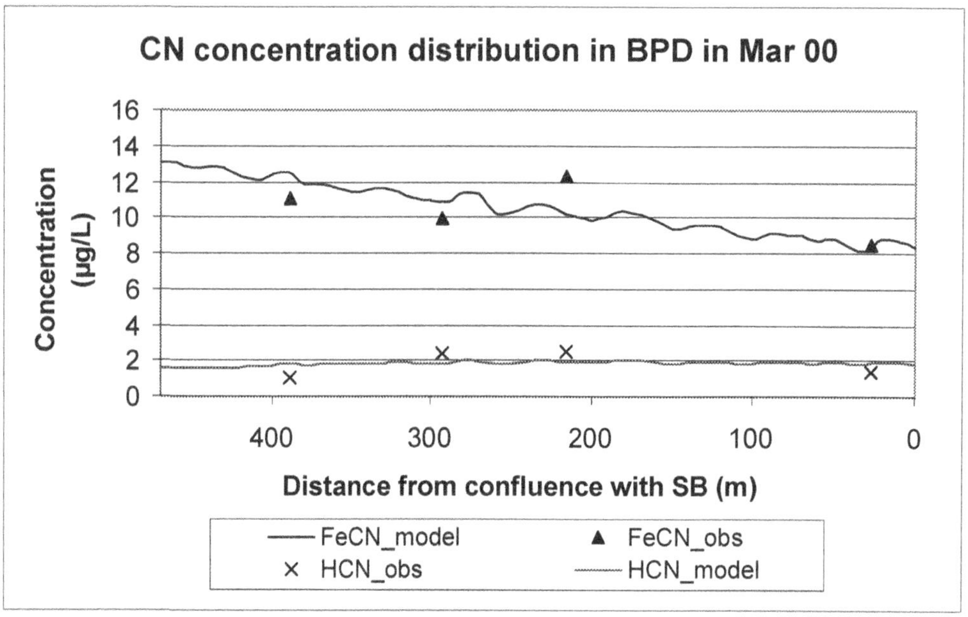

Figure 8-26. Longitudinal Distribution of Measured Cyanide Concentration in Blue Pond Drainage During July 1999 and March 2000 Sampling Events and Model Calibration Results.

Table 8-12. Input Parameters for Model Used to Calibrate Blue Pond Drainage.

Sampling Event	Temperature (°C)	Discharge (m^3/s)	Depth (m)	Flow Velocity (m/s)	Wind Speed (m/s)	Photolysis Rate (1/d)	Volatilization Rate (1/d)	Biodecay Rate (1/d)
Jul 1999	25.2	0.067	0.15	0.26	2.06	20	6.8	30.6
Mar 2000	11.8	0.008	0.05	0.24	2.88	20	19.0	56.0

By comparing the two simulation results, it can be seen that the photolysis rate was about the same for the two data sets. This is an interesting result because the depth in July 1999 was greater by a factor of three approximately than those measured during the March 2000 sampling. This may be explained by differences in the turbidity of the water between the sampling events. Besides the water depth, turbidity would be expected to be an important factor affecting the photolysis of complex cyanide. Broderius and Smith (1980) studied the direct photolysis of hexacyanoferrate complexes and found that photolysis rates increased greatly when turbidity was manipulated upward by the addition of either Bentonite clay or silt. This is thought to be caused by redirection or scattering of light when passing through a medium. Thus, in a shallow system, a greater turbidity would result in greater light scattering and longer pathlengths for transmitted light to traverse and donate energy to the system.

As may be seen in the data below, during July 1999, the water turbidity in the Blue Pond Drainage averaged about 14.4 NTU (nephelometric turbidity units), while in March 2000, the average was 6.69 NTU (see Table 8-13). It seemed that the decreasing of photolysis rate in July 1999 (because of deeper water) was balanced by the higher turbidity. This higher turbidity could increase the photolysis rate. The turbidity did not cause the difference of photolysis in the Blue Pond Drainage and Springfield Branch under average conditions because the average water turbidity is very similar in the Blue Pond Drainage and Springfield Branch. The volatilization rate in the two trips in the Blue Pond Drainage changed most likely caused by the difference of water depths between the two trips. Because the water velocity and wind speed are very similar, the increasing degree of mass transfer coefficient by higher temperature in July 1999 is not enough to balance the decrease degree of volatilization rate by shallower water depth.

Table 8-13. Water Turbidity (NTU) Measured in the Blue Pond Drainage (BPD) and Springfield Branch (SB) During Field Sampling.

Sampling Period	November 1998	March 1999	July 1999	November 1999	March 2000	Average
BPD	5.63	32.63	14.4	4.99	6.69	12.9
SB	4.58	11.2	30.6	2.96	11.35	12.1

Figure 8-27 illustrates the calibration results and observed longitudinal distribution of cyanide concentration in Springfield Branch during the November 1999 and March 2000 sampling events, respectively. The parameters used in the model calibration (photolysis and biodecay) are listed in Table 8-14.

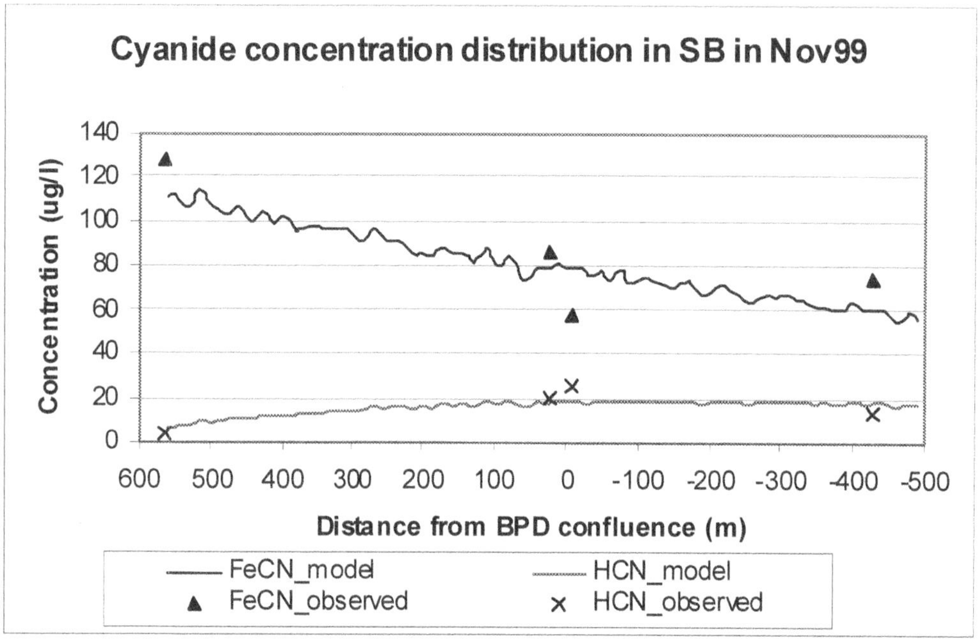

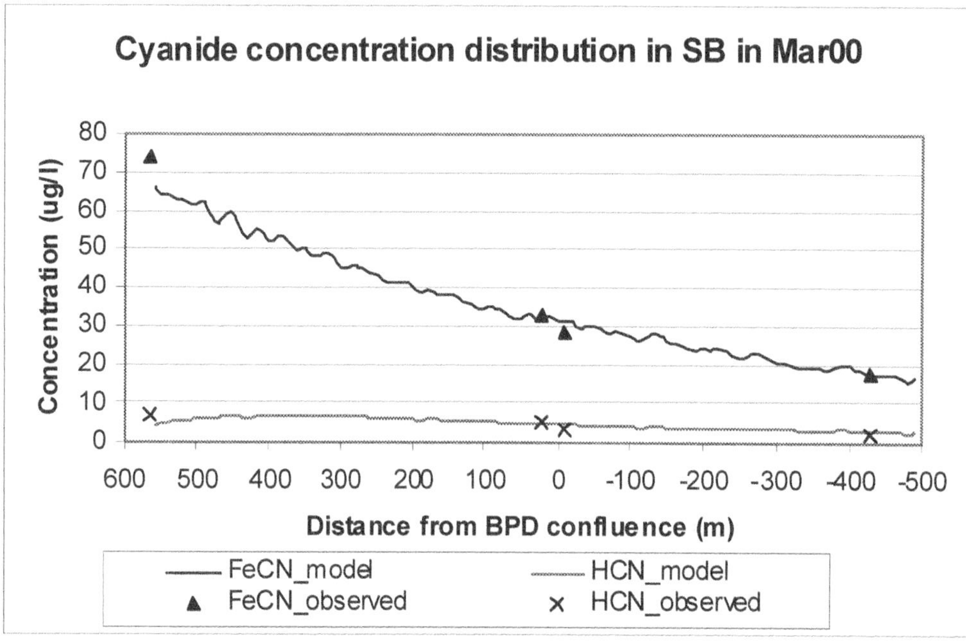

Figure 8-27. Longitudinal Distribution of Measured Cyanide Concentrations in Springfield Branch During November 1999 and March 2000 Sampling Events and Model Calibration Results.

Table 8-14. Model Clibration for Springfield Branch During November 1999 and March 2000 Sampling Eents.

	Temperature (ºC)	Discharge (m^3/s)	Depth (m)	Flow Velocity (m/s)	Wind speed (m/s)	Photolysis rRate (1/d)	Volatilization rate (1/d)	Biodecay rate (1/d)
November 1999	12.0	0.102	0.117	0.170	2.22	10	5.9	24.0
March 2000	11.8	0.325	0.351	0.166	2.88	20	2.2	109.9

It can be seen from thc calibration results for CN in the Springfield Branch that the photolysis rate in March 2000 was greater than during the November 1999 sampling. That difference may be a result of turbidity having a greater effect on the photolysis rate than the depth of the water. The turbidity in March 2000 was about four times greater than it was when measured in November 1999. Another factor that was operating to differentiate the rates of photolysis between November and March was the fact that the vegetative canopy was largely intact during November whereas new leaves and other shading vegetation had not developed by the time of the March sampling. Hence, the amount of light available for photolysis during November was less than during March, which would lead to greater photolysis rates during March.

The volatilization rate found for Springfield Branch during March 2000 was less than that found for November 1999 because the water was deeper during March 2000. The temperature, water velocity and wind speed were very similar between the two sampling events. Finally, the biodecay rate of HCN for Springfield Branch during March 2000, estimated by calibration, was larger than that estimated for any other event for Springfield Branch or for the Blue Pond Drainage calibrations. No reason was apparent to explain this large value, which points toward the need to investigate further the biological utilization of HCN in natural receiving waters.

8.6 Summary and Conclusions

This investigation identified and provided a successful numerical simulation of three processes that simultaneously appear to determine the fate of essentially all cyanide species in an industrially-impacted receiving water. The processes include photolysis of hexacyanoferrate [i.e., $Fe(CN)_6^{-4}$ and $Fe(CN)_6^{-3}$], volatilization of HCN, and biotic utilization of HCN. To accomplish this required effort from four interdependent activities: literature review and data gap determination, site selection and periodic monitoring for relevant physical and chemical environmental variables in system, data gap elimination by experimental observation of fate-determining processes, and numerical model development and calibration.

The principal data gaps identified during the literature search included a lack of knowledge pertaining to the volatilization and biotic utilization of HCN that would be sufficient to permit adequate representation of these processes in a numerical modeling framework. To remedy these deficiencies, a modest program of field (*in situ*) and laboratory (*in vitro*) observation and experimentation was undertaken. The program included the use of a flux chamber to directly observe the flux of HCN from the water surface to air at two field sites, which were affected by hexacyanoferrate contaminants in surface water discharges. The results demonstrated a range of HCN volatilization fluxes from ~ 2 to 15 μg HCN m^{-2} hr^{-1}.

Volatilization flux values depended on both the water phase concentration (i.e., concentration gradient for volatilization) and the temperature. In addition, the work demonstrated that compensation of the measured fluxes for wind speed effects on the gas film mass transfer coefficient could be accomplished successfully using methods previously described (Chapra, 1997, Mills et al., 1982).

The monitoring data on cyanide concentrations in water samples collected from the field site demonstrated a significant decline in total cyanide concentrations—greater than would result from measured volatilization losses—during continuous loading by cyanide-contaminated inflowing water. The specific monitoring site that showed this behavior was a pilot-scale constructed wetland, which was installed to test feasibility as a treatment technology for removal of cyanide and other contaminants of concern. The magnitude of cyanide losses was viewed as presumptive evidence for biotic utilization of HCN in the system inasmuch as other potential fate-determining processes (e.g., volatilization, precipitation, or sorption and sediment accumulation) did not appear to function at a magnitude that could account for the losses.

Accordingly, laboratory studies were undertaken to assess the aerobic bioutilization of HCN by microflora obtained from the constructed wetland. Preliminary experiments on KCN-spiked slurries of wetland sediments and water to assess nutritional needs showed that amendments such as municipal wastewater and standard heterotrophic microbiological growth media (nutrient broth) did not enhance utilization rates. The results demonstrated several features that were consistent with microbial utilization of cyanide when it was the only nitrogen source available in the medium. These included a time lag prior to HCN utilization that depended in a systematic way on incubation temperature and initial HCN concentration, an HCN utilization rate that depended on HCN concentration [directly dependent at low concentrations (<50 μg HCN L^{-1}) and inversely dependent at high concentrations (>50 μg HCN L^{-1})]; and an HCN utilization rate that depended on temperature in a manner that obeyed a customary Arrhenius analysis (E_a ~ 46.6 kJ $mole^{-1}$). It was noted that the observed *in vitro* HCN utilization was strong presumptive evidence for bioutilization of HCN in the constructed wetland and remaining receiving water system. Nevertheless, it was noted also that the rates *in vitro* could not be used in the simulation model directly because the laboratory assessment was undertaken using microbiota in suspended culture whereas the field situation would be a combination of suspended and fixed film organisms. Accordingly, calibration of the model to estimate the biodecay rate was still necessary.

Model development began with an Eulerian framework applied to the state variables hexacyanoferrate and HCN and the fate-determining processes photolysis, volatilization, and biotic utilization. A conventional temporal and spatial discretization was applied to formulate a finite difference scheme for solution of the governing equations, but early results suggested a Lagrangian framework would be better suited to the advection-dominated conditions that prevailed at the field site. Indeed, the calibrated Lagrangian model reproduced concentration magnitudes and spatial trends in the coupled cyanide species.

The results from the modeling work led to the following general conclusion: it would appear that cyanide fate process knowledge is such that site-specific numerical modeling of cyanide species fate can be accomplished when aided by the collection of routine monitoring data and simple availability bioassays. More specific conclusions also may be drawn; these are stated here. The receiving water system did not experience "unchecked" accumulation of HCN from hexacyanoferrate photolysis. Rather, the combination of biotic utilization and volatilization prevented HCN from accumulating. Indeed, the modeling results suggested that exposure to HCN *never* began to approach the initial analytical total of cyanide species.

Stated differently, the characteristics of a longitudinal concentration profile for HCN downstream from a point source would involve the development of a maximum exposure concentration, a "critical point" for biotic exposure that is reminiscent of the behavior of oxygen deficit as modeled by the classical Streeter-Phelps approach. This means numerical experiments could be undertaken with the model to identify conditions that would be necessary for HCN accumulation to become an issue at a specific site including providing an estimate of the maximum likely exposure to HCN. With a small amount of speculation, the success of the modeling approach indicates that this conclusion may be carried forward an additional step. That is to suggest that the potential exists to develop and apply a waste load allocation strategy for cyanide (i.e., WLA/TMDL) much as is the current practice for managing the discharge of more conventional pollutants to receiving waters (e.g., oxygen demand and phosphorus). Such an application would appear possible because the underlying model described here is essentially generic in that the cyanide fate-determining processes therein not only are the most significant ones for the studied system, but the processes themselves are expected to be functional and likely of greatest importance in the vast majority of all surface waters systems where cyanide contamination is an issue at least for systems where HCN/CN^- and hexacyanoferrates predominate the cyanide speciation.

The most significant source of uncertainty to consider for planning future work on the fate of cyanide in surface waters and the application of the results from this investigation is the small scale of the field site used for the present study. The small size made the system "flashy"; or prone to respond rapidly to meteorological events, and difficult to characterize in average terms. Other sources of uncertainty included limited resources for performing both the *in situ* HCN flux measurements and the *in vitro* bioutilization characterization, and constraints imposed on field sampling and monitoring by time and distance.

CHAPTER 9.0

EVALUATION OF WATER QUALITY CRITERIA AND DISCHARGE STANDARDS FOR CYANIDE

9.1 Introduction

The primary objectives of this task were to critically evaluate existing federal and state water quality criteria and discharge standards for cyanide and to develop a more scientifically based framework for setting ambient CN water quality criteria and discharge standards based on cyanide chemical speciation.

This stage of the study consisted of three subtasks: compilation and evaluation of current cyanide water quality discharge standards and criteria, evaluation of the potential for species-based criteria, and development of objectives for new or modified water quality standards and criteria.

9.2 Compilation and Evaluation of Current Cyanide Water Quality Criteria and Discharge Standards

Available data on water quality-related criteria and standards for cyanide are voluminous and complex. Tables 9-1 through 9-5 present a nonexhaustive summary of this information according to the governing authority (national or state), regulatory focus (surface water, drinking water, wastewater), health effects target (human or aquatic life), cyanide form (total, free, complex, amenable) if specified, and water usage or origin. State criteria are often assigned based on water usage category. Table 9-6 summarizes such categories as they are known. A complete listing of all health-related cyanide standards can be found in the U.S. Department of Health and Human Services report on the Toxicological Profile for Cyanide (ATSDR, 1997). U.S. EPA (1984) has presented in detail the aquatic toxicity database underlying the national ambient water quality criteria for cyanide.

Table 9-1. National Standards for Cyanide in Water (as summarized in ATSDR, 1997)

Agency	Description	Limit	Designation
U.S. EPA	**Proposed rule: Great Lakes system water quality standards**		
	Acute water quality criteria for protecting aquatic life	22 µg/L	free cyanide
	Chronic water quality criteria for protection of aquatic life	5.2 µg/L	free cyanide
	Water quality criteria for protection of human health		
	Drinking	800 µg/L	
	Nondrinking	60,000 µg/L	
	MCLs	200 µg/L	free cyanide
	National wastewater pretreatment standards		
	Pretreatment standards for existing sources		
	Discharging < 38,000 L/day		
	1-Day maximum	5,000 µg/L	amenable cyanide
	4-Day average maximum	2,700 µg/L	amenable cyanide
	Discharging > 38,000 L/day		
	1-Day maximum	1,900 µg/L	total cyanide
	4-Day average maximum	1,000 µg/L	total cyanide
	Organic chemicals, plastics, and synthetic fibers category		
	Direct discharge point sources with and without end-of-pipe biological treatment		
	1-day maximum	1,200 µg/L	total cyanide
	Maximum monthly average	420 µg/L	total cyanide
	Ferroalloy manufacturing: covered		
	Metal finishing		
	BPT, BAT, PSES, and PSNS[a]		
	1-Day maximum	1,200 µg/L	total cyanide
	30-Day average	650 µg/L	total cyanide
	BPT, BAT, PSES, and PSNS		
	1-Day maximum	860 µg/L	amenable cyanide
	30-Day average	320 µg/L	amenable cyanide

[a] BPT, best practical treatment; BAT, best available technology; PSES, pretreatment standard for existing sources; PSNS, pretreatment standard for new sources.

In reviewing the information contained in Tables 9-1 through 9-5, several features are evident. First, criteria for cyanide vary significantly depending on the use to which the water is put. These uses, in turn, reflect the large differential toxicities that cyanide compounds can exhibit among target organisms. For instance, standards related to human consumption and exposure are generally considerably higher than those for sensitive aquatic organisms. Further, the most sensitive organism can vary depending on the specific aquatic environment (for example marine vs. fresh waters, or cold vs. warm waters). Second, in recognition of these toxicity differences among cyanide compounds (regardless of target organism), many criteria attempt to differentiate among chemical forms of cyanide, sometimes by specifically listing them (e.g., free cyanide, copper cyanide, potassium cyanide), or by specifying an operational class of cyanide compounds as measurable by an analytical procedure (for example cyanide amenable to chlorination). Third, and perhaps most critically, the degree of specificity among cyanide compounds and chemical forms listed among the various standards, when viewed collectively across national and state criteria, is inconsistent with available approved analytical methods (i.e., many more water uses and chemical forms are recognized as being of importance than current analytical methods can accommodate).

Table 9-2. National Guidelines for Cyanide (as summarized in ATSDR, 1997).

Agency	Description	Limit	Designation
U.S. EPA	**1-Day health advisory**		
	Child	220 µg/L	cyanide
	10-Day health advisory		
	Child	220 µg/L	cyanide
	Lifetime health advisory		
	Adult	200 µg/L	cyanide
	Longer-term health advisory		
	Child	200 µg/L	cyanide
	Adult	800 µg/L	cyanide
	Maximum contaminant level		
	Copper cyanide, cyanide, potassium silver cyanide, sodium cyanide	200 µg/L	
	Maximum contaminant level goal		
	Cyanide, potassium silver cyanide, sodium cyanide, potassium cyanide	200 µg/L	
	Copper cyanide	1,300 µg/L	
	Potassium cyanide	200 µg/L	cyanide
	Ambient water quality criteria for human health		
	Potassium silver cyanide, sodium cyanide, potassium cyanide, copper cyanide (water and fish)	200 µg/L	
	Ambient water quality criteria for aquatic organisms		
	Sodium cyanide, potassium cyanide		
	Freshwater acute	22 µg/L	
	Freshwater chronic	5.2 µg/L	
	Marine acute	1 µg/L	
	Copper cyanide		
	Freshwater acute	9.2 µg/L	
	Freshwater chronic	6.5 µg/L	
	Marine acute	2.9 µg/L	

Such a state of affairs lends itself to the oft-noted appearance that there is little consistency among various regulatory agencies regarding water quality criteria and discharge standards for cyanide. As McFarland (1997) has noted, despite the stated intention on the part of the U.S. EPA to recognize the difference between toxic and relatively nontoxic forms of cyanide, there is considerable confusion within the regulated community over how current standards for cyanide are to be interpreted, particularly with respect to the analytical methods that are to be used. Indeed it is not uncommon for unofficial, but more appropriate, methods to be approved for use in specific instances (e.g., for the specific discharge of a less toxic form of cyanide), or for variances from permitted discharge levels to be granted in cases where the critical organism upon which the standard is based can be demonstrated to not be present in the receiving water (Brix et al., 2000; Elserod et al., 1994). This is further exacerbated by uncertainties in the fate of cyanide in wastewaters and receiving waters: the role of photolysis of some complex cyanides, of biodegradation and volatilization of hydrogen cyanide, of synthesis of cyanides by bacteria and other organisms, of the partitioning of cyanide to solid phases, and of the generation of cyanide by some inorganic and organic compounds upon chlorination. Many of these issues have been addressed in this study. Present water quality criteria and discharge standards for cyanide explicitly or implicitly recognize these complexities, but are limited in their applicability by the lack of approved analytical methods capable of distinguishing among cyanide chemical species.

Table 9-3. State Human Health Standards for Cyanide (as summarized in ATSDR, 1997).

State	Description	Limit	Designation
Arizona	Drinking water guideline	220 µg/L	
	Domestic water source (DWS)	140 µg/L	total cyanide
	Fish consumption	210,000 µg/L	total cyanide
	Full body contact	3,100 µg/L	total cyanide
	Partial body contact	3,100 µg/L	total cyanide
Colorado		200 µg/L	
Connecticut	Degree of treatment		
	Disinfection and chemical	10 µg/L	
	Complete	200 µg/L	
	Maximum permissible level	200 µg/L	
District of Columbia	Class C	3 µg/L	
	Class D	200 µg/L	
Florida	Domestic/drinking	200 µg/L	
	MCL	200 µg/L	
	Criteria for surface waters, Class I-V	5.0 µg/L	
Hawaii	Freshwater: acute	22 µg/L	
	Freshwater: chronic	5.2 µg/L	
	Saltwater: acute	1.0 µg/L	
	Saltwater: chronic	1.0 µg/L	
	Fish consumption	No standard (NS)	
Iowa	MCL		
	Class B waters	5 µg/L	
	Class C waters	20 µg/L	
Illinois	MCL	200 µg/L	
Idaho	MCL	200 µg/L	
Indiana	Continuous (4-day average) point of water intake	200 µg/L	
Kansas	Drinking water guideline	154 µg/L	
Kentucky	Maximum allowable in-stream concentration		
	Chronic	5 µg/L	free cyanide
	Acute	22 µg/L	free cyanide
	MCL: Domestic water supply	200 µg/L	free cyanide
Massachusetts	Drinking water guideline	140 µg/L	
Maine	Drinking water guideline	154 µg/L	
Michigan	Domestic/drinking	150 µg/L	free cyanide
Minnesota	Drinking water guideline	154 µg/L	
	Class A & B waters	10 µg/L	(CN)
	Class D waters	200 µg/L	(CN)
Mississippi	Freshwater: acute	22 µg/L	
	Freshwater: chronic	5.2 µg/L	
	Saltwater: acute	1.0 µg/L	
	Saltwater: chronic	1.0 µg/L	
New Hampshire	Drinking water guideline	154 µg/L	
	MCL	10 µg/L	(CN)
	Municipal/domestic	200 µg/L	
New Jersey	Domestic/drinking	200 µg/L	
	Groundwater quality	200 µg/L	
New Mexico	Groundwater	200 µg/L	(CN)
New York	Domestic/drinking	100 µg/L	
	Groundwater effluent standards: maximum allowable concentration	400 µg/L	

Table 9-3. State Human Health Standards for Cyanide (as summarized in ATSDR, 1997) (continued).

State	Description	Limit	Designation
	Surface waters and groundwater		
	A, A-S, AA, AA-S	100 µg/L	
	GA	100 µg/L	
	Freshwater	5.2– 22 µg/L	
	Saltwater	1.0 µg/L	
North Carolina	Class GSA groundwater	154 µg/L	
Ohio	30-Day average	200 µg/L	
Oklahoma	Maximum allowable levels	200 µg/L	
Oregon	Domestic/drinking	200 µg/L	
Rhode Island	Drinking water guideline	150 µg/L	
Tennessee	Domestic/drinking	200 µg/L	
Utah	MCL	200 µg/L	free cyanide
Virginia	Groundwater	5 µg/L	
Vermont	Drinking water standard	154 µg/L	
	Class A or B waters	200 µg/L	
Wyoming	MCL - groundwaters	200 µg/L	
Wisconsin	Public water supplier		
	Warm water sport fish communities	600 µg/L	total cyanide
	Cold water communities	600 µg/L	total cyanide
	Great Lakes communities	600 µg/L	total cyanide
	Nonpublic water supplier		
	Warm water sport fish communities	40,000 µg/L	total cyanide
	Cold water communities	40,000 µg/L	total cyanide
	Warm water forage and limited forage	120,000 µg/L	total cyanide
	Groundwater		
	Enforcement standard	200 µg/L	
	Prevention action limit	40 µg/L	
West Virginia	Water quality criteria		
	Warm water fishery streams	5 µg/L	free cyanide
	Trout waters	5 µg/L	free cyanide
	Small nonfishable streams	5 µg/L	free cyanide
	Water contact, recreation	5 µg/L	free cyanide
	Water supply, public	5 µg/L	free cyanide

9.3 Evaluation of the Potential for Cyanide Species-Based Water Quality Criteria and Discharge Standards

As indicated by the results of Task 1 in this study (Chapters 2.0 and 3.0), the U.S. EPA considers cyanide to be a method-defined analyte, thus the availability of analytical methods capable of distinguishing among cyanide chemical species at environmentally relevant concentrations is an essential part of the development of water quality criteria that are based on the differential toxicity of cyanide chemical species. Data presented in Task 1 demonstrate that the application of such methods yields satisfactory results: method detection limits are sufficiently low, and bias, precision and percent recovery of target analytes or classes of cyanide compounds are within acceptable ranges for most types of samples and conditions. Some interference problems were encountered for the available cyanide, ion chromatography, and thin film distillation methods when applied to raw municipal wastewater, and recovery of some

Table 9-4. State Aquatic Life Standards for Cyanide (as summarized in ATSDR, 1997).

State	Description	Limit	Designation
Alabama	Freshwater: acute	22.0 µg/L	
	Freshwater: chronic	5.2 µg/L	
	Marine: acute	1.0 µg/L	
	Marine: chronic	--	
Arizona	Acute criteria for aquatic & wildlife		
	Cold water fishery (A&Wc)	22.0 µg/L	T(total recoverable)
	Warm water fishery (A&Ws)	41.0 µg/L	T
	Effluent dominated water (A&Wedw)	41.0 µg/L	T
	Ephemeral (A&We)	84.0 µg/L	T
	Chronic criteria for aquatic & wildlife		
	A&Wc	5.2 µg/L	T
	A&Ws	9.7 µg/L	T
	A&Wedw	9.7 µg/L	T
	A&We	19.0 µg/L	T
Indiana	Acute aquatic criterion	22.0 µg/L	
	Continuous (4-day average) outside of mixing zone: Chronic aquatic criterion	5.2 µg/L	
Maryland	Ambient surface waters		
	Freshwater: acute	22.0 µg/L	
	Freshwater: chronic	5.2 µg/L	
	Estuarine: acute	--	
	Estuarine: chronic	--	
	Saltwater: acute	1.0 µg/L	
	Saltwater: chronic	--	
Minnesota	Class A, B, C waters	20.0 µg/L	(CN)
Mississippi	Freshwater: acute	22.0 µg/L	
	Freshwater: chronic	5.2 µg/L	
	Saltwater: acute	1.0 µg/L	
	Saltwater: chronic	1.0 µg/L	
Missouri	Chronic toxicity	5.0 µg/L	amenable to chlorination
	Acute toxicity	22.0 µg/L	amenable to chlorination
Nevada	Single value	52.0 µg/L	
	24-hour average	3.5 µg/L	
	Propagation of wildlife	5.0 µg/L	
New York	Surface waters & groundwaters		
	A, A-S, AA, AA-S, B, C	5.2 µg/L	
	D	22.0 µg/L	
	SA, SB, SC	1.0 µg/L	
	SD	1.0 µg/L	
North Dakota	Class I streams	5.0 µg/L	total cyanides
North Carolina	Freshwater	5.0 µg/L	
Puerto Rico	Coastal estuarine waters	20.0 µg/L	
	Surface waters	20.0 µg/L	
Oklahoma	Acute	45.93 µg/L	
	Chronic	10.72 µg/L	
Ohio	Outside mixing zone (maximum)		
	Cold water	46.0 µg/L	free cyanide
	Limited resource warm water	22.0 µg/L	free cyanide
	30-day average cold water	5.2 µg/L	free cyanide
	Inside mixing zone (maximum)		

Table 9-4. State Aquatic Life Standards for Cyanide (as summarized in ATSDR, 1997) (continued).

State	Description	Limit	Designation
	Cold water	45.0 µg/L	
	Limited resource warm water	92.0 µg/L	
Virginia	Freshwater	5.2 µg/L	total cyanide
	Saltwater	1.0 µg/L	total cyanide
Vermont	Acute	22.0 µg/L	
	Chronic	5.2 µg/L	
Wyoming	Special A waters	5.0 µg/L	
Wisconsin	Great Lakes	22.4 µg/L	free cyanide
	Cold water	22.4 µg/L	free cyanide
	Warm water sport fish	46.2 µg/L	free cyanide
	All others	46.2 µg/L	free cyanide

diffusible cyanide in the microdiffusion test of the nickel–cyanide complex. Further refinement of these methods would be needed for application under these specific conditions. Nevertheless, it seems clear that in the majority of cases it is possible to distinguish unambiguously among inorganic cyanide species with acceptable accuracy.

The results of monitoring and species-specific analysis of cyanide in a receiving water, reported in Task 2 of this study (Chapters 4.0 through 8.0), illustrates the dynamics of cyanide transformations under natural conditions. While not exhaustive of all possible scenarios, the results indicate convincingly that photolysis of strongly complexed iron–cyanide—although much slower than under ideal laboratory conditions (e.g., Broderius and Smith, 1980)—can indeed occur within an environmentally relevant timeframe. Further, this study shows that the subsequent fate of the free cyanide generated is subject to additional reaction, either biodegradation or volatilization, the former mechanism being dominant in the system studied. Other observations have suggested that toxic forms of cyanide may accumulate in aquatic systems (Burdick and Lipschuetz, 1950). Given anticipated variability of such cyanide transformation mechanisms across the spectrum of receiving waters, the potential for multiple species of cyanide, including free cyanide, to coexist, interconvert, and degrade in a system must be considered, irrespective of the forms introduced or initially present. Thus in a given receiving water, the toxicological profile of cyanide can be expected to vary both temporally and spatially and, as noted above, with the target cyanide-sensitive organism.

Table 9-5. Miscellaneous State Standards for Cyanide (as summarized in ATSDR, 1997).

State	Description	Limit	Designation
	Water quality: agricultural uses		
Arizona	Agricultural irrigation (AgI)	No numerical standard	
	Livestock watering (Ag I)	200 µg/L	total recoverable
Nevada	Ag L	200 µ/L	
	Hazardous constituents		
Indiana	Allowable concentration using leaching test method:		
	Class IV	200 µg/L	
	Class III	2,000 µg/L	
	Class II	5,000 µg/L	

The potential for environmental conditions and the various fate mechanisms to alter the speciation and concentration of cyanide in water leads to questions about the adequacy of chemistry and environmental controls in the aquatic toxicity experiments underlying the existing U.S. EPA and state ambient water quality criteria for cyanide. WERF has initiated a new study focused explicitly on this issue. A project has been undertaken to reassess cyanide criteria for the protection of aquatic life in both the water column and benthos.

9.4 Development of Objectives for New or Modified Water Quality Criteria and Discharge Standards for Cyanide

This research has demonstrated both the availability of robust analytical methods capable of measuring cyanide chemical species, and the need to apply them to receiving waters in which cyanide is introduced or found. In addition, analysis of existing water quality criteria for cyanide reveals a complex set of standards that contain considerable variability according to jurisdiction, water use, and target organism, and that are clearly hampered by the lack of approved methods of analysis that can assist in interpreting more accurately the true toxicological profile that cyanide and its compounds present. Although toxicological studies were not conducted as part of this study, it is clear from the available literature that the toxicity of cyanide, irrespective of target organism, is strongly related to the generation and availability of free cyanide (i.e., HCN or CN^-) (Doudoroff, 1976; Doudoroff et al., 1966). Accordingly, it is recommended that water quality criteria and discharge standards for cyanide be revised to achieve the following objectives:

Standards should be developed that are based on and inclusive of available methods of analysis that distinguish among cyanide chemical species or groups of species. The methods with greatest potential for widespread use, identified in Task 1, should receive priority for federal and state approval. The recently introduced Alternate Test Procedure and New Method Approval Program of the U.S. EPA Office of Water (U.S. EPA, 1999a,b) are flexible programs that can help with expansion of approved analysis tools.

Numerical values of standards, although reflective of the specific susceptibilities of local organisms, should also reflect the ability of cyanide compounds to undergo transformations under some environmental conditions that can either increase or decrease its toxicological impact. It is important to recognize that in some regulatory contexts involving cyanide discharge into receiving waters, an opportunity exists to take into consideration these transformations through a water quality modeling approach as illustrated, for instance, in Task 2 of this report and the work of Elserod et al. (1994).

Although the results of this research cannot be used to suggest specific numerical criteria for all possible aquatic systems, a potentially useful approach would consist of developing a "multi-tiered" set of standards in which specific chemical species or classes of cyanide, as measured by appropriate analytical methods, are identified and applied. For example criteria might be set for total cyanide, as measured by method U.S. EPA 335.2 (U.S. EPA, 1983), and free cyanide as determined by the diffusible cyanide method ASTM D4282-95 (ASTM, 1998a). Alternatively, classes of cyanide might be regulated such as total cyanide, weak acid dissociable cyanide (as determined by APHA/AWWA/WEF, 1998 – Method 4500-CN-I), and diffusible cyanide. A third, more complex approach that could be warranted in some cases would be to set criteria for several cyanide species and achieve inorganic cyanide mass balance closure in which all species are directly measured using total cyanide, free cyanide by microdiffusion, and metal–cyanide complexes as measured by ion chromatography (Dionex Application Note No.55; 1988).

These objectives are consistent with the U.S. EPA's stated intention to affirm the distinction between toxic and comparatively nontoxic forms of cyanide (U.S. EPA, 1992a), as further illustrated by the subsequent approval of method OIA-1677 for "available" cyanide (U.S. EPA, 1999c). Furthermore, they reflect previous approaches to water quality standards that incorporate distinctions between different chemical forms, such as for particulate versus dissolved forms of metals, and criteria that distinguish between chemical species such as total dissolved chromium and chromium(VI) (U.S. EPA, 1997b; 62 FR 42159, 1997), and total dissolved arsenic and arsenic(III) (U.S. EPA, 1995), both of which are analyte defined. The evolution of such criteria that explicitly incorporate distinctions among chemical forms and species is based on three critical factors: generation and review of the most recent toxicological information, the development of necessary analytical methods, and knowledge of relevant environmental chemistry. It can be argued that such a state of affairs, augmented by research that has improved understanding of cyanide speciation and by the findings of this research project, now exists for cyanide as well. The achievement of these objectives would result in significantly improved regulation of cyanide in receiving waters that is based on the most current capability to distinguish among cyanide chemical species, and that reflects the available data on cyanide toxicity and transformation potential.

Table 9-6. Selected State Water Classifications.

Florida
Class I – Potable water supplies
Class II – Shellfish propagation or harvesting
Class III – Recreation, propagating and maintenance of healthy, well-balanced population of fish and wildlife
Class IV – Agricultural water supplies
Class V – Navigation, utility, and industrial use
Minnesota
Class 1– Domestic consumption
Class 2– Aquatic life and recreation
New Jersey
GW1 – Ground water of special ecological significance
GW2 – Ground water for potable water supply
GW3 – Ground water with uses other than potable water supply
New York
Class N – fresh surface waters.
Class A, AA, Special (AA-S) – fresh surface waters (drinking)
Class B – fresh surface waters (primary and secondary contact)
Class C, D – fresh surface waters (fishing)
Class SA – saline surface waters (fish propagation and survival)
Class SB – saline surface waters (primary and secondary contact)
Class SC – saline surface waters (fishing)
Class I – saline surface waters (secondary contact recreation)
Class SD – saline surface waters (fish survival)
Class GA – fresh groundwaters (drinking)
Class GSA – saline groundwaters (potable mineral waters)
Class GSB – saline groundwaters (receiving waters)
North Carolina
Class GA – groundwaters (drinking water).
Class GSA – groundwaters; usage and occurrence (potable mineral waters)
Class GC – groundwaters; usage and occurrence (nondrinking uses)
Vermont
Class A(1) – Ecological waters
Class A(2) – Public water supplies
Class B – Cold and warm water fish habitats
Wyoming
Special A waters – Suitable for fish and aquatic life.

CHAPTER 10.0

FUTURE RESEARCH NEEDS AND STRATEGIC APPROACHES TO MEETING COMPLIANCE STANDARDS FOR CYANIDE DISCHARGE

10.1 Conclusions

10.1.1 Task 1 - Evaluation and Testing of Analytical Methods for Cyanide Species in Municipal and Industrial Waste Waters

10.1.1.1 Analytical Method Testing

An evaluation and testing of seven methods for the analysis of cyanide in reagent water and different contaminated water matrices demonstrated that various species-specific cyanide analytical techniques can be employed reliably for low-concentration analysis of their target cyanide analytes or analyte groups with a range of contaminated water types. The only exceptions were the available cyanide, the ion chromatography and the thin film distillation methods that, when applied to raw municipal wastewater, resulted in significant interference problems and/or low recoveries. Also, there was recovery of some diffusible cyanide in microdiffusion tests with nickel-cyanide-spiked samples apparently reflecting dissociation of this weak metal-cyanide complex during the test. The conventional total cyanide analytical method (APHA/AWWA/WEF, 1998 – Method 4500-CN-C), the WAD cyanide method (APHA/AWWA/WEF, 1998 – Method 4500-CN-I), the free cyanide by microdiffusion method (ASTM, 1998a – Method D4282-95), and some automated total cyanide analytical methods appeared to be the most suitable methods tested for routine WWTP monitoring. These methods were used in a one-year WWTP monitoring program as well as in a laboratory study to investigate cyanide production during wastewater treatment.

10.1.1.2 Organic Cyanide Method Testing

The presence of organocyanide compounds in wastewater could potentially contribute to elevated cyanide levels in the chlorinated effluents of WWTPs. Four model compounds—acetonitrile, amygdalin, cyanocobalamin, and 2-acetoxy-3-butenenitrile—were selected and tested in this study with varying chlorine dosages for release of free cyanide by total and diffusible cyanide procedures. The coenzyme form of vitamin B_{12}, which does not contain cyanide, was also tested. It was found that acetonitrile and amygdalin do not yield cyanide with or without chlorination, cyanocobalamin had increased release with increased chlorine dosages and 2-acetoxy-3-butenenitrile had generally decreased release of cyanide with increased chlorine dosages. Both cyanocobalamin and coenzyme vitamin B_{12} gave evidence of the formation of the cobalt cyanide complex after chlorination, even though the molecular structure of the latter compound contains no cyanide.

10.1.2 Task 2 – Source, Transport, and Fate of Cyanide Species of WWTPs in Receiving Waters

10.1.2.1 Fate of Cyanide in WWTPs

Cyanide Occurrence and Formation in WWTPs. An intensive WWTP monitoring program was implemented at six full-scale treatment facilities of participating municipal utilities. A database was developed to investigate cyanide formation in wastewater secondary treatment. Data from all participating WWTPs showed evidence of cyanide formation in this one-year monitoring effort although cyanide formation patterns varied significantly from one plant to another and from one season to another. Generally, the chlorination of thiocyanate seemed to be the most important mechanism for the formation of cyanide in wastewater treatment processes, especially during chlorination/dechlorination. This observation is consistent with the findings of the laboratory study, which was performed to investigate the mechanisms of cyanide formation in wastewater treatment processes.

Effects of Thiocyanate on the Formation of Free Cyanide. Thiocyanate (SCN^-) is commonly encountered in municipal wastewater and appears to be an important precursor for cyanide formation. Effects of thiocyanate on the formation of free cyanide during chlorination and UV disinfection of WWTP secondary effluent were investigated through a laboratory study with synthetic solutions and WWTP secondary effluents. Results indicated that total cyanide was detected in thiocyanate solutions after chlorination in which the chlorine dose was not sufficient to destroy SCN^- completely thus ensuring no residual chlorine to destroy any free cyanide produced. It was also found that thiocyanate can be broken down to yield cyanide by UV irradiation. In addition, thiocyanate was observed to cause a significant positive error in the conventional total cyanide analysis technique involving acidic distillation and standard colorimetric measurement when nitrate was also present in the sample. Overall, it appears that SCN^- is an important precursor for cyanide formation in wastewaters subjected to oxidative disinfection. Therefore, frequent measurements of thiocyanate in wastewater influent and effluent are strongly recommended for WWTPs that have encountered the problem of cyanide formation in the past or for WWTPs that receive industrial discharges containing thiocyanate in the influent.

Effects of Chloramination on the Formation of Cyanide and Cyanogen Chloride. A laboratory study investigating the role of the chlorination process in the formation of cyanide focused on the potential generation of cyanide species in WWTP wastewater upon chlorination in the presence of residual ammonia (i.e., in the chloramination process). Results demonstrated that low concentrations (5-25 μg/L as CN) of cyanogen chloride (CNCl), a highly toxic cyanide species not measured in total or free cyanide analyses, could be detected as a result of chloramination reactions in WWTP secondary effluents. The potential for chloramination of nitrogen-bearing organic compounds to yield cyanogen chloride and/or free cyanide was demonstrated in experiments with synthetic solutions spiked with selected precursor organics: L-serine, benzene, catechin, and humic acid. The amino acid L-serine yielded the largest concentrations of CNCl upon chloramination. Additionally, detectable cyanide, around 10 μg/L, was observed in solutions of L-serine and in WWTP secondary effluents that were chloraminated followed by dechlorination to prevent destruction of any free cyanide produced. Thus, chlorination of WWTP secondary effluent containing residual ammonia can lead to chloramination of organic compounds and the resulting production of cyanogen chloride and free cyanide.

Effects of Nitrosation on the Formation of Cyanide. The presence of nitrite ion as a common constituent in domestic wastewater effluents may play an important role in the formation of cyanide through reaction with certain kinds of organic compounds (e.g., 1,2,4-trihydroxybenzene). Free cyanide (CN^-) was identified as a product of nitrosation of 1,2,4-trihydroxybenzene at 25°C at low pH. Substantial free cyanide formation was also observed at low pH in experiments with WWTP effluents when reactive organics and nitrite were both spiked. Formation of cyanide through nitrosation was strongly pH dependent, being most significant at low pH and negligible at neutral to high pH. This result points to HNO_2 as being more reactive than the dissociated NO_2^- ion. Therefore, the presence of nitrite ion in domestic wastewater effluents may contribute to the formation of cyanide if low pH conditions are encountered in the treatment process (e.g., in the vicinity of the point of hypochlorous acid addition). The reaction of nitrite with organics also occurs, and at greater rates, in conventional analyses for total cyanide which involve distillation. Sufficient sample pretreatment with sulfamic acid at the time of sampling, not at the time of analysis, is highly recommended.

10.1.2.2 Transport and Fate of Cyanide Species in Surface Waters

In general, this investigation identified and provided a successful numerical simulation of three processes that simultaneously appear to determine the fate of essentially all cyanide species in industrially impacted receiving water. These processes include photolysis of hexacyanoferrate [i.e., $Fe(CN)_6^{-4}$ and $Fe(CN)_6^{-3}$], volatilization of HCN, and biotic utilization of HCN. The investigation required four interdependent activities including 1) literature review and data gap determination, 2) data gap elimination by experimental observation of fate-determining processes, 3) site selection and periodic monitoring for relevant physical and chemical environmental variables in system, and 4) numerical model development and calibration.

Literature Review and Data Gap Determination. The principal data gaps identified during the literature search included a lack of knowledge pertaining to the volatilization and biotic utilization of HCN that would be sufficient to permit adequate representation of these processes in a numerical modeling framework.

Data Gap Elimination by Experimental Observation of Fate-Determining Processes. To remedy the above-mentioned deficiencies, a modest program involving field (*in situ*) and laboratory (*in vitro*) observations and experimentation was undertaken. The program included the use of a flux chamber to directly observe the flux of HCN from the water surface to air at two surface water field sites impacted by hexacyanoferrate. The results demonstrated a range of HCN volatilization fluxes from 2–15 μg HCN m^{-2} hr^{-1}. The specific value of the flux depended on both the water phase concentration (i.e., concentration gradient for volatilization) and the temperature. In addition, the work demonstrated that compensation of the measured fluxes for wind speed effects on the gas-film mass transfer coefficient could be accomplished successfully using previously described methods (Chapra, 1997; Mills et al., 1982).

Site Selection and Periodic Monitoring for Relevant Physical and Chemical environmental Variables in System. The monitoring data on cyanide concentrations in water samples collected from the field site demonstrated a significant decline in total cyanide concentrations—greater than would result from measured volatilization losses—during continuous loading by cyanide-contaminated inflowing water. The specific monitoring site that showed this behavior was a pilot-scale constructed wetland, which was installed to test the feasibility of using wetlands for the removal of cyanide and other contaminants of concern. The magnitude of cyanide losses was viewed as presumptive evidence for biotic utilization of HCN in the system inasmuch as other potential fate-determining processes (e.g., volatilization, precipitation or sorption, and sediment accumulation) did not appear to function at a magnitude that could account for the majority of cyanide losses. Accordingly, laboratory studies were undertaken to assess aerobic bioutilization of HCN by microflora obtained from the constructed wetland. Preliminary experiments using KCN-spiked slurries of wetland sediments and water to assess nutritional needs revealed that the use of amendments such as municipal wastewater and standard heterotrophic microbiological growth media (nutrient broth) did not enhance the bioutilization rate of HCN. These results demonstrated several features that were consistent with microbial utilization of CN when it was the only nitrogen source available in the medium. These features included the following:

- A time lag before HCN utilization that depended in a systematic way on incubation temperature and initial HCN concentration;
- An HCN utilization rate that depended on HCN concentration—directly dependent at low concentrations ($< 50\ \mu g\ HCN\ L^{-1}$) and inversely dependent at high concentrations ($> 50\ \mu g\ HCN\ L^{-1}$); and
- An HCN utilization rate that depended on temperature in a manner that obeyed a customary Arrhenius analysis ($E_a \sim 46.6$ kJ $mole^{-1}$).

It was noted that the observed *in vitro* HCN utilization was strong presumptive evidence for bioutilization of HCN in the constructed wetland and remaining receiving water system. Nevertheless, it was also noted that the rates *in vitro* could not be directly used in the simulation model because the laboratory assessment was undertaken using microbiota in suspended culture whereas the field situation would be a combination of suspended and fixed-film organisms. Accordingly, calibration of the model to estimate the biodecay rate was still necessary.

Numerical Model Development and Calibration. Model development began with a Eulerian framework applied to the state variables (hexacyanoferrate and HCN) and the fate-determining processes (photolysis, volatilization, and biotic utilization). A conventional temporal and spatial discretization was applied to formulate a finite difference scheme for solution of the governing equations. However, early results suggested that a Lagrangian framework would be better suited to the advection-dominated conditions that prevailed at the field site. Indeed, the calibrated Lagrangian model reproduced concentration magnitudes and spatial trends in the coupled cyanide species.

The results from the modeling work led to the overall conclusion that the cyanide fate process knowledge is such that site-specific numerical modeling of cyanide species fate can be accomplished when aided by collection of routine monitoring data and simple availability bioassays. More specific conclusions may also be drawn. For example, the receiving water system did not experience "unchecked" accumulation of HCN from hexacyanoferrate photolysis; rather, the combination of biotic utilization and volatilization prevented HCN from accumulating. Indeed, the modeling results suggested that exposure to HCN never began to approach the initial analytical total of cyanide species.

Stated differently, the characteristics of a longitudinal concentration profile for HCN downstream from a point source would involve development of a maximum exposure concentration, a "critical point" for biotic exposure that is reminiscent of the behavior of oxygen deficit as modeled by the classical Streeter-Phelps approach. This means numerical experiments could be undertaken with the model to identify conditions that would be necessary for HCN accumulation to become an issue at a specific site, including providing an estimate of the maximum likely exposure to HCN. With a small amount of speculation, the success of the modeling approach indicates that this conclusion may be carried forward an additional step. This is to suggest that the potential exists to develop and apply a waste load allocation strategy for cyanide (e.g., WLA/TMDL) much as is the current practice for managing the discharge of more conventional pollutants to receiving waters (e.g., oxygen demand and phosphorus). Such an application would appear possible because the underlying model described here is essentially generic in that the cyanide fate-determining processes therein not only are the most significant ones for the studied system, but the processes themselves are expected to be functional and likely of greatest importance in the vast majority of all surface waters systems where cyanide contamination is an issue, at least for systems where HCN/CN^- and hexacyanoferrates predominate the cyanide speciation.

The most significant source of uncertainty to consider for planning future work on the fate of cyanide in surface waters and the application of the results from this investigation is the small scale of the field site used for the present study. The small size made the system prone to respond rapidly to meteorological events and therefore difficult to characterize in average terms. Other sources of uncertainty included limited resources for performing both the *in situ* HCN flux measurements and the *in vitro* bioutilzation characterization, and constraints imposed on field sampling and monitoring by time and distance.

10.1.3 Task 3 – Relationship of Cyanide Speciation and Cyanide Toxicity

This study was developed to evaluate the current state of understanding of the relationship between cyanide speciation and cyanide toxicity. The original objectives of this task were as follows:

- To provide an annotated review of existing data for toxicity of cyanide species;
- To assess the extent to which speciation was taken into account in the experimental design;
- To summarize the current understanding of metal-cyanide, organocyanide, and free cyanide toxicity with respect to aquatic organisms;
- To determine if any toxicity identification evaluation studies have successfully distinguished cyanide as a source of toxicity through toxicity protocols rather than trial and error;
- To provide recommendations to improve the effectiveness of toxicity identification evaluations at distinguishing cyanide toxicity.

On the basis of the limited data available and because of the budget constraints of this study, it was not possible to draw any detailed or meaningful conclusions regarding cyanide toxicity and the data supporting the current water quality criteria standard. In response, the project subcommittee decided to combine all further toxicity investigations into a future request for proposals to adequately address current concerns related to national criteria. As a result, the scope of Task 3 was modified to include the following activities:

- Compilation of a list of articles/studies in which the toxicity of cyanide species has been reported including search methodology and bibliographic listing;
- Evaluation of species-specific consideration for evaluated toxicity studies;
- Formal recommendations for potential future cyanide toxicity testing protocols including specific test procedures and tested species.

This evaluation of toxicity information was successfully compiled during the course of the study. It includes a brief evaluation of fresh and marine cyanide water quality criteria and chemistry, discussion of cyanide speciation in experimental design, summary of literature review, toxicity mechanisms of cyanide, and recommendations for potential paths for future cyanide-related toxicity studies. It was clear from this study that the toxicity of HCN is well established in fresh water but disputed in marine systems. In addition, the toxic effect of cyanide complexed with metals or organic compounds does not appear to be understood in aquatic environments. Evidence suggests that most cyanide complexes are not readily bioavailable. Iron cyanide complexes have received the most attention, and toxicity or bioavailability of the complexed forms has not been indicated. Other cyanide complexes have not been subjected to rigorous toxicity testing. Therefore, detailed toxicity testing has been proposed as a suggestion for further research to resolve these issues and data gaps.

10.1.4 Task 4 – Evaluation of Water Quality Criteria for Cyanide

The overall objectives of this task were to critically evaluate existing federal and state water quality criteria and discharge standards for cyanide, and develop a more scientifically based framework for setting ambient CN water quality criteria and discharge standards based on cyanide chemical speciation.

The evaluation consisted of three subtasks including the compilation and evaluation of current cyanide water quality discharge standards and criteria, the evaluation of the potential for species-based criteria, and the development of objectives for new or modified water quality standards and criteria.

Evaluation of current federal and state cyanide water quality criteria revealed that they vary significantly depending on the use to which the water is put, and that many criteria attempt, appropriately, to differentiate among the various chemical forms of cyanide because of their differential toxicity. However, it was also determined that the degree of specificity for the cyanide compounds and chemical forms listed in the various state and federal water quality criteria and standards is inconsistent with available approved analytical methodologies. Present water quality criteria and discharge standards for cyanide explicitly or implicitly recognize the complexities of cyanide chemistry and toxicology, but are limited in their applicability by the lack of approved analytical methods capable of distinguishing among cyanide chemical species.

The U.S. EPA considers cyanide to be a method-defined analyte, thus the availability of analytical methods capable of distinguishing among cyanide chemical species at environmentally relevant concentrations is an essential part of the development of water quality criteria that are based on the differential toxicity of cyanide chemical species. Results from the analytical methods evaluation work conducted in this study demonstrate that the application of such methods yields satisfactory results: method detection limits are sufficiently low, and bias, precision and percent recovery of target analytes or classes of cyanide compounds are within acceptable ranges for most types of samples and conditions.

It is recommended that water quality criteria and discharge standards for cyanide be revised to achieve the following objectives:

- Standards should be developed that are based on and inclusive of available methods of analysis that distinguish among cyanide chemical species or groups of species.
- Numerical values of standards, although reflective of the specific susceptibilities of local organisms, should also reflect the ability of cyanide compounds to undergo transformations under some environmental conditions that can either increase or decrease its toxicological effect.
- Although the results of this research cannot be used to suggest specific numerical criteria for all possible aquatic systems, a potentially useful approach would consist of developing a "multitiered" set of standards in which specific chemical species or classes of cyanide, as measured by appropriate analytical methods, are identified and applied.

These objectives are consistent with the U.S. EPA's stated intention to affirm the distinction between toxic and comparatively nontoxic forms of cyanide, and they reflect previous approaches to water quality standards that incorporate distinctions between different chemical forms, such as for particulate versus dissolved forms of metals, and criteria that distinguish between chemical species such as total dissolved chromium and chromium(VI), and total dissolved arsenic and arsenic(III), both of which are analyte-defined. The achievement of these objectives would result in significantly improved regulation of cyanide in receiving waters that is based on the most current capability to distinguish among cyanide chemical species, and that reflects the available data on cyanide toxicity and transformation potential.

10.2 Strategic Approaches to Meeting Compliance with Cyanide Discharge Standards

10.2.1 Compliance Challenges for WWTPs

The presence of cyanide in raw and treated wastewaters poses numerous challenges to both municipal and industrial dischargers. The overall objective of this study was to develop a comprehensive assessment of management options for the control of cyanide in wastewater discharges. This study has demonstrated that alternatives and analytical methodologies can provide accurate characterization of wastewaters with respect to the most toxic cyanide form (i.e., free cyanide). This study has also demonstrated that fate and transport analyses of cyanide species can provide a quantitative assessment of the potential risks from discharging cyanide-containing compounds. Finally, this study has also highlighted the difficulties facing many municipal dischargers when dealing with cyanide levels in wastewaters close to, or slightly above, practical quantification limits of existing analytical methods.

Examples of the types of challenges facing both industrial and municipal dischargers include the following:

- Municipal discharges of wastewater with total cyanide above 1 μg/L, the water quality criterion for marine environments;
- Municipal discharges of wastewater with total cyanide of 5-15 μg/L, which may exceed the local water quality criteria if no mixing credits are allowed;

- Recent regulatory developments as the California Toxic Rule, which places severe fines on any discharger exceeding water quality criteria in any sample taken in the treated water from the discharger;
- Industrial facilities discharging wastewater to a WWTP where pretreatment limits are based on total cyanide as opposed to a measurement of the toxic cyanide species (e.g., free cyanide or weekly dissociable cyanide); and
- Industrial dischargers releasing treated wastewaters to surface waterbodies where standards for compliance are also based on total cyanide rather than on a species-specific analytical method.

Compliance is clearly critical for NPDES permit holders. This is especially true in California following the promulgation of the California Toxic Rule (U.S. EPA, 2000) and the implementation of the state water plan (state implementation plan, or SIP; California Dept. of Water Resources, 1998). Discussions of four key challenges facing WWTPs in California, and in some cases nationwide, are presented below. In addition, a discussion of these challenges has been recently published (Deeb et al., 2003).

10.2.1.1 Permit Limits Below Method Detection Limits (MDLs)

For many of the utilities in California, the first detect could result in noncompliance regardless of the credibility of the data at the MDL (i.e., analytical variability or interferences). Following implementation of the California Toxic Rule in 2000 and the SIP in 2001, NPDES permit holders with "reasonable potential" (established by a formula in the SIP) could have effluent concentration allowances (ECAs or discharge limits) placed in their permit at the water quality criterion. Frequently these ECAs are below the lowest limit of quantification available using the required U.S. EPA-approved analytical methods. In the case of total cyanide, the permit limit for shallow water dischargers was set at 1 μg/L (water quality criterion for marine waters) despite the fact that MDLs for total cyanide using U.S. EPA 335.2 (the only approved method; U.S. EPA, 1983) typically was 3-10 μg/L among laboratories.

10.2.1.2 Permit Limits Just Above MDLs

For example, the San Francisco Bay/Delta Regional Water Quality Control Board (RWQCB) allows deep-water dischargers such as East Bay Municipal Utility District (EBMUD) a 10:1 dilution credit for non-303(d)-listed, permitted contaminants. This is well below the actual 40:1 minimum dilution ratio for EBMUD and other deep-water dischargers determined by dye studies. Even though the dilution credit for such dischargers places the analytical MDL for total cyanide at or below the calculated permit effluent limit of 10 μg/L, the narrow range between the MDL and the effluent limit of 0-7 μg/L is unacceptable to ensure NPDES permit compliance.

10.2.1.3 Reasonable Potential Analysis

For each of the 126 pollutants listed in the California SIP and the additional three pollutants identified by the RWQCB, a reasonable potential analysis is required to determine whether a discharger will receive an NPDES permit limitation for a particular constituent.

According to the SIP, the determination of a priority pollutant requiring a water quality-based effluent limit is based on whether a discharger may (a) cause, (b) have reasonable potential to cause, or (c) contribute to an excursion above the water quality criteria. Setting effluent limits based on reasonable potential is not straightforward. In the SIP, reasonable potential is defined by the following equation:

$$ECA = C + D(C - B) \quad (10\text{-}1)$$

where:

ECA	=	Effluent concentration allowance
C	=	The most stringent water quality criteria for the pollutant in the receiving water
D	=	Dilution credit in the mixing zone
B	=	Ambient background concentration

Because "C" is fixed at 1 µg/L for total cyanide because of the inclusion of *Cancer irroratus* aquatic toxicity data developed by U.S. EPA in 1984, and B is limited by the MDL using current analytical technology allowed by U.S. EPA for total cyanide, the following is true:

ECA = 1 µg/L + 1(1 µg/L – 3 µg/L)
ECA = 1 µg/L + 1(0 µg/L) [note that negative values of (C – B) default to zero]
ECA = 1 µg/L

In other words, the final permit limit is set at the water quality objective for all dischargers regardless of dilution credit.

10.2.1.4 SB 709 – Mandatory Fines for Permit Violations

In California, SB 709 (the Clean Water Enforcement and Pollution Prevention Act of 1999) requires a mandatory $3,000 fine for every exceedance of a discharge permit limitation by 20% (for carcinogens and other priority pollutants) or 40% (for physical and certain noncarcinogens parameters). In the case of cyanide, with a permit limit of 1 µg/L, a concentration of 1.2 µg/L (beyond the method sensitivity) equates to a $3,000 fine.

Additionally, under SB 709 civil penalties may be assessed for repeated failures to comply with the law. Such civil penalties can amount to tens of thousands of dollars in fines within a very short period. In a case where cyanide may be formed as part of the disinfection process, this is a monumental liability.

Finally, third-party litigation is always a threat to WWTPs when dealing with ongoing problems meeting discharge limitations.

10.3 Compliance Strategies for WWTPs

The results of this study suggest that there are several alternative strategies that can be pursued by municipal and industrial dischargers. This is especially true if it can be shown that the standard based on total cyanide measurements is too conservative and does not provide a benefit commensurate with the increase in costs associated with monitoring and/or treatment to meet the total cyanide standard. The goal of any discharger should be to protect human health and the environment, but at a cost commensurate with the benefits obtained. Standards should be established that provide sufficient protection to human health and the environment without requiring expenditures that provide only limited additional benefits and limited additional risk reduction. It must be stressed, however, that each wastewater discharge case involving cyanide species may have site-specific factors that must be accounted for to determine whether an alternative strategy as discussed below should be pursued.

Alternative compliance strategies for WWTPs include the following:

- Reduction of cyanide precursors to sanitary sewers from industrial or commercial dischargers;
- Obtaining a site-specific discharge permit;
- Requesting a change in the point of monitoring where compliance would be evaluated;
- Modification of the activities within a wastewater plant causing production of cyanide, namely the disinfection process; and
- Requesting a change in the permit requirements with specific attention to alternative analytical methods measuring the concentration of the toxic form of cyanide.

10.3.1 Reduction of Cyanide or Cyanide Precursors

Under this strategy, a municipal agency responsible for operating a WWTP could undertake a survey of current industrial dischargers and, through characterization of current and future waste streams, determine the magnitude of cyanide species and cyanide precursors (such as thiocyanate) being discharged to the WWTP. Pretreatment standards and pretreatment limits could be established to reflect the desire to reduce cyanide or cyanide precursor input to the WWTP. Improved monitoring and enforcement may be required if the quantity of cyanide or cyanide precursor cannot be reduced. For example, in some cases, municipal treatment plants generate cyanide internally through practices such as incinerator gas blow-down. In such a case, optimizing the operation of the incinerator can reduce the quantity of cyanide produced in the incinerator off-gas.

Implementing this strategy does not always yield positive results. For example, an initial response by EBMUD to its cyanide problem was to evaluate plant influent and collection system samples for industrial sources of cyanide to implement pollution prevention and control mechanisms. Results of this evaluation revealed that all but one metal plating firm (the principal user of industrial cyanide) had either implemented effective on-site removal systems, moved out of the service area, or had been shut down. The only remaining problem plater was served with a cease-and-desist order for illegal discharges and was placed under investigation by U.S. EPA. Influent results and pollution prevention actions could not support or correct cyanide found in treatment plant effluent samples.

10.3.2 Request a Site-Specific Discharge Permit

U.S. EPA guidelines (U.S. EPA, 1985) for the derivation of numeric aquatic site-specific water quality criteria acknowledged that the national criteria serve as benchmarks and may require adjustments for site-specific applications.

One of the options for dischargers is to challenge the basis upon which the water quality criteria have been established. Such an example is discussed earlier in this report where a site-specific standard was approved for dischargers in the Puget Sound region in the state of Washington. Water quality criteria previously established were based on a crab species not indigenous to the area. An alternative species was tested and anew water quality criterion was established. Alternatively, one can request a change in the numerical standard based on the mixing credit analysis or a change in the analytical technique. For example, the Municipal Water Reclamation District of Greater Chicago pursued and received approval for an alternative water quality criterion that was consistent with relevant aquatic organisms in the receiving water and the appropriate kinds of cyanide analyses.

The most commonly used option in site-specific criteria development uses the U.S. EPA indicator species approach to determine water effects ratios (WER). To date, toxicity in effluent and/or receiving waters has not been widely evaluated to ascertain whether cyanide exhibits a WER > 1 mg/L.

Bay Area dischargers are seriously considering a site-specific objective as a means for addressing the cyanide issue. Bay Area Clean Water Agencies and Western States Petroleum Association representatives are meeting with San Francisco Bay RWQCB staff on this issue to develop and implement a regional site-specific objective plan for San Francisco Bay.

10.3.3 Change Point of Compliance

It has been demonstrated that following chlorination, dechlorination can often lead to a reduction in the amount of cyanide that is either present or has been measured as an analytical artifact. The EBMUD pursued this approach to resolve their cyanide compliance problem as discussed earlier in this report. The strategy to move the point of compliance following dechlorination was accepted by the regulatory agency and the problem has been essentially resolved.

10.3.4 In-Plant Modification of Oxidation Processes

This study demonstrated that chlorination of wastewater under certain chemical conditions can lead to the formation of cyanide. Additionally, it was demonstrated that UV disinfection can also cause the dissociation of cyanide compounds leading to the release of free cyanide. Consequently, modifications to the disinfection process (either chlorination or UV disinfection) require site-specific studies to demonstrate the effectiveness of one strategy or another. In the case of the chlorine disinfection process, one possible solution is more effective control of the in-process mixing regime to reduce local high concentrations of chlorine and thus minimize the potential formation of cyanide during the complex parallel and sequential reactions that occurs once chlorine is added to the wastewater.

10.3.5 Alter the Method of Compliance

The final strategy that can be pursued is modification to discharge permits by changing the acceptable analytical method from the total cyanide method to an alternative approved method such as the available cyanide method (U.S. EPA Method OIA-1677) or other methods as discussed in this report that provide an accurate measurement of more toxic species of cyanide.

10.3.6 Other

In addition to the above-mentioned options, two others are being considered by WWTPs in the San Francisco Bay Area. These methods are discussed below.

10.3.6.1 Confirm that Cyanide is a Non-Issue in San Francisco Bay Waters by Challenging the Appropriateness of the Water Quality Criteria

This may be accomplished by reevaluation of the original *C. irroratus* data to confirm or reject its validity. Bay Area Clean Water Agencies, however, has determined that this option provides only partial relief to Bay area dischargers. Removal of the *C. irroratus* data from the database will not adjust the water quality criteria to a point where shallow water dischargers receive regulatory relief. This is because the recalculation produces a water quality criterion between 2-3 μg/L (insignificantly different from current 1 μg/L), and shallow water dischargers are not entitled to a 10:1 dilution credit.

This can also be accomplished by attempting to understand the environmental persistence and affects of cyanide by better understanding the acute and chronic toxic effects of the various cyanide forms (e.g. hydrogen cyanide, cyanate, thiocyanate, cyanogen chloride, organocyanide compounds, metal cyanide complexes, etc.). There is significant potential in undertaking this effort only if there is some assurance that the outcome will result in a reevaluation of the water quality criteria in the context of the determining that certain cyanide complexes have little or no adverse environmental impact and that analysis of the toxic components is possible.

10.3.6.2 Cyanide Translator Number

An unexplored option for possible resolution to the cyanide dilemma may be to look at this compound in the same way metals are viewed. Fore example, copper was considered to be a problem in South San Francisco Bay. A site-specific study was conducted that resulted in the development of a translator between total copper (the regulated form) and free copper (the presumed bioavailable form). This translator permitted the analysis of total copper than by applying a translator factor converted this number into a value that more appropriately represented "toxic concentration." A similar approach could be used for cyanide where a translator value applied to total cyanide concentrations links these data to the acutely toxic free cyanide form.

Developing cyanide translators would require more certain knowledge than currently exists that certain organic and metal complexes of cyanide do not contribute to either acute or chronic toxicity and result in their exclusion. Though little has been done in this area in either research or practice, it remains an option should the other approaches prove to be unsuccessful.

10.4 Recommendations for Future Research

Many questions remain regarding the fate and transport of cyanide species in municipal wastewater treatment processes as well as the accuracy of analytical methods for measuring cyanide species, especially at low concentration. Issues that need to be addressed that came to light as a result of this research are summarized below.

10.4.1 Analytical Methods for Cyanide Species

The cyanide analytical speciation evaluation task demonstrated that the capability exists to analyze specific classes of cyanide species or in some cases single metal cyanide complexes as opposed to just the total cyanide. Many of the methods evaluated such as the ion chromatography (IC) method for metal cyanide complexes and the WAD cyanide and the free cyanide by microdiffusion analysis method exhibited relatively high accuracy and precision in the majority of the matrices tested. The application of such methods affords the opportunity to conduct a complete cyanide mass balance thus accounting for all cyanide species present and yielding an improved determination of the potential toxicity of wastewaters.

This work is just the beginning, however. To use any of these methods for regulatory monitoring, they need to receive official approval from the U.S. EPA Office of Water. U.S. EPA approval is achieved through the completion of a method validation study. The U.S. EPA Office of Water has published specific requirements for carrying out an interlaboratory comparison study for method validation in their Streamlining Proposal for new test procedures (U.S. EPA 1992b; FR 14976, Volume 62, No. 60, March 28, 1997). Depending on the ultimate application of a new test method, data requirements for method validation can be quite significant. These requirements can translate into significant efforts and costs to the entity seeking approval. Future

WERF research should focus on assisting in the effort to move selected cyanide speciation methods forward through the U.S. EPA method approval process. Such an effort would involve collaborative research between industry, commercial laboratories, and U.S. EPA research laboratories to generate the necessary data towards final agency approval.

The continuing problem of interpretation of the meaning of cyanide species data near method detection limits, which are often close to or even higher than cyanide water quality criteria, could be addressed by further method development to reduce detection limits. For example, the total and WAD methods can be altered to distill larger sample volumes into a smaller NaOH trap volumes, thus reducing detection limits by a factor of 10 or more (Hansen and Associates, 1990).

False cyanide positives are still a concern for several cyanide analysis methods because many factors could contribute to the formation of cyanide during the analytical process [e.g., a combination of certain organic compounds with oxidized nitrogen species (such as nitrite); or a combination of thiocyanate with oxidants (such as nitrate and/or chlorine)]. The following issues need to be thoroughly addressed to ascertain that the cyanide analytical method is reliable and every cyanide analytical result is valid.

10.4.1.1 Interference from Pyridine-Barbituric Acid

Almost all the commonly used cyanide analytical methods utilize a two-step process: a high-temperature, acidic distillation and a colorimetric finish with pyridine-barbituric acid reagent. Interferences from the distillation step have been adequately addressed (APHA/AWWA/WEF, 1998) while interferences from pyridine-barbituric acid (i.e., a red color with absorbance at 578 nm), resulting from reaction of pyridine-bartituric acid with compounds other than free cyanide, are still unknown. Although this kind of interference may be eliminated or reduced to a minimum by distillation, it is possible that some substances in wastewaters may also distill and react with pyridine-bartituric to form a red color, thus causing a false positive.

10.4.1.2 Interference from Thiocyanate Reacting with Nitrate

It was revealed in this study that free cyanide was generated in the total cyanide analysis of nitrate-bearing wastewater effluents spiked with a certain amount of thiocyanate. This apparent interference could not be eliminated by adding sulfamic acid, a chemical reagent effective for removing nitrite interference in samples for total cyanide analysis. A sample preliminary treatment protocol should be established to eliminate the interference. This is a critical issue for total cyanide analysis of thiocyanate-bearing wastewaters that are fully nitrified.

10.4.1.3 Analysis of Thiocyanate in WWTPs

Thiocyanate, a cyanogen compound commonly encountered in municipal wastewater, tends to decompose to free cyanide in oxidative processes such as chlorination and UV irradiation. However, frequent and accurate monitoring of thiocyanate in wastewater is not commonly practiced in most WWTPs. This has caused many thiocyanate-related issues to be complicated and unsolved. Several attempts were made to measure thiocyanate in wastewater influents, postchlorination effluents, and post-dechlorination effluents in some of the WWTPs participating in the one-year monitoring program. This proved to be very beneficial to a better understanding of cyanide–thiocyanate interactions in wastewater treatment processes. Therefore, it is very important to keep track of the thiocyanate profile in the wastewater treatment process, especially for WWTPs with fairly high influent concentrations of thiocyanate (mg/L level).

10.4.2 Fate of Thiocyanate in Wastewater Treatment Process

As demonstrated in this work, thiocyanate (SCN^-) plays an important role in the formation of cyanide in wastewater treatment processes, especially in chlorination/dechlorination processes as well as UV processes, and potentially in all oxidative processes. Although biological degradation of thiocyanate has been extensively studied in the context of treatment of thiocyanate-laden industrial wastes, the overall fate of thiocyanate during WWTP treatment processes is not well understood. Some issues have been investigated in this study, such as the probable fate of thiocyanate in chlorination and UV irradiation. However, many still need to be addressed before a complete and conclusive assessment of environmental fate of thiocyanate in wastewater treatment processes can be made.

10.4.2.1 Photolysis

The photodecomposition of thiocyanate by UV irradiation and the resultant cyanide formation in test solutions and WWTP effluents were observed in the laboratory study. No systematic studies have been performed to investigate the photodecomposition of thiocyanate in WWTP treatment, nor have any rate constants been calculated. Such a study could help assess the significance of photodecomposition to the environmental fate of SCN^- and the importance of photodecomposition as a cyanide formation mechanism.

10.4.2.2 Volatilization

Because of the similarities in chemical and physical properties with the halides, thiocyanate is widely considered a pseudo-halide, with the corresponding pseudo-halogen being thiocyanogen, $(SCN)_2$. No data are available regarding the vapor pressure of thiocyanogen or thiocyanic acid (HSCN). Investigation of the volatilization of thiocyanate in WWTP treatment would be beneficial to a better understanding of the fate of thiocyanate.

10.4.2.3 Sorption

Cyanide can be sorbed by a variety of materials, including clays, biological solids, activated carbon and sediments (U.S. EPA, 1979). In contrast, there are few resources in the literature about the sorption of thiocyanate in WWTPs. Data are required to assess the importance of sorption on the fate of thiocyanate in WWTPs.

10.4.3 Effects of Oxidized Nitrogen on the Formation of Cyanide

Oxidized nitrogen species (i.e., nitrite and nitrate) have been demonstrated to be reactive precursors for the formation of cyanide either in some circumstances in WWTPs or in cyanide analytical procedures. A rigorous study should be performed to assess the overall effects of oxidized nitrogen on the formation of cyanide.

10.4.3.1 Nitrite

One reaction that has been largely unrecognized, yet potentially very important in determining the formation of cyanide in wastewater treatment process is the nitrosation of organics. Although some WWTPs observed a significant cyanide formation when the plant was fully nitrifying, especially in summer, direct evidence of correlation between cyanide and nitrite/nitrate has not been obtained in this study.

10.4.3.2 Nitrate

Nitrate is the ultimate product of a nitrification process and can be denitrified/reduced to nitrite. The effects of nitrate are two fold. Once reduced to nitrite, nitrate can potentially lead to the formation of cyanide by the nitrosation process. On the other hand, nitrate itself can react with thiocyanate to generate free cyanide. Consideration of the combination of both effects may yield important information on the overall effects of nitrate on the formation of cyanide.

10.4.4 Fate of Organic Cyanide in Wastewater Treatment Process

The fate of organic cyanides (e.g., nitriles, cyanohydrins, etc.) has received little attention in the literature. It has been estimated that the fate of low molecular weight organic cyanides (nitriles) may parallel the fate of hydrogen cyanide (U.S. EPA, 1979). However, very few data exist to support this hypothesis and no systematic work has been conducted. Thus, it is difficult to justify the parallelism in the environmental fate of nitrile and cyanide.

10.4.4.1 Nitrile

Although most of the cyanide analytical techniques do not intend to measure nitrile, the possibility of cyanide formation from nitrile merits a careful investigation. It is evident from the literature that nitriles can be biodegraded and some of the low molecular weight nitriles can be used by microbes as their sole source of carbon and nitrogen (DiGeronimo and Antoine, 1976). It is unlikely that free cyanide would be the intermediate product or the end product of the biodegradation of nitriles, but more work needs to be done before the quantitative aspects of this process can be determined.

10.4.4.2 Cyanohydrin

It is well known that aldehydes convert cyanide to cyanohydrin, which forms nitrile under the distillation conditions (APHA/AWWA/WEF, 1998). Therefore, investigation of the fate of nitrile can help to explore the fate of cyanohydrin, which is largely unknown.

10.4.5 Fate of Cyanogen Chloride in Wastewater Treatment Process

As demonstrated in the laboratory study, cyanogen chloride (CNCl), a highly toxic cyanide species not measured in total or free cyanide analyses, could be detected as a result of chloramination reactions in WWTP secondary effluents. However, the conditions for this process to occur vary widely in WWTPs, compared to the laboratory study, and it is difficult to generalize these laboratory results to the issue of environmental fate. The overall fate of CNCl in wastewater and surface water, if determined, would serve as a significant contribution to a better understanding of the environmental fate and chemistry of cyanide.

10.4.6 Fate and Transport of Cyanide in Biosolids

Because of its highly toxic nature, the behavior of cyanide in soils (i.e., the mobility and/or attenuation in soils) has been studied (Alesii and Fuller, 1976; Meeussen, et al., 1992; Ghosh et al., 1999). Because cyanide is readily biodegraded, some residual cyanide is expected to transfer from wastewaters into activated sludge (biosolids). There are, however, very few cyanide sludge concentrations available in the literature other than those provided by Kelada (1989). The fate and transport of cyanide in activated sludge and the potential inhibition of sludge digestion has not been thoroughly investigated. This could be a critical step to estimate accurately the overall fate of cyanide species within WWTP treatment process, especially for those WWTPs discharging the biosolids to the headworks of the treatment plant.

10.4.7 Cyanide Fate and Transport in Surface Receiving Waters

The modeling work conducted at the Alcoa Tennessee site suggests that site-specific factors such as water color and turbity, water column depth, and the presence of cyanide degrading microflora can significantly reduce the free cyanide concentration and prevent toxic accumulations in the environment. Future research should evaluate the affect of these and other site-specific factors on the fate and transport of cyanide. The knowledge gained would improve the predictive power of fate and transport models for more technically sound risk-based management of cyanide in the environment.

APPENDIX A

WORK PLAN FOR TASK 1 – CYANIDE ANALYTICAL METHOD EVALUATION AND TESTING

WORK PLAN FOR TASK 1 – CYANIDE ANALYTICAL METHOD EVALUATION AND TESTING

Anping Zheng, David A. Dzombak, Richard G. Luthy
Carnegie Mellon University

1. OVERVIEW

The performance of analytical methods for cyanide species will be evaluated during the first year of the WERF project. A total of seven test methods will be evaluated by conducting intralaboratory studies at six laboratories with each laboratory evaluating one or two specific methods. The analytical methods to be evaluated and laboratory assignments are listed in Table 1. The method for measurement of organic cyanide is under development at Clarkson University, but will not be subjected to the same range of evaluation tests as the existing methods. During the study, the accuracy, precision, reliability, and ease of implementation for each of the methods will be assessed via testing. This will employ a common set of synthetic and actual wastewaters and receiving waters from participating facilities. This work plan provides detailed procedures for the intralaboratory studies to obtain the method detection limit, bias, and precision of each test method for cyanide species in different contaminated water matrices.

2. TASK 1 ORGANIZATION

Task 1 activities will be coordinated by Carnegie Mellon University with David Dzombak as the lead contact and overall coordinator for conducting the study. Each participating laboratory will have a supervisor to oversee the conduct of the study within the laboratory and to communicate with the coordinator. These supervisors and their contact information are listed in Table 2.

All data obtained will be transmitted to David Dzombak of Carnegie Mellon for compilation. Data will be forwarded from Carnegie Mellon to Malcolm Pirnie, Inc.

Table 1. Cyanide Analytical Methods to be Evaluated.

Methodology	Species Measured	Reference	Laboratory
Automated Total by Thin Film Distillation	Total CN	ASTM D4374-93	Chicago
Automated Total by UV Digestion, Distillation	Total CN	EPA Method 335.3	Boston
Automated Weak Acid Dissociable by Thin Film Distillation	WAD CN	ASTM D4374-93	Chicago
Metal-cyanides by Ion Chromatography	Metal-cyanides	Dionex, 1988	Alcoa
Free Cyanide by Micro Diffusion	Free CN	ASTM D4282-95	Carnegie Mellon
Weak Acid Dissociable Cyanide	WAD CN	SM 4500-CN-I	Carnegie Mellon
Available Cyanide	WAD CN + Hg-CN	ASTM, 1998	Bayer Corp.

Organic Cyanide	Organic CN	under development	Clarkson

Table 2. Task 1 Organization.

Code	Laboratory	Lab. Supervisor	Telephone	Email
1	Chicago	Prakash Tata	708-222-4059	prakasam.tata@mwrdgc.dst.il.us
2	Boston	Michael Delaney	617-539-4302	mdelaney@mwra.state.ma.us
3	Alcoa	Sharon Drop	724-337-2160	sharon.drop@alcoa.com
4	Clarkson	Thomas Theis	315-268-3853	tlt0@sun.soe.clarkson.edu
5	Carnegie Mellon	David Dzombak	412-268-2946	dzombak@cmu.edu
6	Bayer	John Sebroski	304-451-2227	john.sebroski.b@bayer.com

3. TASK 1 WORK PLAN

The Task 1 studies will begin with analysis of cyanide species spiked in a clean water matrix in order to determine the detection limit for each method. Precision and bias will also be assessed for the clean water matrix tests. These tests will not only enable determination of the method detection limit, but will provide baseline information about performance of the analytical methods under ideal conditions.

After completion of the clean water matrix tests, the performance of the methods will be assessed by testing a total of five matrices comprising both wastewater and contaminated groundwater samples. The samples will be chosen to represent a range of contaminated water matrix types. For samples with sufficiently low cyanide concentrations, matrix-dependent method detection limits will be determined. This may not be possible for all samples. At a minimum, however, method precision and bias will be assessed for each sample.

3.1 Reagents and Sample Handling

At a minimum, reagent grade chemicals shall be used in all operations. Reagents of the highest available purity shall be used for spike analytes and demonstrated to be free of interfering substances for the subsequent tests to be performed. It is recommended that all laboratories use chemical reagents from Aldrich, Inc. References to water shall be understood to mean reagent water as defined by Type II of Specification in ASTM D 1193 (ASTM, 1991) and demonstrated to be as free of cyanide as possible.

Table 3 lists four reagents of cyanide species recommended for use in this study. Ferrocyanide is a representative and commonly occurring strong metal-cyanide complex, while the nickel cyanide is a representative weak metal-cyanide complex and can be recovered completely in the WAD test for the concentration range of interest in this study. Potassium cyanide dissociates completely in water to yield free cyanide. Thiocyanate normally is not detected in conventional cyanide analyses and can interfere with measurements under some conditions. The degree to which it is detected by the cyanide speciation methods listed in Table 1 will be evaluated.

With regard to potassium tetracyanonickelate (II) hydrate, the uncertainty in the number of hydrated waters will affect the calculation and accurate addition of this reagent to achieve the expected concentration. Although the Aldrich catalog gives the estimation of attached water molecules (x) as 5, several determinations of x by different laboratories resulted in different values of x and it may change with time. So it is important that each laboratory perform Atomic

Absorption or Inductively Coupled Plasma analysis of Ni in stock and standard solutions to check the x value.

All samples containing cyanide species shall be stored in closed, amber bottles in a cool place (4°C). Strong NaOH solution should be added to raise sample pH to 12.5. Samples must be analyzed within the maximum holding time of 14 days (U.S. EPA, 1997). Because most cyanides are very reactive and unstable, it is strongly recommended that samples be analyzed as quickly as possible. The exact holding time prior to analysis should be noted and recorded for each sample analyzed.

Note that special safety precautions are absolutely necessary when handling and storing cyanide-containing material and reagents. Such precautions are described in Standard Methods (APHA, 1995) and at a minimum all sample preparation should be performed in a laboratory hood. Samples should under no circumstances be acidified prior to analysis, this promotes formation of hydrocyanic acid (HCN) which is volatile and highly toxic. Hydrogen Cyanide gas monitoring is recommended.

Table 3. Cyanide Chemical Reagents to Be Used in the Study.

Product Name	CAS Registry No.	Chemical Purity	Molecular Formula	Aldrich Catalog No.
Potassium ferrocyanide, trihydrate	14459-95-1	99%	$K_4Fe(CN)_6{\cdot}3H_2O$	22,768-4
Sodium cyanide	143-33-9	99.99%	NaCN	43,159-1
Potassium tetra-cyanonickelate, hydrate	238-082-7 (ELINCS/EINECS No.)	—	$K_2Ni(CN)_4{\cdot}xH_2O$	41,515-4
Potassium thiocyanate	333-20-0	99%	KSCN	20,779-9

3.2 Methods Validation Study in Reagent Water Matrix

The purpose of the methods validation study with reagent water systems is to measure the performance of each testing method for cyanide species under ideal conditions with respect to method detection limit, bias and precision in the context of a single operator. This portion of the Task 1 effort is expected to take two months to complete.

Each laboratory will develop its standard curve (calibration curve) for their species-based test methods at the beginning of this study. his calibration curve should be specific to each analyte and should cover the entire working range of the method. Construct the curve using at least three points, including one near the upper point of the concentration range and one near the lower limit with an equitable distribution of the remaining points. The actual minimum number of calibration points depends upon the width of analytical range and the slope of the calibration curve. Sodium cyanide will serve as the standard for total cyanide, WAD cyanide and free cyanide testing methods, while potassium tetracyanonickelate and potassium ferrocyanide will serve as the standards for metal-cyanides by ion chromatography and Atomic Absorption or Inductively Coupled Plasma analyses of metals in standard solutions should be conducted to check the actual concentration of metal-cyanides. All laboratories are expected to utilize the cyanide chemicals from Aldrich, Inc. (please refer to Table 3) for standard cyanide solution and standardize against standard silver nitrate ($AgNO_3$) titrant as described in Section 4500-CN- D.4 (APHA, 1995). As an alternative, each laboratory may purchase standard cyanide 1000 ppm (1 mL = 1 mg CN^-) from a certified chemical reagent supplier such as Fisher Scientific (Catalog No. LC13545-1). If cyanide standards are provided by an independent unit other than made in laboratory, the report should so note, including the concentrations and any specific instructions for their use.

3.2.1 Method Detection Limit

Each laboratory will determine the MDL of its species-based test method using reagent water spiked with each of four cyanide species listed in Table 3. Separate spiking experiments will be conducted for each species. The general procedure for conducting an MDL test is as follows:

(a) Each laboratory should prepare reagent water as specified in Section 3.1.

(b) Prepare a laboratory standard sample (one cyanide species spiked in reagent water) at a concentration that is at least equal to or in the same concentration range as the estimated method detection limit. (Recommend between 1 and 5 times the estimated MDL). Detailed guidance for spiking into aqueous samples is presented in Section 3.2.2.

(c) Take seven aliquots of the sample and process each subsample through the entire analytical method.

(d) Calculate the variance (S^2) and the standard deviation (S) of the replicate measurements.

(e) Compute the MDL as follows: MDL = 3.143 * S (U.S. EPA, 1994a)

Each laboratory will determine four MDLs corresponding to each of the four spiked species for its specific testing method(s). For each MDL testing, seven replicate samples should be prepared. Therefore, a total of 28 samples will be analyzed by each laboratory for each method.

3.2.2 Precision and bias

The concept of accuracy encompasses both precision and bias. The accuracy of a measurement process is affected by both random and systematic sources of error, with the former measured as precision and the latter as bias (ASTM, 1990). The precision of a test method is expressed quantitatively as the standard deviation computed from the results of a series of controlled determinations (APHA, 1995). In this study, precision will be measured as the single-operator standard deviation for analyses in reagent water matrix. The difference between the true value (known concentration) and the related mean of replicates is the bias of the test method which reflects the average systematic error (ASTM, 1990). Bias will be assessed in this study through analysis of samples of known concentrations and by recovery of spikes.

3.2.2.1 Spiking Procedure

Cyanide species selected for spiking in each test method are listed in Table 4. Please refer to Table 3 for detailed information about each chemical reagent. Each cyanide species should be spiked separately into the reagent water by each laboratory following the procedures listed below. The order of compound spiking tests should be the same with every new water sample tested, in an attempt to make the holding time approximately equal for each cyanide species from sample to sample.

(a) Determine the volume of the portion of sample to be spiked, depending on such factors as the sample volume required by specific analytical method to be used, convenience of dilution factors, and amount of sample for replicate analysis.

(b) Prepare a spiking solution of suitable concentration. Where solubilities permit, stock spiking solutions may be prepared 25 to 100 times as concentrated as the working spike solution and diluted volumetrically to produce the working spike solution at the time of use.

(c) Determine the appropriate amount of spike and spiking solution to be added to the reagent water. The spiking solution will generally constitute less than 2% of the total volume of the sample, so the matrix is not altered appreciably and the carrier solvent must not interfere in the test method.

(d) Add the desired volume or mass of spiking solution or spike to the reagent water. Cap the sample and mix well.

Table 4. Cyanide Species Selected for Spiking Purpose.

Cyanide Type	Reagent Water Matrix	Wastewater Matrix
Total cyanide	$K_4Fe(CN)_6{\cdot}3H_2O$	$K_4Fe(CN)_6{\cdot}3H_2O$
Free cyanide	NaCN	NaCN
WAD cyanide	$K_2Ni(CN)_4{\cdot}xH_2O$	$K_2Ni(CN)_4{\cdot}xH_2O$
-	KSCN	no spike

3.2.2.2 Measurement of precision and bias

The study concentrations should generally be uniformly distributed over the range of the test method. At least three concentrations covering the range of the test method should be included with one concentration near the MDL determined in Section 3.2.1, one concentration in the middle of the calibration range and one near the upper limit of the concentration range or a concentration of interest to the analyst. It is recommended that three representative concentrations, i.e. the MDL (or 5ppb), 30ppb and 100ppb, be adopted by each laboratory in the study. It is up to each laboratory to choose whether or not more concentrations should be used. The concentration level used for determination of the MDL may be used as one of the three concentration levels.

Analyze seven replicates at each of three concentrations spiked by each of four cyanide species following the standard operating procedure for the method. A total of 84 samples will be analyzed for each method.

3.2.2.3 Data Reporting

The study results shall be provided to the coordinator in the form of a table, which will usually include the true concentration added to the matrix and the concentration measured. Each participating laboratory should submit for each analyte a data reporting form, an example of which is provided in Section 4.

Precision and bias for a single concentration in a single matrix are determined from seven replicate analyses of a standard with a known concentration of c as follows (ASTM, 1996a; APHA, 1995):

where x_i = result of each analysis (7 replicates),

$$\bar{x} = \sum_{i=1}^{7} x_i \Big/ 7 \qquad (1)$$

$$\text{precision} = \sqrt{\sum_{i=1}^{7} (x_i - \bar{x})^2 \Big/ 6} \qquad (2)$$

$$\text{bias} = \sum_{i=1}^{7} (x_i - c)^2 \Big/ 7 \qquad (3)$$

$\bar{x}$ = mean of 7 replicates,

c = true (spiked) concentration of the sample.

3.2.3 Analysis of Unknown Samples

This step in the method validation procedure requires analysis of independently prepared standards for which the value is unknown to the analyst.

An independent entity, Environmental Resource Associates (ERA) of Arvada, Colorado, will prepare the samples for this study and ship them to the laboratory participants. Upon completion of aforementioned method validation study in reagent water, each laboratory will receive one blind sample and will analyze triplicates of the sample by its species-based methods. Samples are available in 1000mL size.

Standard products available from ERA are listed in Table 5. For this study, ERA will be requested to prepare an aqueous solution with a total cyanide concentration of approximately 50 ppb which comprises of potassium cyanide as an example of free cyanide, potassium tetracyanonickelate (II) as WAD cyanide, and potassium ferrocyanide. The concentrations of each species will be specified by ERA and will be compared to the results from each test method.

Table 5. ERA Products Description.

ERA Product Name	Cyanide Species	Conc. (mg/L)
1000mg/l Metal & Ion Standards Cyanide (free)	Metal & Ion Standards Cyanide (free)	1000
1000mg/l Metal & Ion Standards Cyanide (complex)	Metal & Ion Standards Cyanide (complex)	1000
Free Cyanide	Free Cyanide	0.05-1.0
Ready-to-Use WasteWatR – Cyanide	Total Cyanide	not specified

3.3 Method Validation Study for Wastewater and Contaminated Groundwater Samples

This subtask of the validation procedure focuses on determination of the method detection limit (MDL) and the bias and precision of each specific method performed by a single operator for samples of municipal wastewater and contaminated groundwater. This step in the study is expected to last for five months with the analysis of one sample matrix conducted per month.

3.3.1 Sample Acquisition

Samples of municipal wastewater and contaminated groundwater representing a range of matrix types will be collected at each of five locations shown in Table 6. Twenty-six liter samples will be acquired. From the 26 liters, two liters will be taken for characterization analyses and the remaining 24 liters will be adjusted to pH=12.5 with NaOH and then distributed into one-liter amber bottles. The large sample should be kept cool and in the dark after acquisition, and should be preserved and subdivided into the amber bottles as soon as possible. Groups of four one-liter samples should be packed in coolers and shipped to each of the six participant labs for next day delivery.

Table 6. Wastewater and Groundwater Samples for Cyanide Analytical Method Evaluation and Testing.

Matrix ID	Source	Location
1	MGP site contaminated groundwater	MGP site
2	Effluent from municipal wastewater treatment plant (MWWTP) before chlorination	Chicago, IL
3	Effluent from municipal wastewater treatment plant (MWWTP) after chlorination	Chicago, IL
4	95% MWWTP primary clarifier effluent mixed with 5% MGP water	Boston, MA
5	Spent potlining site contaminated groundwater	Alcoa smelting plant

3.3.2 Sample Pretreatment

Preliminary treatment of wastewater samples prior to preservation will be conducted if interfering substances are present. Sulfides and oxidizing agents such as residual chlorine are the most important interferences for the test methods of interest and will be removed by special procedures.

3.3.2.1 Quick Test of Sulfides

At the time of sample collection, a quick test for sulfide will be performed by placing a drop of sample on lead acetate test paper previously moistened with acetic acid buffer solution at pH 4 (APHA, 1995, Section 4500-Cl.C.3e). Darkening of the paper indicates the presence of sulfide. If sulfide is present, lead acetate will be added to precipitate it. If the sulfide concentration is high, indicated by significant darkening of the test paper, powdered lead carbonate $PbCO_3$ will be added to avoid reducing pH. The quick test will be repeated until a drop of treated sample no longer darkens the acidified lead acetate test paper. If the test indicates presence of sulfide, oxidizing compounds would not be expected and the following test is not necessary. Through preliminary discussions with individuals at the source facilities, every attempt will be made to obtain waters that do not have excessive amount of sulfide. After sulfide removal by precipitation of lead sulfide, samples should be filtered immediately before raising pH for stabilization.

3.3.2.2 Quick Test of Oxidizing Agents

Testing for oxidizing agents will be performed by placing a drop of sample on a strip of potassium iodide (KI)-starch paper previously moistened with acetate buffer solution at pH 4 (APHA, 1995, Section 4500-Cl.C.3e). If a bluish discoloration is noted, sodium thiosulfate ($Na_2S_2O_3$) will be added and the sample will be retested, with care to avoid dosing to a concentration greater than 0.1g $Na_2S_2O_3$/L sample.

3.3.3 Sample Characterization

Contaminated water samples collected from the five locations in Table 6 will be characterized after necessary pretreatment. Characterization will include analysis of major ions, alkalinity, organic content (BOD, TOC), and total cyanide. The characterization analyses will also permit further evaluation of potential interferences for the various methods. If a major interference is detected, the sample should be appropriately pretreated prior to analysis or re-sampling will be conducted. Table 7 lists the constituents to be analyzed in sample characterization. Characterization analyses will be performed by the organizations that provide the samples. Upon receipt, each participating laboratory should analyze (in triplicate) total cyanide in the sample by a U.S. EPA-approved method. These data will serve to demonstrate uniformity of the samples analyzed and in aggregate will provide a statistically strong estimate of the total cyanide in the sample.

3.3.4 Method Detection Limit

All laboratories will receive one of the samples listed in Table 6 every month for five months. Identical sub-samples will be provided in four one-liter bottles. Each laboratory will determine matrix-dependent MDLs for its species-based test methods based on analysis of the various cyanide-bearing field samples. The general procedure for conducting an MDL test in reagent water should be followed, as described in Section 3.2.1.

If the concentration of cyanide species in a contaminated water sample is less than the detection limit obtained from the reagent water matrix tests, each of three cyanide species listed in Table 4 should be added to bring the level of analyte between one and five times the estimated detection limit. Thus, a total of 21 samples will be prepared and analyzed. If the measured level of analyte is greater than five times the estimated detection limit, it will not be possible to determine a matrix-based MDL for the sample. If the measured analyte level is in the range of one to five times the estimated MDL, then no spiking is needed for MDL determination.

3.3.5 Precision

For assessment of precision at different concentration levels, each laboratory will prepare three spiked samples of different concentrations for each contaminated water matrix. Three relevant cyanide species listed in Table 4 will be spiked separately. Precision will be assessed at the three different concentration levels (the MDL or 5ppb, 30ppb and 100ppb recommended). The concentration level used for determination of the MDL (if such a determination is possible) can be used as one of the three concentration levels.

Upon spiking particular cyanide species in contaminated waters, they will react with various solution constituents, thus making it impossible to know the actual concentration of the species achieved by spiking. For this reason, it will not be possible to assess accuracy (bias) for use of the method in the contaminated water matrices. The only possible exception is in the analysis of iron-cyanide-spiked samples with total cyanide techniques. Iron-cyanide species are generally quite stable and nonreactive in the dark over time periods as short as those involved with the planned experiments.

Each laboratory will analyze seven replicates of a sample at each of three concentrations following the standard operating procedure for the method. A total of 63 samples will be analyzed for each method (if the MDL data are available, only 42 additional samples are required). The calculation of precision and bias is the same as described in Section 3.2.2.

Table 7. Analyses for Sample Characterization.

Constitute	Method	Test Method	Reference
pH	Electrometric	EPA 150.1	U.S. EPA, 1983
Ca, total	ICP	EPA 200.7	U.S. EPA, 1994b
Mg, total	ICP	EPA 200.7	U.S. EPA,, 1994b
Na, total	ICP	EPA 200.7	U.S. EPA,, 1994b
K, total	ICP	EPA 200.7	U.S. EPA, 1994b
Cl^-	Argentometric	SM 4500-Cl^- B	APHA, 1995
SO_4^{2-}	Turbidimetric	SM 4500-SO_4^{2-}-B	APHA, 1995
Alkalinity	Titration	SM 2320-B	APHA, 1995
NO_2^- N	Colorimetric	SM 4500-NO_2^--B	APHA, 1995
NO_3^- N	Colorimetric	SM 4500-NO_3^--B	APHA, 1995
NH_3-N	Titrimetric	SM 4500-NH_3-C	APHA, 1995
CN, total	Colorimetric	SM 4500-CN-C	APHA, 1995
TOC	Combustion-Infrared	SM 5310-B	APHA, 1995
BOD	5-Day BOD Test	SM 5210-B	APHA, 1995
SCN^-	Colorimetric	SM 4500-CN-M	APHA, 1995
S^{2-}	Iodometric	SM 4500-S^{2-} F	APHA, 1995
Residual Cl_2	Iodometric	SM 4500-Cl-C	APHA, 1995
ORP	Potentiometric	SM 2580-B	APHA, 1995

4. MANAGEMENT

4.1 Sample Preparation and Distribution

Samples will be collected and stored in closed amber bottles without any headspace. Contaminated water samples will be shipped to each participant laboratory unfiltered except if precipitation of lead sulfide is conducted for sulfide removal. It is the responsibility of each laboratory to determine whether or not sample filtering is necessary prior to analysis.

Label each 1-liter sample bottle with an identification code for the source, i.e., CHI-EFF, and a sequential number for six laboratories and four bottles for each laboratory, i.e., the sample of Chicago effluent would be numbered from CHI-EFF-1 to CHI-EFF-24. Samples should be packed securely in a cooler and shipped for next day delivery to every participating laboratory. Samples will be delivered regularly at mid-month and a notice for sample delivery will be sent to each laboratory by the coordinator at least two weeks in advance. Once the samples have been shipped, the lab supervisors will email the coordinator within a week to confirm that all samples have arrived safely.

4.2 Laboratory Quality Control (QC)

All laboratories will implement standard quality control procedures. As a minimum, each laboratory will analyze laboratory control samples whenever a batch of over 20 samples will be analyzed or once each day, whichever is more frequent. Laboratory control standards (LCS) for different cyanide species at approximately 20 ppb as CN^- will be prepared by each laboratory using chemical reagents from Aldrich, Inc. (Please refer to Table 4 for chemicals designated for each cyanide species). Results from the LCS sample analysis must be within the acceptable limits established by U.S.EPA, i.e. ±25% for cyanide (U.S. EPA, 1997). If errors exceed this limit, corrective action will be taken and documented, and a follow-up LCS analyzed as soon as possible to demonstrate the problem has been corrected. Result of single blind samples analysis described in Section 3.2.3 will also serve as part of quality control purposes. In addition, laboratory blank samples will be analyzed on a routine basis for each analytical method under evaluation, in order to identify sources of contamination arising from the reagents or handling procedures used in performing the analysis. Determine reagent water, reagents, and solvent blanks for each set of samples analyzed, when there is a change in the reagent water system, or whenever a new source (newly prepared reagent or solvents) is introduced into the analytic system that may cause change of reagent purity.

4.3 Data Reporting and Management

An example of the data reporting forms that should be submitted by each participating laboratory for each analyte and matrix is provided as Table 8. Exact sample holding time for each sample is required in the data reporting form. Each laboratory should provide its own reporting forms.

Table 8. An Example of Data Reporting Form.[1]

Lab Code			**Date**				
MDL Test by Spiking NaCN		**Precision and Bias Test by Spiking NaCN**					
		Conc.1 (≈5ppb)		Conc.2 (≈30ppb)		Conc.3 (≈100ppb)	
		True conc.		True conc.		True conc.	
Sple No.	Result	Sple No.	Result	Sple No.	Result	Sple No.	Result
1		1		1		1	
2		2		2		2	
3		3		3		3	
4		4		4		4	
5		5		5		5	
6		6		6		6	
7		7		7		7	
s		s		s		s	
MDL= 3.143*s		p(%)		p(%)		p(%)	
Holding time (hr)		Holding time (hr)		Holding time (hr)		Holding time (hr)	

[1] s is the standard deviation of 7 replicates and p is the mean recovery.

The completed data sheets will be examined by the coordinator immediately upon receipt in order to detect unusual results or issues that require clarification.

Each laboratory should also keep a record (or log) of experience with the analytical techniques employed during the study. This record should include information about operational requirements in implementing the test technique, such as average time to prepare a sample, average time to run a sample, time spent in maintenance of the instrument, and time spent in preparing the instrument for the day. This record should also describe average training time for an analyst to be familiar with the specific technique, and a summary of experiences obtained from running sample tests. Such a record will provide a valuable source of information both in dealing with unusual data and in evaluating the performance of the test method.

4.4 Schedule

The schedule for the Task 1 work plan is presented in Table 9.

Table 9. Schedule for Task 1.1 WERF Project
1999 Plan for Cyanide Analytical Method Evaluation and Testing

ID	Task Name
1	**Method Validation Study in Reagent Water**
2	
3	**Method Validation Study in Wastewater**
4	Matrix 1
5	Matrix 2
6	Matrix 3
7	Matrix 4
8	Matrix 5
9	
10	**Data Arrangement and Reporting**

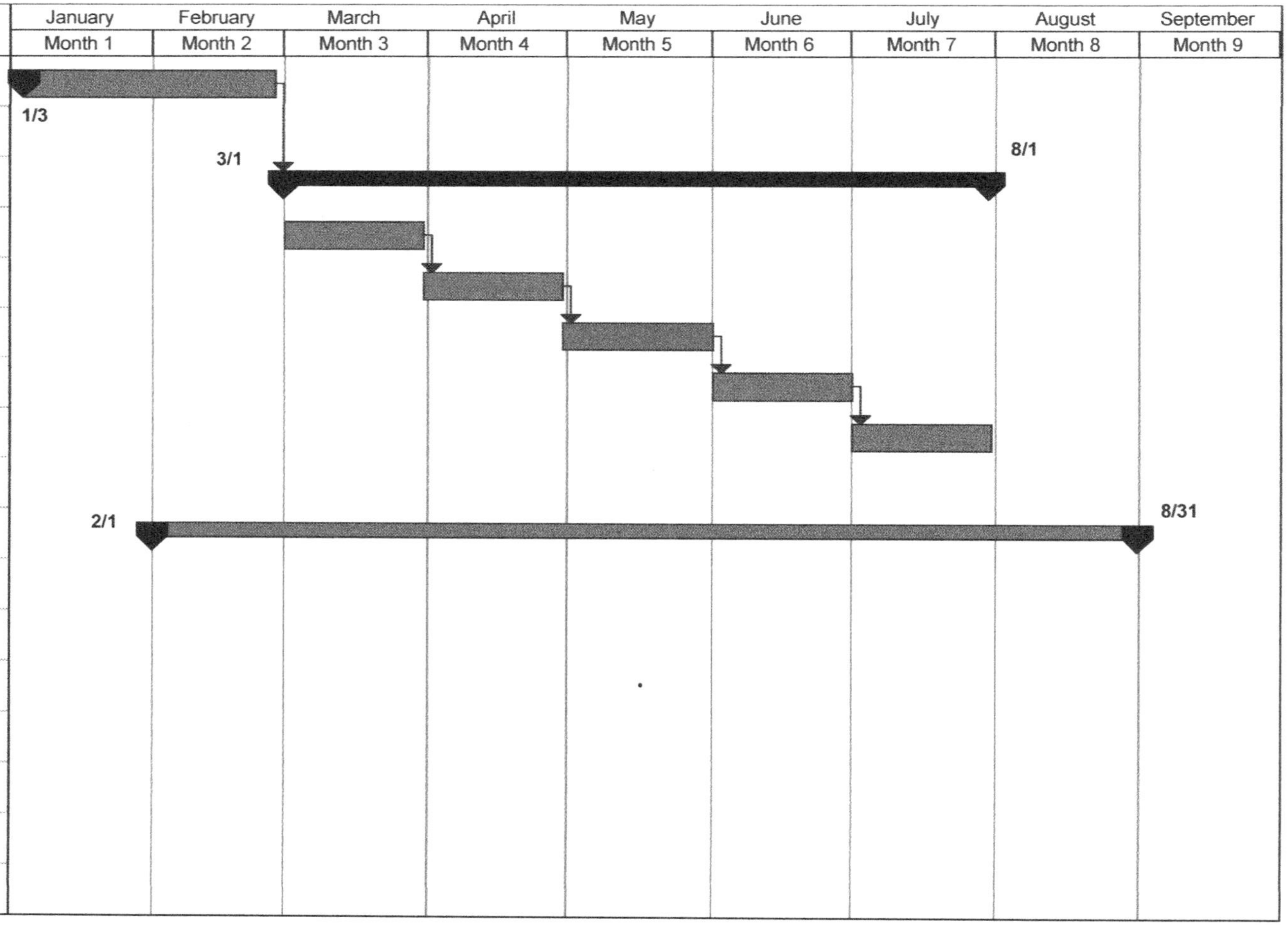

5. REFERENCES

(1) APHA/AWWA/WEF, 1995. Standard Methods for the Examination of Water and Wastewater, 19th Edition, American Public Health Association, American Water Works Association and Water Environment Federation. Washington, D.C.

(2) ASTM, 1989. "Standard Practice for Intralaboratory Quality Control Procedures and a Discussion on Reporting Low-Level Data". Designation D4210-89. 1998 Annual Book of ASTM Standards, Vol.11.01. American Society for Testing and Materials, Philadelphia, PA.

(3) ASTM, 1990. "Standard Practice for Use of the Terms Precision and Bias in ASTM Test Methods". Designation E 177-90a. 1998 Annual Book of ASTM Standards, Vol.14.02. American Society for Testing and Materials, Philadelphia, PA.

(4) ASTM, 1991. "Specification for Reagent Water". Designation D 1193-91. 1998 Annual Book of ASTM Standards, Vol.11.01. American Society for Testing and Materials, Philadelphia, PA.

(5) ASTM, 1992. "Standard Practice for Conducting an Interlaboratory Study to Determine the Precision of a Test Method". Designation E691-92. 1998 Annual Book of ASTM Standards, Vol.14.02. American Society for Testing and Materials, Philadelphia, PA.

(6) ASTM, 1993. "Standard Test Methods for Cyanides in Water-Automated Methods for Total Cyanide, Dissociable Cyanide, and Thiocyanate". Designation D4374-93. 1998 Annual Book of ASTM Standards, Vol.14.02. American Society for Testing and Materials, Philadelphia, PA.

(7) ASTM, 1995. "Standard Guide for Good Laboratory Practices in Laboratories Engaged in Sampling and Analysis of Water". Designation D3856-95. 1998 Annual Book of ASTM Standards, Vol.11.01. American Society for Testing and Materials, Philadelphia, PA.

(8) ASTM, 1996a. "Standard Practice for Determination of Precision and Bias of Applicable Test Methods of Committee D-19 on Water". Designation D2777-96. 1998 Annual Book of ASTM Standards, Vol.11.01. American Society for Testing and Materials, Philadelphia, PA.

(9) ASTM, 1996b. "Standard Guide for Spiking into Aqueous Samples". Designation D5810-96. 1998 Annual Book of ASTM Standards, Vol.11.01. American Society for Testing and Materials, Philadelphia, PA.

(10) ASTM, 1997. "Practice for Preparation, Standardization, and Storage of Standard and Reagent Solutions for Chemical Analysis". Designation E200-97. 1998 Annual Book of ASTM Standards, Vol.15.05. American Society for Testing and Materials, Philadelphia, PA.

(11) ASTM, 1998. "Standard Test Method for Available Cyanide with Ligand Displacement and Flow Injection Analysis (FIA) Utilizing Gas Diffusion Separation and Amperometric Detection". Proposed Draft Method, ASTM Committee D-19.06.

(12) Dionex, 1988. "Determination of Metal Cyanides". Application Note No.55, Dionex Corporation, Sunnyvale, CA.

(13) U.S. EPA, 1983. Methods for Chemical Analysis of Water and Wastes. EPA-600/4-79-020. US Environmental Protection Agency, Cincinnati, OH.

(14) U.S. EPA, 1994a. "Definition and Procedure for the Determination of the Method Detection Limit", Appendix B, Part 136, 40CFR.

(15) U.S. EPA, 1994b. Methods for the Determination of Metals in Environmental Samples – Supplement I. EPA-600/R-94-111. US Environmental Protection Agency, Cincinnati, OH.

(16) U.S. EPA, 1997. Manual for the Certification of Laboratories Analyzing Drinking Water - Criteria and Procedures Quality Assurance, 4th Edition. U.S. Environmental Protection Agency, Cincinnati, OH.

Attachment A
Analytical Chemistry Definitions

The following definitions are from APHA(1995), EPA(1992) and ASTM(1989).

Accuracy – combination of bias and precision of an analytical procedure, which reflects the closeness of a measured value to a true value.
Bias – consistent deviation of measured values from the true value, caused by systematic errors in a procedure.
Detection limits – please refer to Table A.1 for the definition of different detection limits issued by U.S. EPA, ASTM and Standard Methods.
Matrix spike – the quantity (mass) of a component (analyte) of interest that is added to a sample (matrix) in order to test the bias as measured by recovery (of that component under specific analytical conditions) and reported as percent recovery (P).
Precision – measure of the degree of agreement among replicate analyses of a sample, usually expressed as the standard deviation.
Quality Control – set of measures within a sample analysis methodology to assure that the process is in control.
Replicate – repeated operation occurring within an analytical procedure, two or more analyses for the same constituent in an extract of a single sample constitute replicate extract analyses.
Spiking solution – the solution in which one or more spikes are dissolved (along with any necessary preservatives). This solution acts as a carrier to provide ease of measurement and more rapid and thorough mixing of the spike into the sample, as compared to adding the spike as a pure compound.

Table A-1. Detection Limits Defined by EPA, Standard Methods (SM) and ASTM.

Org.	Detection Limits	Definition	Reference
US EPA	Method Detection Limit (MDL)	MDL is the minimum concentration of a substance that can be measured and reported with 99% confidence that the analyte concentration is greater than zero and is determined from analysis of a sample in a given matrix containing the analyte.	U.S. EPA. 40 CFR 136, Appendix B, para.1
	Practical Quantitation Limit (PQL)	PQL is the lowest level that can be reliably achieved within specified limits of precision and accuracy during routine laboratory operating conditions. PQL is determined through inter-laboratory studies, such as the Performance Evaluation (PE) studies. Generally, PQL=EQL.	U.S. EPA. Federal Register, Vol.50, No.219, Nov.13, 1985. p46906, para.11
	Estimated Quantitation Limit (EQL)	The lowest concentration that can be reliably achieved within specified limits of precision and accuracy during routine laboratory operating conditions. The EQL is generally 5 to 10 times the MDL. However, it may be nominally chosen within these guidelines to simplify data reporting. For many analytes the EQL analyte concentration is selected as the lowest non-zero standard in the calibration curve. Sample EQLs are highly matrix-dependent.	U.S. EPA. SW-846, "Test Methods for Evaluating Solid Waste, Physical/ Chemical Methods", Revision 1, July 1992. p1-25, para.1
Standard Methods	Instrumental Detection Level (IDL)	The constituent concentration that produces a signal greater than five times the signal/noise ratio of the instrument. This is similar, in many respects, to "critical level" and "Criterion of detection." The latter level is stated as 1.645 times the s of blank analyses.	APHA/AWWA/WEF. Standard Methods for the Examination of Water and Wastewater, 19th Edition, 1995. p1-3, para.13
	Lower Level of Detection (LLD)	The constituent concentration in reagent water that produces a signal (2)(1.645)s above the mean of blank analyses. This sets both Type I and Type II errors at 5%. Other names for this level are "detection level" and "Level of Detection" (LOD).	APHA/AWWA/WEF. Standard Methods for the Examination of Water and Wastewater, 19th Edition, 1995. p1-3, para.14
	Method Detection Level (MDL)	The constituent concentration that produces a signal with a 99% probability that it is different from the blank. For seven replicates of the sample, the mean must be 3.14s above the blank where s is the standard deviation of the seven replicates. Compute MDL from replicates measurements one to five times the actual MDL. The MDL will be larger than the LLD because of the few replications and the sample processing steps and may vary with constituent and matrix.	APHA/AWWA/WEF. Standard Methods for the Examination of Water and Wastewater, 19th Edition, 1995. p1-3, para.15
	Level of Quantitation (LOQ)	The constituent concentration that produces a signal sufficiently greater than the blank that it can be detected within specified levels by good laboratories during routine operating conditions. Typically it is the concentration that produces a signal 10s above the reagent water blank signal.	APHA/AWWA/WEF. Standard Methods for the Examination of Water and Wastewater, 19th Edition, 1995. p1-3, para.16
ASTM	Criterion of Detection (COD)	The minimum quantity (analytical result) which must be observed before it can be stated that a substance has been discerned with a n acceptable probability that the statement is true. The criterion of detection must always be accompanied by the stated probability.	ASTM D4210-89. "Standard Practice for Intralaboratory Quality Control Procedures and a Discussion on Reporting Low Level Data". 1998 Annual Book of ASTM Standards, Vol.11.01. American Society for Testing and Materials, Philadelphia, PA.
	Limit of Detection (LOD)	A concentration of twice the criterion of detection when it has been decided that the risk of making a Type II error is to be equal to a Type I error.	Limit of Detection (LOD)

APPENDIX C

Work Plan for WWTP Monitoring Program

WORK PLAN FOR WWTP MONITORING PROGRAM

Kara J. Forman[1], Anping Zheng[2], David A. Dzombak[2], Richard G. Luthy[2]
[1] Malcolm Pirnie, Inc.
[2] Carnegie Mellon University

1. PURPOSE AND OBJECTIVES

A growing number of publicly owned treatment works (WWTPs) across the U.S. have reported detecting cyanide in chlorinated effluents at levels exceeding the influent concentration of cyanide. Since water quality criteria and related discharge limits are typically low, in some cases as low as 5 μg/L, some of these WWTPs have periodic exceedances of effluent cyanide standards. Based on preliminary studies conducted in municipal wastewater treatment plants of the City of Sunnyvale, CA (Malcolm Pirnie, 1997), the East Bay Municipal Utility District, IL (Environ, 1996), and the City of Boston, MA (Delaney, 1999), it appears that this phenomenon is attributable to formation of cyanide during the wastewater chlorination process. In Year 2 of the WERF project, a comprehensive process monitoring program will be implemented at several full-scale WWTPs to assess the fate and effects of cyanide species in chlorination and dechlorination processes at WWTPs. This monitoring program will involve sampling and analysis for individual cyanide species before and after chlorination and dechlorination processes at treatment facilities of the municipal utilities participating in the project. Monitoring results will be verified and utilized to investigate the potential formation of cyanide. These results will also be incorporated in a project database with subsequent availability to all investigators via electronic media.

The aim of the WWTP monitoring is to obtain a better understanding of the cyanide formation phenomenon as it occurs in the field. A separate laboratory investigation by Carnegie Mellon will explore the chlorination issues in depth. The primary objectives of the monitoring program are to:

(1) monitor the removal or production of cyanide species in chlorination/dechlorination processes of a WWTP over a cycle of seasons (one year);
(2) evaluate the utility of the various cyanide species measurement techniques for routine application in WWTP operation and monitoring; and
(3) for periods when post-chlorination effluent cyanide concentrations are elevated, monitor intensively the fate of cyanide in the chlorination process.

The monitoring program presented in this work plan provides for periodic measurement of several cyanide species in the chlorination/dechlorination processes at WWTPs. Cyanide measurements to be performed routinely on samples will include total cyanide, WAD cyanide, and free cyanide by microdiffusion. These techniques can be performed with instrumentation already available at WWTPs (total cyanide, WAD cyanide), or that is inexpensive to acquire (free cyanide by microdiffusion). Other cyanide speciation measurements that require expensive instrumentation (available cyanide, automated analysis of total cyanide with in-line UV digestion, automated analysis of total and WAD cyanide by thin film distillation, ion chromatography) will be employed in the monitoring programs at selected facilities. Sampling will be performed routinely at designated locations in the WWTPs participating in the project.

2. FATE AND EFFECT OF CYANIDE DURING CHLORINATION PROCESS

Although treatment of cyanide-bearing industrial wastewater has been extensively studied, the fate of cyanide and its species during wastewater treatment at conventional activated sludge process has received less attention in the literature. Industrial wastewaters containing cyanide traditionally have been treated by the two-stage process of alkaline-chlorination that converts cyanide to cyanate and cyanate to bicarbonate and nitrogen gas:

Theoretically, chlorination of cyanide-containing wastewater in WWTPs should eliminate any free cyanide present. However, numerous plants throughout the country have observed that cyanide concentrations can increase slightly following chlorination compared to pre-chlorination concentrations. Results from Task 1 of WERF project revealed that analytical artifacts are unlikely to be the cause of this phenomenon (Malcolm Pirnie, 1999).

Based on a review of published literature, the processes of potential significance with regard to formation of CN species in chlorination include, but are not limited to:

- Cyanide formation from thiocyanate reacting with hypochlorous acid;
- Cyanide formation from thiocyanate reacting with nitrate;
- Cyanogen chloride formation from chloramines reacting with various organic compounds including: aromatic and aliphatic amino acids, humic acids, aromatic hydrocarbons, aromatic amines, phenolic compounds and formaldehyde.

3. SELECTION OF ANALYTICAL TECHNIQUES

During the Task 1 of this study, a total of seven cyanide species testing methods were systematically evaluated including: metal cyanide complex by ion chromatography, available cyanide, automated analysis of total cyanide with in-line UV digestion, automated analysis of total and weak acid dissociable (WAD) cyanide by thin film distillation, WAD cyanide, and free cyanide by microdiffusion. A subset of these methods will be implemented in the WWTP monitoring program. Criteria for selection of methods to be employed are: 1) method performance with wastewater samples, including percent recovery and method detection limit; 2) cost of implementation; and 3) ease of implementation.

The Task 1 study (Malcolm Pirnie, 1999) indicated that all of the methods achieved very low method detection limits (MDLs) and met the U.S. EPA detection limit quantity requirement for cyanide (CFR, 1997), which is 0.04 ppm. It was also shown that each test method yielded reproducible results in all water matrices in terms of precision and bias (except for the available cyanide technique when applied to raw wastewater). Therefore, in the development of the monitoring program, primary consideration has been given to the trade-off between cost and ease of implementation of each test method as seen in Table 1.

Table 1. Analysis Time and Relative Cost of Cyanide Species Analytical Techniques.

Analytical Technique	Time for Analysis	Ease of Implementation	Instrument Cost
Ion Chromatography	Minutes	Not easy	Expensive
Available Cyanide	Minutes	Easy	Expensive
WAD Cyanide	3 hr	Easy	Less expensive
Free CN Microdiffusion	5 hr	Easy	Inexpensive
Total Cyanide by UV In-Line Digestion	Minutes	Easy	Expensive
Thin Film Distillation	Minutes	Easy	Expensive
Total Cyanide by Distillation (SM)	3 hr	Easy	Less expensive

Cyanide measurements to be performed routinely on samples will include total cyanide, WAD cyanide, and free cyanide by microdiffusion. These techniques can be performed with instrumentation already available at WWTPs (total cyanide, WAD cyanide), or that is inexpensive to acquire (free cyanide by microdiffusion). Other cyanide speciation measurements that require expensive instrumentation (available cyanide, automated analysis of total cyanide with in-line UV digestion, automated analysis of total and WAD cyanide by thin film distillation, ion chromatography) will be employed in the monitoring programs at selected facilities that have or can acquire such instrumentation.

All samples will be analyzed for total cyanide, WAD cyanide, and free cyanide by microdiffusion. The standard analytical methods to be used for cyanide species analyses are listed in Table 2.

Table 2a. Standard Cyanide Analytical Methods for Wastewater.

Methodology	Species Measured	Reference	Preservation
Total Cyanide after Distillation	Total CN	SM 4500-CN-C	NaOH, pH>12
Free Cyanide by Micro Diffusion	Free CN	ASTM D4282-95	NaOH, pH>12
Weak Acid Dissociable Cyanide	WAD CN	SM 4500-CN-I	NaOH, pH>12

Table 2b. Additional Cyanide Analytical Methods.

Methodology	Species Measured	Reference	Preservation
Available Cyanide	WAD CN + Hg-CN	ASTM, 1998	NaOH, pH>12
Automated Total by UV Digestion, Distillation	Total CN	EPA Method 335.3	NaOH, pH>12
Automated Total by Thin Film Distillation	Total CN	ASTM D4374-93	NaOH, pH>12
Automated Weak Acid Dissociable by Thin Film Distillation	WAD CN	ASTM D4374-93	NaOH, pH>12
Metal-cyanides by Ion Chromatography	Metal-cyanides	Dionex, 1988	NaOH, pH>12

4. SAMPLING WORK PLAN

The proposed monitoring plan is intended to provide data needed to investigate the probable cause of cyanide formation in chlorination and dechlorination. This work plan contains procedures and protocols for field sampling and laboratory analyses. A schematic of a WWTP showing the planned monitoring points is provided in Figure 1.

4.1 Sampling Schedule

This monitoring program is planned to last for 12 months; Table 3 summarizes the sampling schedule in terms of sampling duration, frequency, sample type, sample locations, and total samples for the project. Cost and resource availability were considered in a general way in establishing this schedule. Detailed sampling plans will be formulated by the WWTPs in consultation with the project team.

Table 3. Cyanide Sampling Schedule for WWTP Monitoring Program.

Monitoring Type	Duration	Frequency	Sample Type	Sampling Locations a,b	Number of CN Samples per Event
Incident-Detections	Jan. – Dec. 2000	1 day/week	24-hr composite, autosampler (every 3 hours)	1 • Post-chlorination	One per event (52 per year)
Incident-Driven	5 days subsequent to an incident	5 days/incident	Four 6-hr composites, Autosampler (every 1 hour)	2-4 • Prechlorination • Post-chlorination • Pre-dechlorination (as applicable) • Post-dechlorination (as applicable)	40-80 per event

a Please refer to Figure 1 in Attachment A for sampling locations.

b In the event that the residence time between the Post-Chlorination and Pre-Dechlorination sample locations is insignificant, utilities may disregard the Post-Chlorination sample location and sample from only the Pre-Chlorination, Pre-Dechlorination and Post-Dechlorination sample locations.

Incident-Detections monitoring will be performed using 24-hour composite sampling once per week for one year. The composite samples collected from the post-chlorination sample location will be analyzed for total cyanide. An "incident" will be defined as either:

1. a total cyanide concentration at the post-chlorination point greater than 5ppb, or
2. an analytical result 50% greater than the average cyanide concentration (e.g. 15 ppb spike versus a typical effluent cyanide concentration of 10 ppb), or
3. an analytical result 100% greater than the method detection limit.

Each utility should use some discretion in identifying cyanide formation incidents based upon their knowledge of the utility's treatment process and previous cyanide formation history.

The sampling locations will include: pre-chlorination, post-chlorination, pre-dechlorination and post-dechlorination points. For each sampling event, eight three-hour composite samples will be collected at the post-chlorination sample location. Once an incident is detected, Incident-Driven monitoring will be performed intensively for five subsequent days following the incident. Wastewater characterization analyses should be conducted in conjunction with the Incident-Driven sampling as outlined in Section 4.2.2.2.

4.2 Monitoring Program and Sampling Protocol

The program will involve use of a 24-hour composite autosampler to sample chlorinator effluent one day per week for 24 hours. This weekly sampling will be performed for one year. If an elevated total cyanide event is detected in the 24-hour composite sample, then intensive Incident-Driven sampling will be initiated and conducted for five subsequent days as described in Section 4.2.2. The intensive sampling will involve collection of 6-hr composite samples (minimum are grab sample per **hour**) using autosampling equipment. Each participating WWTP will conduct intensive sampling for at least three elevated-cyanide incidents within the year.

Where feasible, participating utilities will ***continue to run the automatic samplers and collect the Incident-Driven samples***, as outlined in Section 4.2.2, while the "incident detection" sample is being analyzed. If the weekly sample does not represent an "incident" the collected samples are simple disposed of properly. If the weekly sample identifies a cyanide "hit" the utility has already started collecting the Incident-Driven samples without having potentially missed the cyanide formation event.

The intensive sampling will involve collection of 3-hr composite samples (minimum are grab sample per **hour**), again with autosampling equipment. Each participating WWTP will conduct intensive sampling for at least three elevated-cyanide incidents within the year.

4.2.1 Incident-Detections Monitoring

The composite sampling program will involve collection of 24-hour, flowrate-weighted composite samples of post-chlorination effluent. Conventional composite automatic samplers will be employed. These devices typically take individual samples for eight three-hour periods, and where feasible these samples are subsequently composited with proportioning according to flow during the sampling periods.

Each individual container will be pre-preserved with sodium hydroxide (NaOH) and, for the post-chlorination samples, sodium thiosulfate[1] ($Na_2S_2O_3$) will also be added. Only total cyanide concentration will be measured for the composite sample. If total cyanide is a) greater than 5 ppb, or b) 50% greater than the average cyanide concentration, an incident will be considered in progress and Incident-Driven monitoring will be initiated immediately the next day.

4.2.1.1 Sampling Protocol for Incident-Detections Monitoring

Wastewater samples for cyanide analysis will be collected as composite samples over 24 hours. Eight, three-hour composite samples will be drawn by an autosampler and composited at the end of the 24-hr period. The 24-hr compositing will be performed in proportion to flow rates during the 6-hr subsampling periods. All three-hour samples will be stored in a container pre-preserved with NaOH and, for the post-chlorination samples, $Na_2S_2O_3$. The total volume of composite sample should not be less than 500mL. NaOH should be added as pellets or liquid to ensure that the pH of the final sample volume is greater than 12. $Na_2S_2O_3$ will be needed to dechlorinate the post-chlorination samples. The $Na_2S_2O_3$ dose should be such that the concentration of $Na_2S_2O_3$ in the final sample does not exceed 0.1 mg per liter of wastewater. All samples should be collected and stored in brown plastic or amber glass containers.

4.2.2 Incident-Driven Monitoring

Once the utility has identified an excursion, the utility will proceed to implement an intensive five-day monitoring program. Separate samples will be collected during the Incident-Driven Monitoring for cyanide analysis and for general characterization of the wastewater.

4.2.2.1 Sampling Protocol for Incident-Driven Monitoring

Wastewater samples for cyanide species and characterization analysis will be collected as four, six-hour composite samples over 24 hours, with sampling of one gallon (3785 mL) every six hours at each sampling location. The samples will be collected according to the procedure listed below.

1) Collect one gallon of wastewater in a 1-gallon container at each sampling location every six hours. Split this sample into two different size containers, i.e., Container 1 of size 500 mL with approximately 500 mL wastewater for cyanide analyses and Container 2 of size one gallon with approximately 3200 mL sample for characterization analyses. Follow step 2 for the Container 1 sample. Follow step 3 for the Container 2 sample.
2) Container 1 should be pre-preserved with NaOH which should be added as pellets or liquid to ensure that the pH of the final sample volume is greater than 12. For the post-

[1] When collecting samples using an autosampler and pre-preserving the bottles for dechlorination interferences, samples should be preserved by adding enough thiosulfate to react with completely with the chlorine dose applied. In the laboratory, samples should then be back-titrated to remove the additional thiosulfate.

chlorination sample, $Na_2S_2O_3$ will be needed to dechlorinate the sample. The $Na_2S_2O_3$ dose should be such that the concentration of $Na_2S_2O_3$ in the final sample volume does not exceed 0.1 mg per liter of wastewater. All samples should be collected and stored in brown plastic or amber glass containers.

3) The Container 2 sample needs to be subdivided into five groups of bottles for characterization analyses:

- two one-liter bottles unpreserved/unfiltered,
- one 500 mL bottle (pre-preserved with base),
- one 100 mL bottle pre-preserved with acid,
- one 100 mL bottle pre-preserved with base, and
- one 50 mL bottle, pre-baked.

From the 3200 mL sample, place appropriate volume of wastewater into each bottle. Dispose of any extra water.

4) Follow Steps 1) to 3) to collect the next six-hour sample.
5) Transport the samples to the designated analytical laboratory.

To mitigate the impact of volatilization during holding times, extraction and analysis for cyanide species is to be ***performed within nine days of collection***. Reporting schedules are outlined in Section 4.5.1.

4.2.2.2 Required Characterization Analysis

The set of constituents to be analyzed for characterization purposes are listed in Table 4. The frequency of analysis is provided in Table 5. When sample frequency is listed as less than daily, the utility may decide on which day they will collect samples for that specific constituent.

Table 4. Analyses for Wastewater Characterization in Incident-driven Monitoring.

Constituent	Method	Test Method	Reference
pH	Electrometric	EPA 150.1	U.S. EPA, 1983
Ca, total	ICP	EPA 200.7	U.S. EPA, 1994
Mg, total	ICP	EPA 200.7	U.S. EPA, 1994
Cd, total	ICP	EPA 200.7	U.S. EPA, 1994
Co, total	ICP	EPA 200.7	U.S. EPA, 1994
Cu, total	ICP	EPA 200.7	U.S. EPA, 1994
Fe, total	ICP	EPA 200.7	U.S. EPA, 1994
Ni, total	ICP	EPA 200.7	U.S. EPA, 1994
Pb, total	ICP	EPA 200.7	U.S. EPA, 1994
Zn, total	ICP	EPA 200.7	U.S. EPA, 1994
Cl^-	Iodometric	SM 4500-Cl-B	APHA, 1995
SO_4^{2-}	IC	SM 4500-SO_4^{2-}-B	APHA, 1995
Alkalinity	Titration	SM 2320-B	APHA, 1995
NO_2^--N	Colorimetric	SM 4500-NO_2^--B	APHA, 1995
NO_3^--N	Colorimetric	SM 4500-NO_3^--B	APHA, 1995
NH_3-N	Titrimetric	SM 4500-NH_3-C	APHA, 1995
TKN	Macro-Kjeldahl	SM 4500-N_{org} -B	APHA, 1995
TSS	Dried at 103-105 °C	SM 2540-D	APHA, 1995
TOC	Combustion-Infrared	SM 5310-B	APHA, 1995
SCN^- [b]	Colorimetric	SM 4500-CN-M	APHA, 1995
S^{2-}	Iodometric	SM 4500-S^2-F	APHA, 1995
Residual Cl_2	DPD Colorimetric	SM 4500-Cl-G	APHA, 1995
ORP	Potentiometric	SM 2580-B	APHA, 1995

Table 5. Sample Frequency for Wastewater Characterization in Incident-Driven Monitoring.

Constituent	Sample Locations[a]	Frequency Per 5-Day Incident Event	Total Number of Samples per Event[a]
pH	All	5 days	10-20
Ca, total	All	1 day	2-4
Mg, total	All	1 day	2-4
Cd, total	All	1 day	2-4
Co, total	All	1 day	2-4
Cu, total	All	1 day	2-4
Fe, total	All	1 day	2-4
Ni, total	All	1 day	2-4
Pb, total	All	1 day	2-4
Zn, total	All	1 day	2-4
Cl^-	All	5 days	10-20
SO_4^{2-}	All	5 days	10-20
Alkalinity	All	5 days	10-20
NO_2^-N	All	5 days	10-20
NO_3^-N	All	5 days	10-20
NH_3-N	Prechlor & Postchlor	5 days	10
TKN	Prechlor, Postchlor, PostDechlor	3 days	9
TSS	All	5 days	10-20
TOC	All	5 days	10-20
SCN^-[b]	Prechlor, Postchlor, PostDechlor	5 days	15
S^{2-}	All	5 days	10-20
Residual Cl_2	All	5 days	10-20
ORP	All	5 days	10-20

Note:

[a] Number of samples depends on the number of sample locations at each utility. Utilities that dechlorinate will have four (4) sample locations. Utilities that do not dechlorinate will have two sample locations.

[b] Thiocyanate should be analyzed for, at a minimum, one of the three Incident Driven sample events.

4.3 Quality Assurance/ Quality Control (QA/QC)

The QA/QC program implemented in this plan includes the following items:

- Field Standard Operation Procedure (FSOP) will be developed by the project team and distributed to each member of the sampling team.
- A *Chain-of-Custody* form will be used for all sampling events, which ensures that the samples are collected, transferred, stored and analyzed only by authorized personnel.
- A field log book will be used to record: sample times at each location, sampler's name, weather conditions, pH of sample, chlorine residual, additional preservative used and other observations.
- Field QC checks will be employed including field blanks.
- Field blanks will be collected at the end of each sampling event day or five-day event. Fill a sample container with analyte-free water and then preserve and document in the same manner as the collected samples.

4.4 Participants

4.4.1 WWTP Participants

The participating WWTPs are listed in Table 5. The cyanide analyses to be performed at each facility for the Incident-Detections and Incident-Driven monitoring are indicated in the table.

Table 6. WWTP Cyanide Monitoring Study Participants and Analytical Methods.

WWTP	Cyanide Analytical Methods	Facility	Chlorinate	Dechlor
MWRDGC, Chicago	Total CN by Thin-Film Distillation WAD CN by Thin-Film Distillation	Hanover Park	Yes	Yes
Gulf Coast Waste Authority [a]	Total, Free Cyanide, Cyanide Amenable to Chlorination, Available Cyanide (OIA 1677)	Washburn Tunnel Facility	No	No
MWRA, Boston	Total CN, Autoanalysis w/ UV Digestion WAD CN, Autoanalysis (no UV Digestion)	Deer Island Treatment Plant	Yes	No
BADA – EBMUD, Oakland	Total CN, WAD CN, Free CN by DIffusion	Oakland Water Pollution Control Plant	Yes	Yes
City of Sunnyvale, CA	Total CN	Sunnyvale Water Pollution Control Plant	Yes	Yes
City of LA	Total CN, WAD Cn	Glendale Water Reclamation Plant Donald C. Tillman Water Reclamation Plant	Yes	Yes

Note:
[a] Gulf Coast will conduct a separate comparison study of several analytical methods.

4.4.1 Other Participating Entities

Additional participating entities are listed in Table 6. Clarkson will participate be providing analytical services for the experimental organic cyanide analytical method. Clarkson will analyze wastewater samples collected from Metropolitan Water Reclamation District of Greater Chicago and Gulf Coast Waste Authority.

4.5 Data Compilation and Reporting Requirements

4.5.1 Data Compilation

E2 Consulting Engineers will maintain all analytical data in an Access database. Data compilation will be prepared by each WWTP participant. Each participating laboratory will submit their data to E2 electronically using the established Excel form file by the *end of the month* following the sample collection. Copies of the Data Submission form are provided as Attachment C. Electronic copies may be either submitted by email or mailed to E2 on a diskette.
E 2 Consulting Engineers, Inc.
Ms. Carrie Campbell

1900 Powell Street, Suite 250, Emeryville, CA 94608-1812
510-596-9669 Carrie.Campbell@lfr.com 510-652-5604 (fax)

Carnegie Mellon will assist Malcolm Pirnie with the data analysis.

4.5.2 Reporting Requirements

Each facility should maintain a logbook (electronic or paper) documenting sample collection and analysis. Information to be collected includes:

- date and time of sampling
- flow rate (if known) at time of sampling
- weather conditions
- unusual wastestream or process conditions
- pH at time of preservation

4.6 Program Management

Overall program management will be performed by Malcolm Pirnie. Comprehensive monitoring program specifications, including the selection of statistical methods, detailed sampling points, minimum sample numbers, sampling intervals, and appropriate quality assurance/quality control procedures, will be developed by the WWTP personnel in consultation with the project team.

5.0 REFERENCES

(1) APHA/AWWA/WEF, 1995. Standard Methods for the Examination of Water and Wastewater, 19th Edition, American Public Health Association, American Water Works Association and Water Environment Federation. Washington, D.C.

(2) CFR, 1997. "Inorganic Chemical Sampling and Analytical Requirements," 40CFR, Part 141.23.

(3) Delaney, M.F., Zilitinkevitch, L., McSweeney, N.E., Epelman, P.E., 1997. Cyanide formation from chlorinated WWTP effluent, Proceedings of the WEF Environmental Laboratories Conference, 6-17 - 6-26.

(4) Environ, 1996. Evaluation of Cyanide at EBMUD's Water Pollution Control Plant Special District No.1.

(5) Malcolm Pirnie, 1997. Investigation of Cyanide Formation in Sunnyvale Water Pollution Control Plant, draft report.

(6) Malcolm Pirnie, 1999. Annual Summary Report for WERF Project 98-HHE-5 "Cyanide Formation & Fate in Complex Effluents and Its Relation to Water Quality Criteria", submitted to WERF.

(7) U.S. EPA, 1983. Methods for Chemical Analysis of Water and Wastes. EPA-600/4-79-020. US Environmental Protection Agency, Cincinnati, OH.

(8) U.S. EPA, 1994. Methods for the Determination of Metals in Environmental Samples – Supplement I. EPA-600/R-94-111. U.S. Environmental Protection Agency, Cincinnati, OH.

Attachment A

Sample Location Schematic

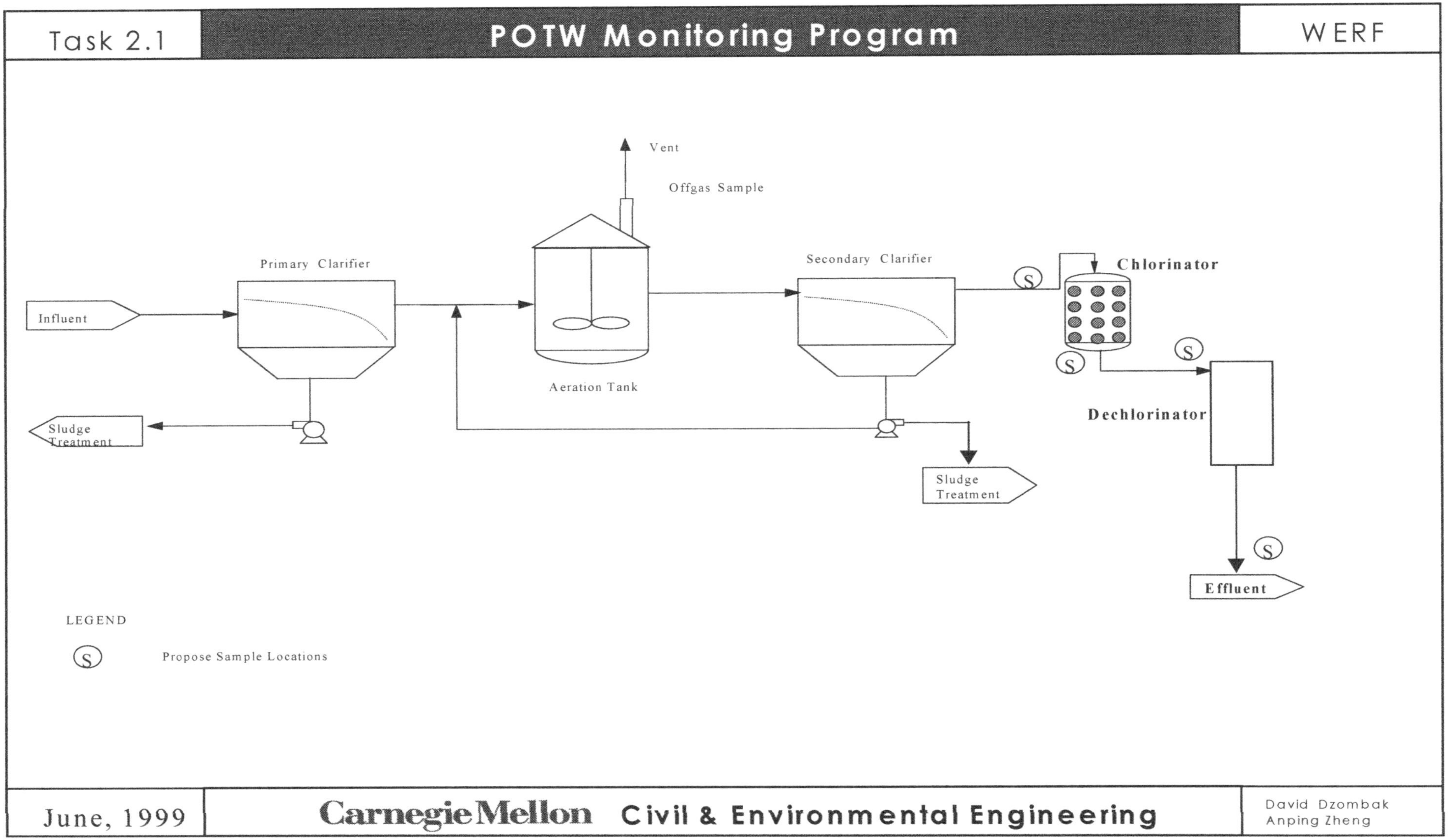
Task 2.1
POTW Monitoring Program
WERF
Vent
Offgas Sample
Primary Clarifier
Secondary Clarifier
Chlorinator
Influent
Aeration Tank
Sludge Treatment
Dechlorinator
Sludge Treatment
Effluent
LEGEND
S
Propose Sample Locations
June, 1999
Carnegie Mellon Civil & Environmental Engineering
David Dzombak
Anping Zheng

Attachment B

Sample Numbering Protocol

WWTP Sample Numbering Protocol

Each sample number should consist of a nine-digit number that represents the utility at which the sample was collected, the sample location, the sample date and the analysis, as follows:

___ ___ ___ - ___ ___ - ___ ___ ___ ___ - ___ ___

Utility ID	Sample Location	Sample Date	Sample Number

Utility ID

WWTP	Utility ID
MWRDGC, Chicago	Chi
MWRA, Boston	Bos
City of Los Angeles	CLA
Gulf Coast	GCW
BADA - EBMUD, Oakland	EBM
City of Sunnyvale, CA	Sun

Sample Location

Sample Location	Sample ID
Raw Influent	Infl
Pre-Chlorination	PrCl
Post-Chlorination	PoCl
Pre-Dechlorination	PrDe
Post-Dechlorination	PoDe
Final Effluent	FiEf

Sample Date

M M D D
e.g. January 15, 2000 written as 0115

Sample Number

e.g. 01, 02, 03, 04, etc.

Sample should be sequentially numbered for each day. During Incident-Detections monitoring there is only one sample (01). During Incident-Driven monitoring there are eight three-hour composite samples that should be label 01, 02, etc. in sequential order.

Attachment C

WWTP Fate and Transport Database Input Sheets

Facility ID: Clearwater Treatment Facility

Name of person entering/submitting data: Jamie Atwood **Date:** 03/14/1999

Sample Location Name	Unique Sample Identifier	Sample Date [(1)]	Sample Time [(2)]	Sample Technique[(3)]	Preservative (if any)	Facility Flow Rate (MGD)	Temp. (deg C)	Chemical Doses (mg/L) Cl	Cl Res	Polymer	Other	Weather Conditions [(4)]	Other comments or notes about this specific sample
Example Data are Shown:													
WT-121	WT-121-A	03/04/1999	04:30 PM	GRAB	NaOH	35.231	19.2					Sunny	Slightly turbid, greenish
WT-121	WT-121-B	03/04/1999	04:30 PM	GRAB		35.231	19.2					Sunny	Slightly turbid, greenish
WT-121	WT-121-C	03/04/1999	04:30 PM	GRAB	HCL	35.231	19.2					Sunny	Slightly turbid, greenish
(leave a space between samples for ease of entry if you prefer)													
EFF-001	EF-001-A	03/04/1999	04:45 PM	GRAB	NaOH	29.780	20.5					Sunny	
EFF-001	EF-001-B	03/04/1999	04:45 PM	GRAB		29.780	20.5					Sunny	
EFF-001	EF-001-C	03/04/1999	04:45 PM	GRAB	HCL	29.780	20.5					Sunny	
EFF-001	EFF-001-C-Dup	03/04/1999	04:48 PM	GRAB	HCL	29.780	20.5					Sunny	

Valid values and entries to use:

(1) Sample Date: Use the format MM/DD/YYYY

(2) Sample Time: Use the format HH:MM am or pm

(3) Sample Technique: Grab or Composite

(4) Weather Conditions at that specific location: Sunny, Partial Sun, Cloudy or Shade

Facility ID: Clearwater Treatment Facility

Name of person entering/submitting data: Jamie Atwood **Date:** 03/14/1999

Unique Sample Identifier	Date Received at Lab (1)	Date Analyzed (1)	Lab Name	Analytical Method No.	Lab QC Batch No.	Sample Type (2)	Analyte Name	Result (3)	Reporting Limit (4)	Units	Qualifiers or Notes	Result Type (5)	Spike Recovery (%)	RPD (%) (6)
WT-121-A	03/04/1999	03/05/1999	Toxscan	SM 4500-CN, E	Q030599-CN	ORIG	Total Cyanide	0.084	0.001	ug/L		TRG		
WT-121-A	03/04/1999	03/05/1999	Toxscan	SM 4500-CN, AM	Q030599-AM	ORIG	Amenable Cyanide	0.028	0.001	ug/L		TRG		
WT-121-B	03/04/1999	03/05/1999	Toxscan	SM 4500 CL F	Q030599-CL	ORIG	Chlorine Residual	ND	0.005	mg/L		TRG		
WT-121-C	03/04/1999	03/05/1999	Toxscan	SM 2550 B	Q030599-TOC	ORIG	Total Organic Carbon	345	10	mg/L		TRG		
(leave a space between sample results for ease of entry if you prefer)														

APPENDIX D

WORK PLAN FOR A LABORATORY STUDY ON THE EFFECT OF CHLORINATION ON CYANIDE SPECIES IN WWTPs

WORK PLAN FOR A LABORATORY STUDY ON THE EFFECT OF CHLORINATION ON CYANIDE SPECIES IN WWTPs

Anping Zheng, David A. Dzombak, Richard G. Luthy
Carnegie Mellon University

1. OVERVIEW

Raw wastewater at Publicly Owned Treatment Works (WWTPs) sometimes contains cyanide at low concentrations (ppb level), primarily from industrial sources. Because cyanide causes acute toxicity to various life forms, including bacteria in activated sludge plants, and because cyanide is a U.S. EPA priority pollutant limited by regulation to low concentrations in discharge, pretreatment for removal of cyanide is required of industrial sources by WWTPs. Thus, influent concentrations of total cyanide at WWTPs typically are very low.

Despite the low influent concentrations, cyanide has been periodically detected of elevated concentrations in WWTP final effluent, especially in chlorinated effluents. Moreover, in a number of such cases, the effluent cyanide concentrations have exceeded the influent cyanide concentration. This has been observed at treatment plants across the United States (Environ Inc. 1996; Delaney et al., 1997; Forman and Kavanaugh, 1999). Monitoring at these plants indicates that cyanide may be formed in the chlorination step, which is surprising in that chlorination is used widely for destruction of free cyanide in wastewater. The phenomenon is of great concern because the water quality criteria and related discharge limits of cyanide are typically very low, in many cases as low as 5 ppb. The mechanism responsible for the apparent cyanide formation in chlorination is not understood.

During the second year of this project, a bench-scale chlorination study will be performed at Carnegie Mellon to study the fate of individual cyanide species in the chlorination process. Data obtained will be used to assess reaction pathways and transformation mechanism for cyanide species in the chlorination process. The objective of this study is to determine the cause of elevated cyanide levels observed in post-chlorination water at WWTPs. Based on a review of available literature, the following potential causes of the apparent production of cyanide in wastewater chlorination have been identified:

- formation of cyanide from reaction with thiocyanate;
- formation of cyanide from reactions not directly involving chlorine (e.g., nitrosation); and
- artifacts of sample handling or interferences in the analytical techniques.

This study will aim to evaluate chemical reactions associated with chlorination of secondary effluent, nitrosation of organic compounds, and analytical interferences due to the presence of ascorbic acid in the wastewater samples containing nitrites. This research will also assess which of these mechanisms is the primary cause of elevated cyanide levels. The specific contributions of the research will be as follows:

- Evaluation of the performance of various cyanide species test methods in wastewater;
- Identification of cyanide species analytical techniques suitable for WWTP monitoring;
- Assessment of the fate of cyanide in chlorination/dechlorination processes;
- Characterization of the major precursors for cyanide formation in chlorination/dechlorination processes;
- Effects of dechlorinating agents on the fate of cyanide species; and
- Recommendation of treatment schemes for the minimization of cyanide formation during WWTP wastewater treatment.

2. BACKGROUND

2.1 Basic Chlorine Chemistry and Chlorination in Water

Chlorine is used widely for the disinfection of drinking water and municipal wastewater. The reactions of aqueous chlorine in potable water and wastewater are very similar, while wastewater has a much greater chlorine demand. The most common chlorine compounds used in wastewater treatment plants are chlorine gas (Cl_2), sodium hypochlorite (NaOCl), calcium hypochlorite [$Ca(OCl)_2$], and chlorine dioxide (ClO_2). NaOCl is often used at large facilities, primarily for reasons of safety.

Sodium hypochlorite (NaOCl) is the chemical reagent to be used in this study as the source of chlorine. NaOCl can ionize in water following the reaction:

$$NaOCl + H_2O \rightarrow HOCl + NaOH \qquad (1)$$

The hypochlorous acid produced is a weak acid with pKa = 7.53; it partially dissociates in water to form the hypochlorite ion as follows:

$$HOCl \leftrightarrows H^+ + OCl^- \qquad (2)$$

As a result, pH is critical in determining the dominant species in solution. The quantity of HOCl and OCl^- present in water is called the free available chlorine. The time to reach equilibrium is very short (less than one second).

Ammonia is commonly present in wastewater. Hypochlorous acid reacts with ammonia and with many of its complex derivatives to produce chloramines in accordance with the following reactions (Metcalf and Eddy, Inc., 1991):

$$NH_3 + HOCl \rightarrow H_2O + NH_2Cl \quad \text{(monochloramine)} \qquad (3)$$

$$NH_2Cl + HOCl \rightarrow H_2O + NHCl_2 \quad \text{(dichloramine)} \qquad (4)$$

$$NHCl_2 + HOCl \rightarrow H_2O + NCl_3 \quad \text{(nitrogen trichloride)} \qquad (5)$$

The extent of chloramine formation and the ratio of NH_2Cl / $NHCl_2$ depend on pH and the ratio of HOCl / NH_3. Monochloramine and dichloramine are two species that predominate in most cases, and the chlorine in these compounds is called combined available chlorine. Continued addition of chlorine will result in a directly proportional increase in the free available chlorine (unreacted hypochlorite). As to disinfection efficiencies, hypochlorous acid is the strongest disinfectant among chlorine species. Chloramine residuals typically are less germicidal than HOCl, but given adequate contact time, monochloramine is nearly as effective as free chlorine in achieving destruction of most bacteria (Metcalf and Eddy, Inc., 1991). The reaction of chloramines with organic carbon sources is one subject of this research since cyanides may be formed during these reactions.

2.2 Basic Cyanide Chemistry

"Cyanide" refers to all of the CN groups in cyanide compounds that can be determined analytically as the cyanide ion, CN^-, which can be protonated to form hydrogen cyanide (HCN). HCN and CN^- are referred to as free cyanide forms. The cyanide compounds can be classified as simple and complex cyanides. Simple cyanides are represented by the formula A(CN)x where A is an alkali or a metal and x is the number of cyanide groups, such as NaCN, KCN, NH_4CN, etc. In aqueous solutions of simple alkali cyanides, the CN group is present as CN^- or molecular HCN, depending on the pH. The acid dissociation constant for hydrogen cyanide is pKa = 9.2 at 25°C. In solutions of simple metal cyanides, the CN group may occur also in the form of complex metal-cyanide ions, e.g., $Cu(CN)_3^-$, $Ni(CN)_4^{2-}$, $ZnCN^+$. Simple metal cyanides, $M(CN)_x^{y-}$, may dissociate further depending on several factors. Metal-cyanide compounds can be further classified as weak acid dissociable (WAD) compounds such as zinc- and cadmium-cyanide complexes, and strongly complexed cyanides such as cobalt- and iron-cyanide complexes. Cyanide in wastewater may be present as free cyanides, WAD metal-cyanides, and strongly-complexed metal-cyanides. The species of cyanide exhibit great differences in their toxicity, reactivity, and environmental fate and transport (Boening et al., 1999; Ghosh et al., 1999; Shifrin et al., 1996).

There are three other cyanide species that may occur in WWTPs. Cyanogen chloride (CNCl), a highly toxic gas of limited solubility, is the first reaction product on chlorination of cyanides. At an alkaline pH, CNCl hydrolyzes to the cyanate ion (CNO^-), which has only limited toxicity. Thiocyanate (SCN^-) is always present in municipal wastewaters as well as industrial influents. It is not very toxic to aquatic life and can be biodegraded to yield ammonium (Boening et al., 1999).

2.3 Cyanide Analytical Methods

The most commonly used cyanide analytical method is the total cyanide method which measures free cyanide and metal-complexed cyanide. The metal cyanide is converted to HCN gas by acid-digestion in a heated reactor, and the liberated HCN is then absorbed in NaOH solution (APHA, 1995; U.S. EPA, 1993; ASTM, 1998). Generally, the total cyanide method does not measure cyanogen chloride, thiocyanate, cyanate, or organo-cyanide complexes. The Standard Methods (APHA, 1995) method 4500-CN-C "Total cyanide after distillation" will be used in this research. The weak acid dissociable cyanide method (APHA, 1995, Method 4500-CN-I) is very similar to the total cyanide method, except that the sample is only slightly acidified. This method does not recover CN^- from strong metal-cyanide complexes (e.g., Fe-CN and Co-CN complexes). It will be used occasionally in this study. Free cyanide by microdiffusion method (ASTM D4282-95) is intended to directly measure diffusible cyanide, i.e., CN^- and HCN, at room temperature from a sample solution at pH 6. It does not measure strong metal-cyanide complexes, and only partially recovers easily dissociable cyanide complexes. This method will also be used in the study. The other cyanide species analytical methods that will be employed in this research include (APHA, 1995): cyanogen chloride (Method 4500-CN-J), thiocyanate (Method 4500-CN-M) and cyanates (Method 4500-CN-L).

2.4 Alkaline Chlorination of Cyanide Species in Water

Industrial wastewaters containing significant concentrations of cyanide are usually treated by the two-stage process of alkaline-chlorination prior to discharge. This is the most popular method for treatment of wastewater containing free cyanide or weakly-complexed cyanide. The first stage is to convert cyanide to cyanate ion (CNO^-) as shown in the following reaction sequence proposed by Chamberlin and Snyder (1955):

$$CN^- + OCl^- + H_2O \rightarrow CNCl + 2OH^- \qquad (6)$$

$$CNCl + 2OH^- \rightarrow CNO^- + Cl^- + H_2O \qquad (7)$$

$$\text{overall reaction:} \quad OCl^- + CN^- \rightarrow CNO^- + Cl^- \qquad (8)$$

The cyanate ion is very much less toxic than cyanide, and, as a result, in many treatment situations the oxidation process is only taken to this point (Teo and Tan, 1987). There is no evidence that cyanates are likely to be reduced to cyanides in wastewater treatment plants (Resnick et al., 1958). If more hypochlorite is applied, the reaction will go to completion. Cyanate is oxidized to bicarbonate and nitrogen gas as the second stage of chlorination:

$$3OCl^- + H_2O + 2CNO^- \rightarrow 2HCO_3^- + N_2 + 3Cl^- \qquad (9)$$

The oxidation of cyanides to cyanogen chloride (CNCl), Equation (6), occurs instantaneously at all pH levels. However, the hydrolysis of CNCl to cyanate, Equation (7), occurs very slowly at pH values around 7.0 but increases rapidly in the pH range of 8.5-9.0 (Teo and Tan, 1987). Since cyanogen chloride is a highly toxic gas, it is essential that it hydrolyze quickly to avoid a buildup of dangerous concentrations in non-basic solution. Consequently, first stage chlorination is always conducted at pH 9-10 in order to ensure rapid completion of reaction (7). The overall reaction for the chlorine oxidation of cyanide to cyanate shows that the hypochlorite ion (OCl^-) is the active chlorine group in the oxidation process.

The reaction of hypochlorite with cyanate has received relatively little attention compared to the reaction of hypochlorite and cyanide. This reaction is likely rather complex, with a stoichiometric equation approximated by Equation (9). However, Teo and Tan (1987) showed that the relative amounts of cyanate and hypochlorite reacted deviated from the stoichiometric ratio given by Equation (9). This discrepancy was attributed to a simultaneous parallel reaction as shown by Equation (10).

$$4OCl^- + H_2O + CNO^- \rightarrow HCO_3^- + NO_3^- + 4Cl^- + H^+ \quad (10)$$

Chlorination of thiocyanates (SCN^-) can lead to the formation of cyanate or cyanogen chloride depending on the pH value of the solution [Equation (11) and (12)], which will be discussed in detail in Section 3.

$$4OCl^- + SCN^- \rightarrow CNCl + SO_4^{2-} + 3Cl^- \quad (11)$$

$$4OCl^- + SCN^- + 2OH^- \rightarrow CNO^- + SO_4^{2-} + 4Cl^- + H_2O \quad (12)$$

2.5 Observation of Cyanide Fate in WWTPs

Numerous wastewater plants across the U.S. have reported detecting formation of cyanide in their chlorinated effluents, at levels exceeding cyanide values in the influent, and in some cases, exceeding the NPDES discharge limits. Three examples are listed below.

2.5.1 East Bay Municipal Utility District (EBMUD), San Francisco, California

From 1989 to 1996, EBMUD periodically experienced elevated cyanide levels in its chlorinated final effluent from the wastewater treatment plant, reaching as high as 50 μg/L, much above the District's discharge limit of 10 μg/L. This phenomenon was inconsistent with the low concentrations observed at the plant in the influent and in the dechlorinated effluent, which are discharged to San Francisco Bay. In 1996, EBMUD conducted a 6-month intensive study intended to isolate and eliminate the source of this apparent in-plant cyanide contamination (ENVIRON Inc., 1996). A single source was not established, but it was determined that the apparent production of cyanide was attributable either to chlorination of the secondary effluent, or to analytical interference of ascorbic acid in wastewater samples containing nitrite. It was not determined which mechanism was more likely to produce a detectable amount of cyanide in a wastewater matrix (ENVIRON Inc., 1996, Ellgas, 1999).

2.5.2 City of Sunnyvale, California

The City of Sunnyvale reported, from 1992 to 1996, periodic episodes of cyanide occurrence in the chlorinated effluent from the water pollution control plant at levels exceeding the daily average discharge limit of 5 μg/L. The cyanide appeared during the summer in the tertiary processes at the plant. There were also indications that cyanide formed when the total organic carbon (TOC) was high, and when nitrification occurred. However, there were numerous samples showing nondetectable cyanide when these conditions occurred. Because cyanide only formed when the plant was fully nitrifying, it appeared that the distribution of nitrogen species played a role in cyanide formation. It is unclear whether chlorination of wastewater or nitrification of wastewater, or even the analytical errors (e.g., false positives) might be responsible for cyanide formation (Malcolm Pirnie Inc., 1997).

2.5.3 Massachusetts Water Resources Authority, Boston, Massachusetts

At times low levels of cyanide have been detected in the plant's chlorinated primary effluent, despite the fact that cyanide is rarely observed in the plant's influent. Experiments conducted by Delaney et al. (1997) suggested that cyanide is formed when hypochlorite oxidizes thiocyanate (SCN^-) present in sewage. By monitoring both cyanide and thiocyanate, it was observed that chlorinating primary or secondary effluent caused thiocyanate levels to decrease and cyanide to form. However, in their study, they used ascorbic acid as the dechlorinating agent which may have falsely elevated cyanide concentrations (Delaney et al, 1997).

2.5.4 Summary

Based on preliminary studies (Carr et al, 1997; Delaney et al, 1997; Malcolm Pirnie Inc., 1997; Environ Inc., 1996) conducted to investigate possible causes of cyanide formation in WWTPs, the following hypotheses have been established:

- Formation during wastewater chlorination/chloramination processes;
- Reactions between thiocyanate and nitrate ion;
- False positives due to an artifact of sample handling or the analytical technique.

These preliminary studies indicated that cyanide is very likely to be formed within the wastewater chlorination process. Therefore, the study of cyanide formation during WWTP wastewater chlorination is the focus of this research. In addition, other potential causes will also be addressed.

2.6 Potential Cyanide Formation During Chlorination/Dechlorination Processes

Based on a review of published literature, 15 reactions or reaction groups were identified as potential means of cyanide formation during municipal wastewater treatment. Among these, 10 reactions were directly related to chlorination/chloramination processes. The summary of these reactions is presented in Table 1, and the reactions are discussed in more detail below.

2.6.1 Cyanide formation from thiocyanate reacting with chlorine/hypochlorous acid

The oxidation of thiocyanates has received considerable attention. The oxidizing agents studied have included hydrogen peroxide, ferric ion, Co(III), Mn(III), chromate, bromate, iodate, ozone, liquid SO_2, etc. In general thiocyanates are oxidized to sulfate and either cyanide or cyanate depending on the pH of the solution. The usual oxidation half-reactions cited are (Newman, 1975):

$$SCN^- + 4H_2O \rightarrow SO_4^{2-} + HCN + 7H^+ + 6e^- \quad (13)$$

$$SCN^- + 5H_2O \rightarrow SO_4^{2-} + CNO^- + 10H^+ + 8e^- \quad (14)$$

Data related to thiocyanate oxidation by chlorine or hypochlorous acid are very scarce. Generally, chlorination of thiocyanate can lead to the formation of cyanate or cyanogen chloride depending on the pH value of the solution (Newman, 1975). Hypochlorite has been used in titration of thiocyanates in basic medium according to following reaction:

$$SCN^- + 4ClO^- + 2OH^- \rightarrow SO_4^{2-} + CNO^- + 4Cl^- + H_2O \quad (15)$$

In acidic medium, the reaction leads to the formation of cyanogen chloride based on the equation:

$$SCN^- + 4ClO^- + H^+ \rightarrow HSO_4^- + CNCl + 3Cl^- \quad (16)$$

A relevant study was conducted by Allen et al. (1948) who observed that the chlorine reacted with the SCN^- to form an extremely toxic compound. Subsequent research indicated that this compound was cyanogen chloride. As discussed in the Section 2.6.2, it is possible that cyanide ion may be formed under some conditions of chlorination process. It has been suggested that solutions containing thiocyanate can form free hydrogen cyanide in acidic media (U.S. EPA, 1979), though no data in support of this statement were provided.

Table 1. Summary of Possible Reactions Relating to Formation of Cyanide or Cyanogen Chloride.

Type	No.	Reaction	Reaction Conditions	Chemical Compounds Tested in Literature	Proposed Mechanism	Related to Chlorination	Research in the Work	Reference
CN^- Formation	1	aromatic primary amine + $HNO_2 \rightarrow CN^-$	weak acidic solution	p-nitroaniline, aniline, p-toluidine, ethyl p-aminobenzoate	aniline → phenol →p-nitrosophenol → CN^-	No	Task 3	Kanno et al., 1978
	2	polyhydroxybenzene + $HNO_2 \rightarrow CN^-$	pH=1-4, room temperature	(di, tri) hydroxybenzene	None	No	Task 3	Ohya and Kanno, 1991
	3	$SCN^- + HOCl \rightarrow CN^-$	acidic	SCN^-, OCl^-, H^+	None	Yes	Task 1	Newman, 1975
	4	$SCN^- + NO_3^- \rightarrow CN^- + SO_4^{2-} + 2NO$	Highly acidic	SCN^-, NO_3^-	None	No	Task 1	APHA, 1995
	5	ascorbic acid + $NO_2^- \rightarrow CN^-$	Highly acidic	ascorbic acid, NO_2^-	None	No	Task 3, 4	Rapean et al., 1980
CNCl Formation	6	aliphatic amino acid + $HNO_2 \rightarrow CNCl$	acidic	glycine, serine, threonine	None	No	Task 3	Matsushita, 1979
	7	aromatic amino acids + chloramine → CNCl	weak acidic	phenylalanine, tyrosine, tryptophan	cleavage of the aromatic ring by chloramines	Yes	Task 2	Ohya and Kanno, 1984
	8	aliphatic amino acids + chloramine → CNCl	pH = 6, room temperature	valine, leucine, isoleucine, serine, threonine	N atom of CNCl was from chloramines	Yes	Task 2	Hirose et al., 1988
	9	humic acid + chloramine → CNCl	room temperature	humic acid	None	Yes	Task 2	Ohya and Kanno, 1985
	10	aromatic hydrocarbons + chloramines → CNCl	pH = 5-8, room temperature	benzene, toluene, xylene (o-,m-,p-), naphthalene	cleavage of the aromatic ring by chloramines	Yes	Task 2	Kanno et al., 1982a
	11	aromatic amines + chloramine → CNCl	pH = 5-8, room temperature	aniline, toluidine , naphthylamine (1-,2-)	cleavage of the aromatic ring by chloramines	Yes	Task 2	Kanno et al., 1982b
	12	phenolic compounds + chloramine → CNCl	pH = 5-8, room temperature	phenol, cresol, dihydroxybenzene, naphthol	cleavage of the aromatic ring by chloramines	Yes	Task 2	Kanno et al., 1982b
	13	chlorinated, nitrated, carboxylated or methylated benzenes, phenols and anilines + chloramines → CNCl	pH = 6, 25°C	[(chloro, nitro) benzene, aniline, phenol], benzoic acid, aminobenzoic acid, hydroxybenzoic acid, (dimethyl, trichloro, pentachloro, dinitro) phenol	None	Yes	Task 2	Maeda et al., 1987
	14	$SCN^- + HOCl \rightarrow CNCl$	weak acidic	SCN^-, HOCl	None	Yes	Task 1	Zilligh, 1972
	15	formaldehyde + chloramine → CNCl	weak acidic	formaldehyde	None	Yes	Task 2	Pedersen et al., 1995

2.6.2 Cyanogen chloride formation from chloramines reacting with organic compounds

Formation of cyanogen chloride is of great concern since it is very toxic itself and it may break down to cyanide under certain conditions. Due to the widespread concern in the U.S. over the presence of organic disinfection by-products (DBPs) in chlorinated waters, many treatment plants are turning to ozone, chloramines and chlorine dioxide as a total or partial replacement of chlorine (McGuire, 1989). Chloramine is of particular interest, because it is inexpensive, provides a lasting chlorine residual, and it produces much lower level of many DBPs. One concern over the use of chloramines is the current level of uncertainty in the nature and health impacts of chloramine disinfection by-products. As shown in Table 1, cyanogen chloride can be formed from chloramines reacting with various types of organic compounds including: aromatic and aliphatic amino acids, humic acids, aromatic hydrocarbons, aromatic amines, phenolic compounds and formaldehyde, etc. The mechanisms of these reactions have not been established.

The fate of cyanogen chloride is rather complicated. Its significance as a cyanide source will depend highly on the environmental conditions such as pH and time. Generally, cyanogen chloride is an unstable compound whose half-life values range from 1 minute 45 secods to 10 h at 5°C in ambient waters (Boening and Chew, 1999). As stated previously, the oxidation of cyanide to cyanogen chloride occurs instantaneously at all pH values. In alkaline solutions with pH >8.5, CNCl may continuously undergo hydrolysis to form cyanate (CNO^-), and the rate of hydrolysis is accelerated by a slight excess of chlorine (Eden et al., 1950b; Chamberlin and Snyder, 1955). With no excess chlorine present, it can persist for 24 hr at pH 9 (Pettet and Ware, 1955). When large residual chlorine is available, it may convert to nitrogen gas and bicarbonate (Eden et al., 1950a). There are very few data pertaining to the potential transformation of CNCl to CN. One report states that "When CNCl is heated to decomposition or on contact with water or steam, it will react to produce highly toxic and corrosive fumes of Cl^-, CN^-, and NO_x." (Lewis, 1992).

2.7 Cyanide Formation From Non-Chlorination Reactions

2.7.1 Cyanide formation from nitrite reacting with organic compounds

It has been reported that cyanide can be formed by nitrite (NO_2^-) reacting with organic compounds (APHA, 1995), although these organic compounds have not been clearly identified, and the reaction mechanisms have not been revealed. Koshimizu and Toshiko (1975) reported that aromatic amines and glycine can react with nitrite to form cyanide. Rapean et al. (1980) have shown that nitrite forms HCN via nitrosation of a number of compounds including methyl isobutyl ketone, phenol, catechol, ethylacetoacetate, pyruvic acid, oxalacetic acid and 2,4,4-trimethyl pentanes. Kanno et al. (1978) reported that aliphatic and aromatic amines were reactive with nitrite and described a study of the formation of HCN from aniline showing that the nitrogen in the cyanide was derived from the nitrite ion.

Overall, nitrite-reactive compounds likely occur in municipal wastewaters and cyanide may be formed during the nitrosation of these organics.

2.7.2 Cyanide formation from thiocyanate reacting with nitrate

Thiocyanate might be converted to cyanide in the presence of oxidants such as nitrate in highly acidic media (Csikai and Barnard, 1983; APHA, 1995). It has been observed that 5-95% of thiocyanate present was converted to free cyanide during the distillation test (Csikai and Barnard, 1983). While nitrate is encountered in most wastewater treatment plants, the experiment of Csikai and Barnard (1983) was performed under low pH conditions unlikely to occur in a WWTP plant. This reaction should be carefully investigated to determine if it is significant in the formation of cyanide ion:

$$SCN^- + 2NO_3^- \rightarrow CN^- + SO_4^{2-} + 2NO \qquad (17)$$

2.7.3 Cyanide formation due to analysis artifacts - from ascorbic acid reacting with nitrite

A number of facilities that used ascorbic acid to remove residual chlorine in samples prior to analysis have detected high concentration of cyanide (Delaney, et al., 1997). It has been revealed by a number of researches (Rapean, et al., 1980; Carr, et al., 1997) that cyanide could be formed when ascorbic acid and nitrite ion are both present in wastewater samples prior to cyanide analyses. Indeed, this reaction is of the type described in Section 2.6.1. The use of ascorbic acid as a dechlorinating agent has been recommended by Standard Methods (APHA, 1995), EPA Method 335.1/335.2 (U.S. EPA, 1983), and SW-846 Method 9010 (U.S. EPA, 1986), and this technique is employed in most WWTP facilities to pretreat samples for cyanide analyses. An intensive study on the effect of the dechlorinating agent on cyanide concentration will be conducted in the laboratory.

3. WORK PLAN

3.1 Objective

The objective of this research is to investigate systematically the potential formation of cyanide in wastewater treatment processes, especially during chlorination/dechlorination units, as well as some major cyanide analysis interferences. The overall goal is to gain a broader understanding of the fate of cyanide species during WWTP operations, and especially in chlorination and dechlorination processes.

3.2 Research Approach

This study will seek to determine the major mechanism of cyanide formation among several possible mechanisms that have been identified. It will comprise four main activities. First, effects of thiocyanates in chlorination process will be investigated by Task 1 study. Second, an intensive study, Task 2, on the chloramination of organic compounds will be implemented in reagent water and wastewater matrices. Third, Task 3 will be performed to investigate the possibility of cyanide formation from nitrosation of organic compounds. Finally, although current knowledge leads to the conclusion that cyanide is being formed in wastewater chlorination, the possible role of analytical interferences needs to be carefully examined, which will be addressed in Task 4. Additionally, this research will aim to recommend treatment schemes or chlorination conditions that will minimize the formation of cyanide.

3.3 Task 1 - Study of the Effects of Thiocyanate in Chlorination Process

This study will investigate the potential for cyanide formation upon chlorination of wastewater containing thiocyanates under a range of conditions. Chlorination experiments will be conducted first in reagent water matrix spiked with different amounts of thiocyanate, nitrate and ammonia as well as various chlorination conditions. Thiocyanate will be spiked in reagent water at concentrations of 1, 5, or 10 mg/L. In some experiments, nitrate and ammonia will be spiked as well. Considering the complexity of wastewater matrix, the same experiments will also be performed in several WWTP wastewater effluents with a high concentration of TOC, especially organic nitrogen compounds, to investigate the effect of chlorination due to the presence of other active materials.

3.3.1 Materials and Analyses

Analytical reagent grade chemicals will be used in all operations for the experimental study. Reagent water will be prepared in the laboratory as defined by Type II of Specification in ASTM D1193 (ASTM, 1991). Total Cyanide, WAD cyanide, thiocyanate, cyanate, and cyanogen chloride analyses will be performed according to Standard Methods (APHA, 1995) while Free Cyanide by Microdiffusion analysis will be conducted based on ASTM method (ASTM, 1995). Analyses of NO_3^-, NO_2^-, and NH_4^+ will be performed by ion chromatographic methods (APHA, 1995); Total Kjeldahl Nitrogen (TKN) will be conducted according to Standard Methods 4500-N_{org}-C. Residual chlorine will be measured by DPD colorimetric method (APHA, 1995). All samples will be analyzed in duplicate.

3.3.2 Laboratory Chlorination Procedure

Laboratory chlorination will be carried out in 1-liter beakers. The appropriate amount of sodium hypochlorite solution will be added from a prepared and calibrated solution. A 5 mg/mL stock solution of NaOCl will be prepared in distilled water. The pH of the chlorinated water will then be immediately adjusted with 1N HCl or 1N NaOH to the desired pH. In cases where ammonia is added, the addition will be done before pH adjustment. The ammonia solution will be prepared as 70 mg N-NH_4/mL from NH_4Cl in distilled water. The chlorinated solution will be stirred for the appropriate time and incubated in the dark at ambient temperature for a desired period prior to analysis.

3.3.3 Wastewater Effluent Acquisition and Characterization

Samples of unchlorinated WWTP effluent to be used in this work will be collected at one or several of the following tertiary treatment WWTPs:

- Hanover Park Water Reclamation Plant, Metropolitan Water Reclamation District of Greater Chicago (MWRDGC), IL
- Deer Island Treatment Plant of the Massachusetts Water Resources Authority (MWRA), MA
- Oakland Water Pollution Control Plant, East Bay Municipal Utility District (EBMUD), CA
- Water pollution control plant, City of Sunnyvale, CA

Wastewater samples will be analyzed for TOC, TKN, NH_3, NO_3^-, NO_2^-, residual chlorine, total cyanide, and SCN^- as the general characterization.

3.3.4 Effect of Chlorination Parameters

The effect of chlorine dosage, contact time and pH value on the formation of cyanide will be addressed firstly in reagent water matrix experiments. Typical values of these parameters chosen for the experiments are listed below:

Cl_2 dosage:	2.5, 10, 20 mg/L
Contact time:	2, 6, 18 hr
pH:	3, 7, 11

There are totally nine experiments in this scheme based on the design shown in Table 2.

Table 2. Experiment Design for Effect of Chlorination Parameters.

Run No.	SCN^- Conc. (mg/L)	Cl_2 dosage (mg/L)	Contact time (hr)	pH
1	1	2.5	2	3
2	1	10	6	7
3	1	20	18	11
4	5	2.5	6	11
5	5	10	18	3
6	5	20	2	7
7	10	2.5	18	7
8	10	10	2	11
9	10	20	6	3

Once the conditions most conducive to cyanide formation are identified, these will be used as the base condition for experiments with various wastewater matrices.

3.3.5 Effects of Nitrate (NO_3^-)

Various amounts of sodium nitrate (5, 10 mg/L as NO_3^-) will be added to thiocyanate-spiked solution in reagent water at pH 1,5 and 7. These three pH values were selected to represent the range of environments in analytical procedures and field conditions (pH 1 for total cyanide analysis, pH 5 for WAD CN analysis, and pH 7 as representative of a natural system or wastewater). The experimental design is shown in Table 3. Total cyanide, SCN^-, and NO_3^- concentrations will be measured two hours after the addition, in order to investigate the potential formation of cyanide at presence of nitrate without chlorination. Then, the solution will be subjected to chlorination using 10 mg/L of chlorine for two hours. Samples will be analyzed again for total cyanide, SCN^-, and NO_3^- concentrations after chlorination. The experiment will be repeated in different wastewater matrices using a fixed chlorination condition.

Table 3. Experiment Design for Effect of Nitrate.

Run No.	NO_3^- Conc. (mg/L)	pH	Cl_2 dosage (mg/L)	Contact time (hr)
1	5	1	10	2
2	5	5	10	2
3	5	7	10	2
4	10	1	10	2
5	10	5	10	2
6	10	7	10	2

3.3.6 Effects of Ammonia (NH_3)

Effects of ammonia will be investigated by adding various amounts of ammonium chloride to the thiocyanate-spiked solution at a certain pH level using 10 mg/L of chlorine for two hours. The solution will be spiked with 1, 5, 10 mg/L of NH_3 before chlorination. Total cyanide, SCN^-, NH_4^+ and NO_3^- concentrations will be determined after chlorination.

3.4 Task 2 - Study of the Effects of Organic Compounds in Chloramination Process

As described previously, many types of organic compounds in wastewater have potential to form cyanide or cyanogen chloride upon chlorination. The objective of this study will be to evaluate the significance of the formation of cyanide or cyanogen chloride from the reaction of chloramines with major organic compounds that are frequently present in WWTPs. To make this evaluation, we will measure the yields of cyanide species from the chloramination of model organics and relate these to the concentration of organic compounds found in wastewaters. Additionally, we will also measure the concentrations of target organics after the chloramination process.

The principal groups of organic compounds found in wastewater are proteins (40-60%), carbohydrates (25-50%), and fats and oils (10%) (Metcalf and Eddy, Inc., 1991). In addition, wastewater contains small quantities of a large number of different synthetic organic molecules ranging from simple to extremely complex in structure. These compounds include surfactants, volatile organic compounds, pesticides, etc. Therefore, several typical organic compounds will be chosen to investigate their effects during the chloramination process.

3.4.1 Chloramination of Pure Organic Compounds

Six pure organic compounds which are likely to be present in WWTP wastewater will be chloraminated in the laboratory. These compounds were chosen as representative of organics in WWTPs. Protein and amino acids, humic acids, fulvic acids, carbohydrates, hydrocarbons and plant pigments are chosen and the compound names are listed in Table 4.

Table 4. Organic Compounds Used in the Chloramination Study.

Category	Organic Compound	CAS
Amino Acids	Pepsin	9001-75-6
Humic Substances	Aldrich Humic acid	1415-93-6
Carbohydrates	Cellulose	9004-34-6
Hydrocarbons	Benzene	71-43-2
	Aniline	62-53-3
Plant Pigments	Catechin	154-23-4

10^{-4} mol/L doses of the pure compounds will be used in order to increase the detection probability of any cyanide/cyanogen chloride formation. All chloramination will be done at pH 6 for two hours with approximately 3:2 NaOCl / NH_4Cl ratios. The chloraminated solution will be stirred in a constant temperature (25°C) bath for 2 hours. Immediately after reaction termination, an 800-mL aliquot of each sample will be extracted with 0.015 L of high purity hexanes. Two subsequent extractions will be made, each with 0.015 L of fresh hexanes, and the three extracts will be combined and dried by passing over 20g of anhydrous sodium sulfate. The dried extracts will be brought to 0.050 L volume with hexanes in a volumetric flask, and the target organic compounds will be measured by gas chromatography in a Hewlett Packard 6890 GC System. The remaining 200-mL aliquot in the reaction vessel will be used for cyanide and cyanogen chloride analysis.

3.4.2 Chloramination of Organic Compounds in Wastewater Matrices

Chloramination will also be conducted in different wastewater matrices. First, raw wastewater effluents are chloraminated for two hours. Second, chloramination experiments are conducted in a solution spiked with 10^{-4} mol/L of organic compounds and chloraminated for two hours. The experiment and analysis procedures are similar to those described in Section 3.4.1.

3.5 Task 3 - Study of the Effects of Nitrosation Process of Organic Compounds

Five organic compounds will be selected to investigate potential cyanide formation during the nitrosation process. These compounds include aniline and p-toluidine as models of aromatic amines, 1,2,4-trihydroxybenzene as an example of polyhydroxybenzene, glycine and serine as aliphatic amino acids. The compound names are listed in Table 5.

Table 5. Organic Compounds Used in the Nitrosation Study.

Category	Organic Compound	CAS
Aromatic Amines	Aniline	9001-75-6
	p-toluidine	1415-93-6
Polyhydroxybenzene	1,2,4-trihydroxybenzene	9004-34-6
Aliphatic Amino Acid	Glycine	71-43-2
	Serine	62-53-3

3.5.1 Study on Nitrosation of Organic Compounds in Reagent Water Matrix

Nitrosation will be conducted in a 1-L beaker with 10^{-4} mol/L organic compounds mixed with 10^{-3} mol/L of sodium nitrite. This study will assess the effects of pH values of 3 and 6 and contact times of 12 hr and 24 hr. Therefore, four experiments will be performed with each organic compound.

Each solution will be stirred in a constant temperature bath for a desired time. Immediately after reaction termination, an 800-mL aliquot of each sample will be extracted with 0.015 L of high purity hexane. Two subsequent extractions will be made, each with 0.015 L of fresh hexane, and the three extracts will be combined and dried by passing over 20g of anhydrous sodium sulfate. The dried extracts will be brought to 0.050 L volume with hexane in a volumetric flask, and the target organic compound will be measured by gas chromatography. The remaining 200-mL aliquot in the reaction vessel will be used for total cyanide, nitrite and nitrate analyses.

3.5.2 Study on Nitrosation of Organic Compounds in Wastewater Matrices

Nitrosation studies will also be performed with different wastewater effluents as described below. Firstly, 10^{-3} mol/L dose of sodium nitrite will be added to raw wastewater effluent and the effect of nitrosation on cyanide formation will be addressed. Secondly, 10^{-4} mol/L organic compounds will be spiked in each wastewater effluent in a 1-L beaker; then, 10^{-3} mol/L of sodium nitrite will be added to the solution. The effect of nitrosation in the presence of elevated organic compound concentration will be investigated. Nitrosation will take place in the condition (pH and contact time) pre-determined in Section 3.5.1. The experimental procedures will be the same as those in reagent water.

3.6 Task 4 - Study of the Effects of Dechlorinating Agents in Cyanide Analyses

Among all tasks in this work plan, Task 4 will be conducted first. The conclusions drawn from this study will be applied to other tasks where a dechlorinating agent is employed. Three experiments will be conducted to assess the effects of two dechlorinating agents (sodium thiosulfate and ascorbic acid) on the potential production of cyanides. First, several unchlorinated WWTP effluent wastewater samples will be sent to CMU, and chlorination of these samples will be conducted in the laboratory with NaOCl at an appropriate dosage and a contact time of t hours. Dechlorinating agents will be added to chlorinated effluents prior to the cyanide analyses. Second, the same effluents will be first spiked with 20 mg/L (as NO_2-N) of sodium nitrite and then chlorinated. Samples will then be treated with dechlorinating agents followed by cyanide analyses. Finally, dechlorinating agents will be used in some chlorination experiments in the laboratory with reagent water matrix. Total cyanide is the only cyanide species to be measured in this task and the cyanide analyses will be performed before and after adding appropriate chemicals to the solution. Once a dechlorinating agent shows a significant potential to form cyanide, further study will be conducted to try to identify the major mechanism involved in the formation of cyanide. The other dechlorinating agent will be used consistently in the rest of this research work.

4. SCHEDULE

The schedule for this research is presented in Figure 1.

Figure 1. Schedule for Task 2.2 WERF Project – Year 2000
Plan for Laboratory Study of Cyanide Formation in Chlorination of Municipal Wastewaters.

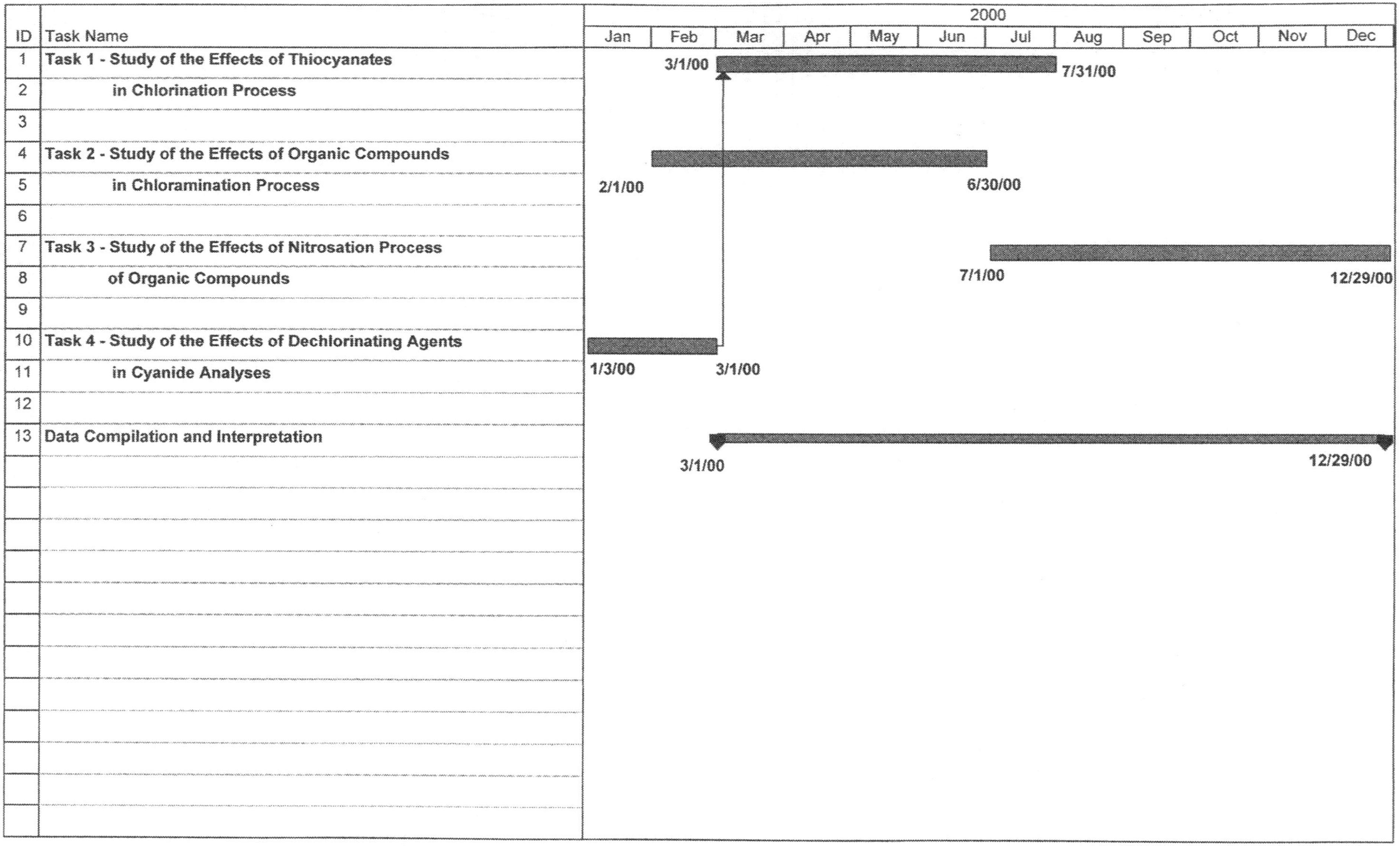

5. REFERENCES

Allen, L.A., et al., 1948. Formation of cyanogen chloride during chlorination of certain liquids and toxicity of such liquids to fish. *The Journal of Hygiene*, 46:184.

APHA/AWWA/WEF, 1995. *Standard Methods for the Examination of Water and Wastewater,* 19th Edition, American Public Health Association, American Water Works Association and Water Environment Federation. Washington DC.

ASTM, 1991. Specification for Reagent Water. Designation D 1193-91. *1998 Annual Book of ASTM Standards,* Vol.11.01. American Society for Testing and Materials, Philadelphia, PA.

ASTM, 1995. Standard Test Method for Determination of Free Cyanide in Water and Wastewater by Microdiffusion. Designation D 4282-95. *1998 Annual Book of ASTM Standards,* Vol.11.01. American Society for Testing and Materials, Philadelphia, PA.

ASTM, 1998. Standard Test Methods for Cyanides in Water. Designation D2036-98. *1998 Annual Book of ASTM Standards,* Vol.11.01. American Society for Testing and Materials, Philadelphia, PA.

Boening, D.W. and Chew, C.M., 1999. A critical review: general toxicity and environmental fate of three aqueous cyanide ions and associated ligands. *Water, Air and Soil Pollution*, 109: 67.

Carr, S.A., R.B., and Lin,B.T., 1997. Wastewater derived interferences in cyanide analysis, *Water Research*, 31 (7): 1543.

Csikai, N.J. and Barnard, A.J.Jr., 1983. Determination of total cyanide in thiocyanate-containing wastewater. *Analytical Chemistry*, 55:1677.

Chamberlin, N.S. and Snyder, H.B., 1955. Technology of treating plating wastes, *Conference Proceedings from the Purdue Industrial Waste Conference*. May 1955.

Delaney, M.F., Zilitinkevitch, L., McSweeney, N. E., and Epelman, P. E., 1997. Cyanide formation from chlorinated WWTP effluent, *Proceedings of WEF Environmental Laboratories Conference*, 6-17.

Eden, G.E., Hampson B.L., and Wheatland, A.B., 1950a. Destruction of cyanide in waste waters by chlorination. *Journal of the Society of Chemical Industry*, 69:244

Eden, G.E. and Wheatland, A.B., 1950b. Effect of temperature and presence of hypochlorite on the rate of hydrolysis of cyanogen chloride alkaline solution. *Journal of the Society of Chemical Industry*, 69:166

Ellgas, W.M., 1999. Cyanide formation in chlorinated wastewater. *WEFTEC '99 Workshop #114 Materials.*

ENVIRON Corp., 1996. *Evaluation of Cyanide at EBMUD's Water Pollution Control Plant Special District No.1.* ENVIRON Corporation, Emeryville, California.

Forman, K.J, and Kavanaugh, M.C., 1999. *Incidents of Cyanide Formation and their Relationship to Water Quality Criteria.* WEFTEC '99 Workshop #114 Materials.

Ghosh, R.S., Dzombak, D.A., Luthy, R.G., and Nakles, D.V., 1999. Subsurface fate and transport of cyanide species at a manufactured-gas plant site. *Water Environment Research*, 71(6): 1205.

Hirose, Y., Maeda, N., Ohya, T., Nojima, K., and Kanno S., 1988. Formation of cyanogen chloride by the reaction of amino acids with hypochlorous acid in the presence of ammonium ion. *Chemosphere*. 17 (5): 865.

Hirose, Y., Maeda, N., Ohya, T., Nojima, K., and Kanno S., 1989. Formation of cyanogen chloride by the reaction of peptides and proteins with hypochlorous in the presence of ammonium ion. *Chemosphere*. 18 (11/12): 2101.

Kanno, S., Nojima, K. and Ohya T., 1978. Formation of cyanide ion by the reaction of organic compounds with nitrous acid or other reagents. *Chemosphere*. 7 (8): 657.

Kanno, S., Nojima, K. and Ohya T., 1982a. Formation of cyanide ion or cyanogen chloride through the cleavage of aromatic rings by nitrous acid or chlorine. IV. *Chemosphere*. 11(7): 663.

Kanno, S., Nojima, K. and Ohya T., 1982b. Formation of cyanide ion or cyanogen chloride through the cleavage of aromatic rings by nitrous acid or chlorine. V. *Chemosphere*. 11(7): 669.

Koshimizu, and Toshiko. 1975. Formation of cyanide ion by the reaction of amines with nitrite ion. II Formation of cyanide ion by the reaction of aromatic amine and nitrite ion. *Eisei Kagaku*, 21(1):1.

Lewis, R.J., Sr., 1992. *Sax's Dangerous Properties of Industrial Materials*, 8th Edition, Van Nostrand Reinhold, New York.

Malcolm Pirnie, Inc., 1997. *Investigation of Cyanide Formation in Sunnyvale Water Pollution Control Plant,* draft report. Malcolm Pirnie, Inc., White Plains, NY 10602.

Maeda, K., Ohya, T., Nojima K., and Kanno S., 1987. Formation of cyanide ion or cyanogen chloride through the cleavage of aromatic rings by nitrous acid or chlorine. IX. *Chemosphere*. 16 (10-12): 2249.

McGuire, M.J., 1989. Preparing for the disinfection by-products rule: a water industry status report. *Journal of American Water Works Association*, 81:35.

Metcalf and Eddy, Inc., 1991. *Wastewater Engineering: Treatment, Disposal and Reuse*, 3rd Edition. McGraw-Hill, Inc., New York.

Newman, A.A., 1975. *Chemistry and biochemistry of thiocyanic acid and its derivatives*. Academic Press Inc. Ltd. London.

Ohya, T. and Kanno S., 1984. *Chemical & Pharmaceutical Bulletin*, 32:280.

Ohya, T. and Kanno S., 1985. Formation of cyanide ion or cyanogen chloride through the cleavage of aromatic rings by nitrous acid or chlorine. VIII. *Chemosphere*. 14 (11-12): 1717.

Ohya, T. and Kanno S., 1991. Formation of cyanide ion or cyanogen chloride through the cleavage of aromatic rings by nitrous acid or chlorine. XII. *Chemosphere*. 22 (3-4): 267.

Rapean, J.C., Johnson, R.A., and Hanson, T.P., 1980. Biodegradation of cyanide: nitrite interference in the standard test for total cyanide. *Conference Proceedings from the Purdue Industrial Waste Conference*. May 13-15, 1980.

Pederson, E.J. III, Urbansky, E.T., Marinas B.J., and Margerum, D.W., 1995. Formation of cyanogen chloride from the reaction of monochloramine and formaldehyde. Extended Abstract, Division of Environmental Chemistry, American Chemical Society 210th National Meeting, Chicago, IL, Aug. 20-24, 1995.

Pettet, A.E., and Ware, G.C., 1955. Disposal of cyanide wastes. *Chemistry and Industry*, 10:1232.

Resnick, J.D., Moore, W.A., and Ettinger, M.B., 1958. Behavior of cyanides in polluted water. *Industrial and Engineering Chemistry*, 50 (1):71

Shifrin, N.S., Beck, B.D., Gauthier, T.D., Chapnick, S.D., and Goodman, G., 1996. Chemistry, toxicology, and human health risk of cyanide compounds in soils at Former Manufactured Gas Plant Sites. *Regulatory Toxicology and Pharmacology*, 23:106.

Teo, W.K. and Tan, T.C., 1987. Hypochlorite oxidation of cyanate under mildly alkaline conditions. *Water Research*, 21(6): 677.

U.S. EPA, 1979. *Water-related environmental fate of 129 priority pollutants, Vol. I.* EPA-440/4-79-029a, United States Environmental Protection Agency, Washington, D.C. 20460.

U.S. EPA, 1983. *Methods for chemical analysis of water and wastes.* Environmental Protection Agency Method 335.2. United States Environmental Protection Agency, Washington, D.C. 20460.

U.S. EPA, 1986. *Test methods for evaluating solid waste.* SW-846, Physical/Chemical Methods, 3rd Edition. United States Environmental Protection Agency, Washington, D.C. 20460.

U.S. EPA, 1993. *Methods for the Determination of Inorganic Substances in Environmental Samples.* EPA/600/R-93/100. United States Environmental Protection Agency, Washington, D.C. 20460.

Zilligh, J.A., 1972. Toxicity of combined chlorine residuals to freshwater fish. *Journal of Water Pollution Control Federation*, 44:212.

APPENDIX E

RELATIONSHIP OF CYANIDE SPECIATION AND CYANIDE TOXICITY

1.0 INTRODUCTION

Toxicity testing provides one method to develop the supporting science for establishing appropriate water quality standards. In cases where measured toxicity provides consistent and reproducible effects on a range of test organisms, there is little debate about the appropriateness of a water quality criterion. The Clean Water Act requires that standards be based on the most sensitive species that could be exposed to a pollutant. Most standards promulgated by U.S. EPA and the individual states are based on toxicity testing that was performed by U.S. EPA and their contractors during the 1970s and early 1980s.

In cases where standards may not be appropriate for local conditions dischargers, can perform water effects ratio tests to determine if the effect of a pollutant is less severe in local waters. A discharger can also perform tests on local species if the species upon which the standard is based is not present in local waters.

The quality assurance quality control protocols for standardized toxicity testing have advanced dramatically in the intervening years. Despite these changes, many toxicity based standards are still shown to be appropriate and well supported by current and historic tests. Standard toxicity tests will clearly measure toxicity caused by hydrogen cyanide (HCN). HCN readily enters organisms and blocks cellular respiration. Much less is known about toxicity of other cyanide species to aquatic organisms, primarily because the methods for distinguishing those species are not well established. In addition, no rigorous testing has been performed on complex cyanide species. Early studies suggested that the equilibrium equations between complex metal cyanides and HCN and the resultant HCN concentrations in receiving waters is well correlated with measured toxicity in fish (Douderoff et al., 1966). The bioavailability of other forms is not well understood primarily because analytical methodology has not been available to assess the presence and concentration of many cyanide complexes in a routine and reproducible manner.

When toxicity is observed in wastewater effluent stream, regulatory agencies typically require Toxicity Identification Evaluation (TIE) and Toxicity Reduction Evaluation (TRE) procedures to be implemented. Currently available analytical methods do not distinguish CN toxicity clearly from other potential sources. There are also no U.S. EPA methods that distinguish toxicity contributions of different cyanide complexes.

This work addresses Task 3 of this study titled "Relationship of Cyanide Speciation and Cyanide Toxicity".

2. OBJECTIVES

Task 3 was developed to evaluate the current state of understanding of the relationship between cyanide speciation and cyanide toxicity. The original objectives of this task were as follows:

- To provide an annotated review of existing data for toxicity of cyanide species;
- To assess the extent to which speciation was taken into account in the experimental design;

- To summarize the current understanding of metal-cyanide, organocyanide, and free cyanide toxicity with respect to aquatic organisms;
- To determine if any TIE studies have successfully distinguished CN as a source of toxicity through toxicity protocols rather than trial and error;
- To provide recommendations to improve TIE effectiveness at distinguishing CN toxicity.

Based on the limited data available and because of the budget constraints of this study, it was not possible to draw any detailed or meaningful conclusions regarding cyanide toxicity and the data supporting the current water quality criteria standard. In response, the Project Sub-Committee decided to combine all further toxicity investigations into a future RFP to adequately address current concerns related to national criteria. As a result, the scope of Task 3 was modified to include the following activities:

A list of articles/studies in which the toxicity of cyanide species has been reported, complete with bibliographic listing;

- Search methodology and complete bibliography;
- Evaluation of species-specific consideration for evaluated toxicity studies;
- Formal recommendations for potential future cyanide toxicity testing protocols including specific test procedures and tested species.

By the end of Year 2, the Project Team finalized the evaluation of the toxicity information compiled during the course of this study. The final evaluation includes a brief evaluation of fresh and marine cyanide water quality criteria and chemistry, discussion of cyanide speciation in experimental design, summary of literature review, toxicity mechanisms of cyanide, and recommendations for potential paths for future WERF studies.

3. SUMMARY OF LITERATURE REVIEW

A literature review was conducted during 1998 and 1999 primarily through electronic search tools with additional manual review of literature cited sections of retrieved references. The literature contains many good examples of toxicity studies performed within the context of available analytical methods. Most studies are based on toxicity derived from spiked samples with known concentrations of simple cyanide salts, which readily form HCN in water. The QA/QC protocols of toxicity tests for cyanide mirror the development of the industry and have improved dramatically in the last decade. The Dialog database was searched under the chemical name cyanide in the category of wastewater pollutant sources, with the keyword toxicity to retrieve references between 1995 and the present.

The literature review suggests that there is a lack of detailed toxicity testing on cyanide species. The range of information available about toxicity related to specific cyanide complexes is almost nonexistent. The acute toxicity of cyanide in marine waters is not as well accepted or understood. The focus of the literature reviewed to date has been on the relationship of cyanide toxicity tests to marine water quality criteria.

4. THE ROLE OF CYANIDE SPECIATION IN EXPERIMENTAL DESIGN

Most toxicity testing has been performed in samples spiked with known concentrations of simple cyanides (i.e., sodium cyanide) that readily disassociate into HCN in water. Furthermore, detailed laboratory QA/QC and raw data are generally unavailable for many of the published reports.

4.1 Total Cyanide

Toxicity studies on complex effluents typically measure Total cyanide by the distillation method. Most toxicity studies where concentrations of sample are spiked to achieve a desired range of cyanide use NaCN or other simple cyanide salts. Given the evolving state of the science of cyanide measurement and the uncertainties of various techniques, few studies have linked cyanide speciation with any analytical precision. The suggestions presented in Task 2 are that the Weak Acid Diffusible method represents the best currently approved method for estimating HCN in wastewater samples.

4.2 Metallic Cyanides

Most studies suggest that metallic cyanide complexes are not readily absorbed by organisms and that toxicity is directly related to the rate at which cyanide complexes (Pablo et al., 1997) or species convert to HCN in solution. A recent series of tests performed in Australia measured toxicity of a variety of iron cyanide complexes. The conclusions of these studies substantiate earlier studies that the toxicity of cyanide is attributable to HCN rather than free cyanide ion. No studies have been located where an attempt was made to quantify the toxicity of organo-cyanide compounds.

4.3 Organic Cyanides

In a 1996 study by Chen et al., the Microtox™ test was used to study synergistic effects between cyanogenic compounds such as acetonitrile and other compounds. Microtox™ testing indicated that greater than additive or synergistic effects were not typical except when these compounds were in mixtures with other reactive toxicants. All of the tests performed in this study used Microtox™ testing as the sole determinant of toxicity.

4.4 Other Cyanide Species

Rainbow trout were used to compare the sublethal effects of cyanide and thiocyanate in another study (Lanno et al., 1996). No mortality was evident, but significant bioconcentration of SCN^- in plasma occurred in all trout exposed to SCN^- or CN^-. Mild chronic SCN^- toxicity, as characterized by diffuse colloid goitre, slightly reduced growth, decreased hepatosomatic and splenosomatic indices and hematocrit, was observed in trout exposed to 720 μmol SCN^-/L (~ 42 mg/L). Exposure of rainbow trout to CN^- at concentrations up to 0.98 mu mol/L (~ 25 μg/L) for 16 weeks does not result in toxicity either due to the direct effects of CN- or to the indirect effects of endogenously produced SCN^-.

5. CURRENT TOXICITY-BASED CYANIDE STANDARDS

5.1 Freshwater Criteria

There is little debate about the validity of cyanide toxicity and associated standards in freshwater. The toxicity testing that supports the development of the freshwater criteria is consistent with current tests. Tests have been reproduced that show similar ranges for acute toxicity.

5.2 Marine Water Criteria

The U.S. EPA marine cyanide water quality standard has been set based on tests with *Cancer irroratus.* In these studies, toxicity was measured at concentrations six times lower than in tests with any other species (Brix et al., 1997). Figure 9-1 illustrates the range of acute values reported by USEPA for *Cancer irroratus* and other marine species compared to the acute value reported by Brix et al. for four species of *Cancer* tested.

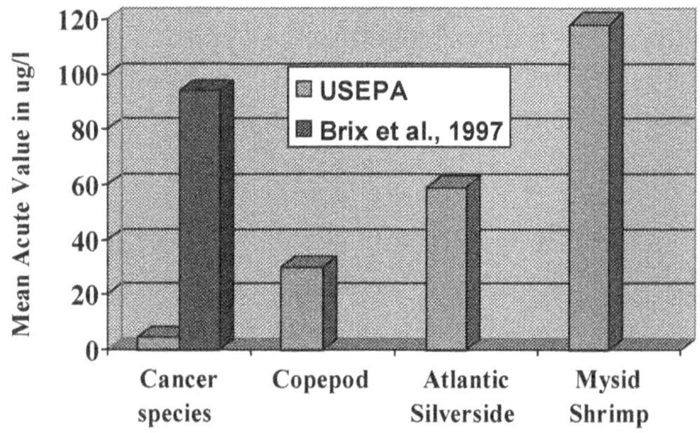

Figure 9-1: Marine Water Quality Criterion Database for Cyanide.

Tests performed by Paramatrix Inc. (Brix et al., 1997) on west coast species of the genus *Cancer* have documented acute toxic responses within the ranges reported by U.S. EPA for marine crab species other than *Cancer Irroratus*. Based on these toxicity studies, the U.S. EPA data on *Cancer irroratus* appears to be a statistical outlier.

The only available method for evaluating the USEPA tests would be replication. An array of tests could be performed on zoea (larva) from *Cancer irroratus* to establish whether or not the original tests were representative of the real toxic response of these crabs to cyanide. However, it is not clear that USEPA would accept the results of such a series of tests as sufficient to modify the current marine water quality standard for cyanide.

Several studies have demonstrated a need to revisit the selection of representative testing species. The acute toxic concentrations of various marine fish and crustaceans suggest that the mysid shrimp *Mysidopsis bahia* is a representative sensitive crustacean species (Suter et al., 1986). In fact the series of tests reviewed by NOAA shows the mysid is more sensitive in all known cases except the toxicity to *Cancer irroratus* measured for cyanide. The toxicity of cyanide to mysid shrimp is in the same range as the toxicity to West Coast *Cancer* species reported by Brix et al. (1997). These data indicate that the mysid may be an ideal candidate for additional toxicity studies. Pablo et al. (1997) have reported a similar range of toxicity for the banana prawn *Penaeus monodon* providing further support for the consideration of the *Cancer irroratus* data as an outlier.

6. SITE-SPECIFIC MODIFICATIONS

One means of easing the burden of stringent cyanide water quality criteria is site specific toxicity tests. Site specific criterion for Puget Sound have been proposed based on the weight of evidence for "local" species versus the east coast based standard. Currently other facilities that are having difficulty attaining the marine water quality criteria have the option to pursue a site-specific cyanide criterion based on toxicity to locally present species. Eastern facilities where *Cancer irroratus* is present will not have this option without redefining the toxicity to that species.

7. POTENTIAL PATHS FORWARD

Several options exist for further defining appropriate standards for discharge of cyanide into both fresh and marine waters. The analytical methods defined in Task 2 of this study can be employed to develop appropriate discharge criteria for organic and cyanide complexes independent of standards based on HCN. The stability and bioavailibility of various cyanide species can be more accurately documented with better analytical methodology.

Water effects ratios can be employed more broadly to determine if the conservative assumptions in the standards development process are appropriate for cyanide and cyanide species. Use of water effects ratios and potential site-specific standards will help to determine if the nature of complex municipal or industrial effluents alters the toxicity either in whole effluent or in combination with receiving water.

The approach developed at length for this task has been to re-examine the toxicity of cyanide to *Cancer irroratus*. This species drives the value of the current acute toxicity standard in marine systems. The reported acute toxicity of cyanide to this species appears to be a significant outlier. The earlier U.S. EPA studies may not have adequately characterized the response of this species to cyanide exposure. These studies should be repeated.

The following discussion describes three avenues of investigation to provide additional information for supporting an alternative toxicity basis for the marine water quality standard. The three levels are chosen to represent a range of cost and quality assurance. Level 1 may not be sufficient to adequately support review and reconsideration of the standard. The Level 2 effort would provide a valid analysis of the response of the rock crab to concentrations of what is measured as weak acid disassociable cyanide. The Level 3 effort would provide information about the sensitivity of *Cancer irroratus* compared to the sensitivity of a more commonly tested marine species (*Mysidopsis bahia*). The Level 3 effort would also provide a better definition of forms of cyanide present in the test solutions that might be sources or sinks of cyanide toxicity.

A single toxicity test has been suggested as a method to define the benefit to be derived from a more comprehensive program of testing. The problem with performing a single analysis for toxicity is the inherent variability of the response of individual organisms to toxicants. Even if a single test were developed around the offspring from several parent crabs, it would be susceptible to error if one of the parent individuals were slightly more or less sensitive than might be typical. Typically three replicate toxicity tests are used when a species that has not been extensively tested is the test organism. Effluent monitoring programs are based on single replicate tests where commonly tested organisms such as the freshwater Cladoceran *Ceriodaphnia dubia,* the fathead minnow *Pimephales promelus* or the marine shrimp *Mysidopsis bahia.*

8. SINGLE TEST SCOPE

The following tasks should constitute the minimum level of effort required to assess the validity of the previous testing performed on *Cancer irroratus*:

9. COLLECTION OF ADULT CRABS

Adult female *Cancer irroratus* in gravid condition would be collected from the field. To insure that appropriate individuals are collected, it is estimated that a two-person crew would require about 18 hours in the field to collect crabs. Additional collection costs would include travel expenses and either scuba diving equipment or boat rental (depending on available sites).

10. REARING

Female crabs would need to be fed and maintained in the laboratory until their eggs hatch and the larvae (zoea) are released. Rearing costs would include preparation of seawater media (natural or artificial or a blend), care of the adults, and record keeping and handling of the newly hatched zoea. The large numbers of zoea released by a single female provide a good source of test organisms. Reliance on zoea from a single crab to perform a test series increases the chance for measurement error. Ideally, test organisms should be blocked so that one replicate at each concentration represents the offspring of a parent. In that manner the variation in response between individuals can be measured.

11. TESTING

Acute static renewal methods would be used to measure mortality at a range of concentrations through a 96-hour test. Historically, many USEPA methods used flow through systems to assess toxicity. Current practice has limited the use of flow through tests because the increased cost and labor involved in conducting flow through test has not been shown to produce proportionately more accurate results. Furthermore the use of flow through tests increases the chance of introduced error that could result from volatilization of the toxic material being tested.

Since data exists that shows LC_{50} concentrations could be between 4 and 200 μg/L, a range of test solutions should be used to bracket that range. Including two controls, ten concentrations should be tested to provide this range. The two controls could be natural and artificial seawater or a blend of natural and artificial seawater. One control should represent the stock water used to make the test concentrations and the other should be made from a different source water. Test solutions would be produced by adding known amounts of cyanide in a readily disassociable form such as sodium cyanide to produce a range of expected concentrations.

The WAD method would be used to test the solutions before, during and after the tests to measure the actual test concentrations. Four replicates of each concentration would be used for the test. Mortality of test organisms would be observed every 24 hours through the 96-hour test. At 48 hours, test solution would be renewed by transferring the surviving zoea into clean chambers of new solutions. Concentrations of WAD-cyanide would be measured in the old and new solutions. General water quality measurements would be measured in each concentration every 24 hours throughout the test. Temperature, pH, dissolved oxygen and salinity would provide the minimum suite of water quality measurements. Chambers would be maintained at $10 \pm 1^{\circ}C$ in environmental chambers or water baths. Chambers should be isolated from the atmosphere to minimize volatilization of cyanide.

12. REPORTING

The results of the tests would be interpreted and analyzed with standard statistical techniques to determine acute values. A report would be prepared to describe the methods used and the results obtained. The report would include a summary of all analytical data, lab records, rearing and collection records and any conditions observed outside of protocol.

13. TRIPLICATE TEST SCOPE

The second highest level of effort would be to perform the testing described above in triplicate. Three individual tests conforming to the description above would be performed. The three tests would provide a greater range of parent organisms and a better measurement of the variability between individuals. This scope does not include any additional cyanide analysis. There is some economy of scale to be realized by completing three tests instead of a single test. The labor estimate for this scope would be slightly less than triple the effort of a single test. This scope assumes that the tests are performed by the same laboratory to take advantage of that economy of scale. Rearing, testing and reporting should be as discussed for a single test.

An alternative would be to hire three independent laboratories to perform each one test. In that event the cost would be more than triple the cost to perform a single test. Additional effort would be required to coordinate between the laboratories and to provide a summary and comparison of the three independent results.

14. THREE REPLICATES AND ADDITIONAL QA/QC

This scope provides an outline of a scientifically defensible battery of tests that will provide very clear definition of the response of *Cancer irroratus* to cyanide. Three additional testing areas are expanded in this scope to provide that extra definition. Standard reference toxicant tests will be used to define response of the crab to other known toxins. Mysid toxicity tests will compare the response of the crab to a more commonly tested species. Additional analytical work will be done to define complexes of cyanide that are formed or degraded during the course of the toxicity test.

15. STANDARD REFERENCE TOXICANTS

Since *Cancer irroratus* is not a frequently tested organism, little or no data is available about the response of this organism to "standard" toxicants. Standard toxicants (generally conservative salts) are used to measure the response of different "batches" of organisms used in standard WET testing. A series of at least three standard reference toxicant tests should ideally be performed to look at the range of responses for these crabs. The protocol for these tests would be identical to the cyanide tests except a range of concentrations of a reference toxicant would be used to replace cyanide.

16. PARALLEL MYSID TESTS

Mysid shrimp have been commonly used as a test species in marine toxicity studies. Various reports suggest that they are representative of a range of sensitive species to many toxic compounds. Since a large database of tests exists to define the toxic response of mysids, it would be useful to directly compare the response of the shrimp to the crabs. A battery of three tests are proposed to be performed alongside the *Cancer irroratus* tests that would be identical except for the different species used in the test chambers.

17. ADDITIONAL CYANIDE CHEMISTRY

It is not entirely clear which methods would be most useful here but it would be valuable to measure the complexes of cyanide present in each of the above tests. Also various improvements to the WAD-cyanide test can be performed in parallel to the standard method to increase the detection limit.

More detailed analytical work and related toxicity testing is clearly needed to assess what role cyanide complexes have on toxicity (Fisher et al., 1998). The use of WET testing to measure acute and chronic effects of effluents streams is well established. Studies show that the use of toxicity tests for effluents have decreased the frequency and severity of toxicity. Cyanide or cyanide complexes are not readily implicated as causes of acute or chronic effects because of the limitations of analytical methods and the absence of Toxicity Identification Evaluation protocols specific for Cyanide.

18. CONCLUSIONS

The toxicity of HCN is well established in freshwater and is disputed in marine systems. Further species-specific testing is indicated as a potential mechanism to clarify the regulatory implications of some past results.

The toxic effect of cyanide complexed with metals or organic compounds is not understood in any aquatic environment. Evidence suggests that most cyanide complexes are not readily bioavailable. Iron cyanide complexes have received the most attention, and toxicity or bioavailability of the complexed forms has not been indicated. Other cyanide complexes have not been subjected to rigorous toxicity testing.

Additional toxicity testing has been proposed in detail as a suggestion for further research to resolve this issue.

REFERENCES

Abulanba, A.; Nieber, J.L. (2000). Modeling the Effects of Nonlinear Equilibrium Sorption on the Transport of Solute Plumes in Saturated Heterogeneous Porous Media. *Adv. Water Res.* 23: 893-905.

Agency for Toxic Substances and Disease Registry (ATSDR) (1997). *Toxicological Profile for Cyanide.* ATSDR, U.S. Department of Health and Human Services, Atlanta, GA.

Allen, L.A. et al. (1948). Formation of Cyanogen Chloride during Chlorination of Certain Liquids and Toxicity of such Liquids to Fish. *J. Hygiene* 46:184.

Allesi, B.A. and Fuller, W.H. (1976) The Mobility of Three Cyanides Forms in Soils. Residual Management by Land Disposal – *Proceedings of the Hazardous Waste Research Symposium. U.S. Environmental Protection Agency*, EPA-600/9-76-015. U.S. EPA, Cincinnati, OH,

Altmayer, F. (1997). Cyanide –ATC Headaches. *Plating Surface Finish.* 2:26-28.

American Public Health Association, American Water Works Association, and Water Environment Federation (APHA/AWWA/WEF) (1995). *Standard Methods for the Examination of Water and Wastewater,* 19th Ed., APHA/AWWA/WEF. Washington, D.C.

APHA/AWWA/WEF (1998). S*tandard Methods for the Examination of Water and Wastewater*, 20th Edition, American Public Health Association, American Water Works Association and Water Environment Federation. Washington, D.C.

American Society for Testing and Materials (ASTM) (1998a). Standard Test Method for Determination of Free Cyanide in Water and Wastewater by Microdiffusion. Designation D4282-95. *1998 Annual Book of ASTM Standards,* Vol.14.02. ASTM, Philadelphia, PA.

ASTM (1998b). *Standard Test Methods for Cyanides in Water. Designation D2036-98. 1998 Annual Book of ASTM Standards*, Vol.11.01. ASTM, Philadelphia, PA.

ASTM (1998c). *Standard Test Method for Available Cyanide with Ligand Displacement and Flow Injection Analysis (FIA) Utilizing Gas Diffusion Separation and Amperometric Detection. Proposed Draft Method*, ASTM Committee D-19.06. ASTM, Philadelphia, PA.

ASTM (1998d). Standard Test Methods for Cyanides in Water-Automated Methods for Total Cyanide, Dissociable Cyanide, and Thiocyanate. Designation D4374-93. *1998 Annual Book of ASTM Standards,* Vol.14.02. ASTM, Philadelphia, PA.

ASTM (1998e). Standard Practice for Intralaboratory Quality Control Procedures and a Discussion on Reporting Low-Level Data. Designation D4210-89. *1998 Annual Book of ASTM Standards*, Vol.11.01. ASTM, Philadelphia, PA.

ASTM (1998f). Standard Practice for Use of the Terms Precision and Bias in ASTM Test Methods. Designation E 177-90a. *1998 Annual Book of ASTM Standards*, Vol.14.02. ASTMPhiladelphia, PA.

ASTM (1998g). Standard Specification for Reagent Water. Designation D1193-91. *1998 Annual Book of ASTM Standards,* Vol.11.01, ASTM, Philadelphia, PA.

ASTM (1998h). Standard Practice for Conducting an Interlaboratory Study to Determine the

Precision of a Test Method. Designation E691-92. 1998 *Annual Book of ASTM Standards*, Vol.14.02. ASTM, Philadelphia, PA.

ASTM (1998i). Standard Guide for Good Laboratory Practices in Laboratories Engaged in Sampling and Analysis of Water. Designation D3856-95. *1998 Annual Book of ASTM Standards*, Vol.11.01. ASTM, Philadelphia, PA.

ASTM (1998j). Standard Practice for Determination of Precision and Bias of Applicable Test Methods of Committee D-19 on Water. Designation D2777-96. *1998 Annual Book of ASTM Standards*, Vol.11.01. ASTM, Philadelphia, PA.

ASTM (1998k). Standard Guide for Spiking into Aqueous Samples. Designation D5810-96. *1998 Annual Book of ASTM Standards*, Vol.11.01. ASTM, Philadelphia, PA.

ASTM (1998l). Practice for Preparation, Standardization, and Storage of Standard and Reagent Solutions for Chemical Analysis. Designation E200-97. *1998 Annual Book of ASTM Standards*, Vol.15.05. American Society for Testing and Materials, Philadelphia, PA.

Arber, R., Speed, M.A.; Scully, F. (1984). In *Water Chlorination: Chemistry, Environmental Impact and Health Effects*, Jolley, R.L.; Bull, R.J.; Davis, W.P.; Katz, S.; Roberts, M.H., Jr.; Jacobs, V.A.; Eds., Vol. 5, Chap. 74. Lewis Publishers: Chelsea, MI.

Banerjee, G. (1996). Phenol- and Thiocyanate-Based Wastewater Treatment in RBC Reactor. *J. Environ. Eng.* 122, 941.

BHE Environmental, Inc. (2000). *ALCOA Blue Pond Demonstration Treatment Wetland.* Final Report. BHE Environmental, Inc., Cincinnati, OH.

Bodek, I.; Lyman, J.L.; Reehl, W.F.; Rosenblatt, D.H. (1988). *Environmental Inorganic Chemistry.* Pergamon Press: New York.

Boening, D.W.; Chew, C.M. (1999). A Critical Review: General Toxicity and Environmental Fate of Three Aqueous Cyanide Ions and Associated Ligands. *Water, Air and Soil Pollut.* 109, 67.

Brezonik, P.L. (1994). *Chemical Kinetics and Process Dynamics in Aquatic System.* Lewis Publisher: Boca Raton, FL.

Brix, K.V.; Cardwell, R.D.; Henderson, D.G.; Marsden, A.R. (2000). *A Site Specific Marine Water Quality Standard for Cyanide*. Manuscript Parametrix, Inc.: Kirkland, WA.

Broderius, S.J.; Smith, L.L. (1980). *Direct Photolysis of Hexacyanoferrate Complexes*, EPA-600/3-80-003. Office of Research and Development, U.S. Environmental Protection Agency, Washington, D.C.

Brown, M.E., Ed. (1957). Water Quality Requirements of Fishes and Effects of Toxic Substances. *The Physiology of Fishes, Vol. 2, Behavior*. New York: Academic Press, Inc., pp. 403-430.

Brown, P.D.; Morra, M.J. (1993). Fate of Ionic Thiocyanate in Soil, *J. Agric. Food Chem*. 41, 978.

Burdick, G.E.; M. Lipshuetz. (1950). Toxicity of Ferro- and Ferricyanide Solutions to Fish, and Determination of the Cause of Mortality. *Trans. Am. Fish. Soc.* 78, 192-202.

California Department of Water Resources (1998) California Water Plan. Bulletin 160-98.

California Department of Water Resources Sacramento, CA.

Carr, S.A.; Baird, R.B.; Lin, B.T. (1997). Wastewater Derived Interferences in Cyanide Analysis, *Water Res.* 31 (7), 1543.

Castric, K.F.; Castric. P.A. (1983). Method for Rapid Detection of Cyanogenic Bacteria. *Appl. Environ. Microbiol.* 45 (2), 701-702.

Chamberlin, N.S.; Snyder, H.B. (1955). Technology of Treating Plating Wastes. In *Conference Proceedings from the Purdue Industrial Waste Conference*. Purdue University, West Lafayette, IN.

Chapra, S.C. (1997). *Surface Water-Quality Modeling.* McGraw-Hill, New York.

Clark, M.M. (1996). *Transport Modeling for Environmental Engineers and Scientists*. Wiley-Interscience, New York.

Csikai, N.J.; Barnard, A.J., Jr. (1983). Determination of Total Cyanide in Thiocyanate-Containing Wastewater. *Anal. Chem.* 55, 1677.

Deeb, R.A.; Dzombak, D.; Theis, T.; Ellgas, W.; Kavanaugh, M. (2003). The Cyanide Challenge: WERF Researchers Discuss Strategies for Managing this Disinfection Byproduct. *Water Environ. Technol.* 15(2), 34-38.

Delaney, M.F.; Zilitinkevitch, L.; McSweeney, N.E.; Epelman, P.E. (1997). Cyanide Formation from Chlorinated POTW Effluent. In *Proceedings of Water Environment Federation Environmental Laboratories Conference*. Water Environment Federation, Alexandra, VA.

Diehl, A.C.; Speitel, G.E., Jr.; Symons; J.M.; Krasner, S.W.; Hwang, C.J.; Barrett, S.E. (2000) DBP Formation During Chloramination. *J. Am. Water Works Assoc.* 92 (6), 76

DiGeronimo, M.J. and Antoine, A.D. (1976) Metabolism of Acetonitrile and Propionitrile by Nocardia-Rhodochrous LL100-21. *Appl. Environ. Microbiol.* 31, 900-906.

Dionex Corporation (1988). Determination of Metal Cyanides. Application Note No.55. Dionex Corporation: Sunnyvale, CA.

Doudoroff, P. (1976). *Toxicity of Fish of Cyanides and Related Compounds: A Review.* Duluth, MN: U.S. Environmental Protection Agency, Office of Research and Development, Environmental Research Laboratory.

Eden, G.E.; Hampson, B.L.; Wheatland, A.B. (1950a) Destruction of Cyanide in Waste Waters by Chlorination. *J. Soc. Chem. Industr.* 69, 244.

Eilbeck, W.J.; Mattock, G. (1987). Chemical Processes in Waste Water Treatment. New York: Ellis Horwood Limited.

Ellgas, W.M. (1999). *Cyanide Formation in Chlorinated Wastewater*. Presented at WEFTEC '99 Workshop, New Orleans, LA.

Elserod, H.J.; Firstenberg, C.E.; Quinn, L.C.; Logan, C.M.; Stine, J.F. (1994). Water Quality-Based Permitting for Cyanide, *Proceedings of WEFTEC94.* Water Environment Federation, Alexandria, VA.

Environ, Inc. (1996). *Evaluation of Cyanide at EBMUD's Water Pollution Control Plant Special District No.1*. ENVIRON Corporation, Emeryville, CA.

Ferguson, D.W.; McGuire, M.J.; Koch, B.; Wolfe, R.L.; Aieta, E.M. (1990). Comparing Peroxone and Ozone for Controlling Taste and Odor Compounds, Disinfection By-products, and Microorganisms. *J. Am. Water Works Assoc.* 82, 181.

Forman, K.J.; Kavanaugh, M.C. (1999). Incidents of Cyanide Formation and their Relationship to Water Quality Criteria. Presented at WEFTEC '99 Workshop #114, New Orleans, LA.

Gerritsen, C.M.; Margerum, D.W. (1990). Non-metal Redox Kinetics: Hypochlorite and Hypochlorous Acid Reactions with Cyanide. *Inorg. Chem.* 29, 2757.

Ghosh, R.S.; Dzombak, D.A.; Luthy, R.G.; Nakles, D.V. (1999). Subsurface Fate and Transport of Cyanide Species at a Manufactured Gas Plant Site. *Water Environ. Res.* 71(6), 1205–1216.

Goncalves, M.M.M.; Pinto, A.F.; Granato, M. (1998). Biodegradation of Free Cyanide, Thiocyanate and Metal Complexed Cyanides in Solutions with Different Compositions. *Environmental Technology* 19, 133-142.

Hansen and Associates. (1990). *Studies in Support of Alternative Cyanide Effluent Limits for Four San Francisco Bay Area Refineries*. Report by S.R. Hansen and Associates, Berkeley, CA, for The Joint Refinery Cyanide Study Group—Tosco Avon, Shell Martinez, Exxon Benicia, and Unocal San Francisco Refineries.

Hirose, Y.; Maeda, N.; Ohya, T.; Nojima, K.; Kanno, S. (1988). Formation of Cyanogen Chloride by the Reaction of Amino Acids with Hypochlorous Acid in the Presence of Ammonium Ion. *Chemosphere*. 17 (5), 865.

Hirose, Y.; Maeda, N.; Ohya, T.; Nojima, K.; Kanno, S. (1989). Formation of Cyanogen Chloride by the Reaction of Peptides and Proteins with Hypochlorous in the Presence of Ammonium Ion. *Chemosphere* 18(11/12), 2101.

Ho, S.L.; Roberts, M.H., Jr. (1984). In *Water Chlorination: Chemistry, Environmental Impact and Health Effects,* Jolley, R.L.; Bull, R.J.; Davis, W.P.; Katz, S.; Roberts, M.H., Jr.; Jacobs, V.A., Eds., Vol. 5, Chap. 43. Lewis Publishers, Chelsea, MI.

Hopke, E. (2000). *Preconcentration on the Dionex IC.* Technical Report, Clarkson University, Department of Civil and Environmental Engineering, Potsdam, NY.

Hung, C.H.; Pavlostathis, S.G. (1997). Aerobic Biodegradation of Thiocyanate. *Water Res.* 31, 2761.

Ingersol, D.; Harris, W.R.; Bomberger, D.C.; Coulson, D.M (1983). *Development and Evaluation of Procedures for the Analysis of Simple Cyanides, Total Cyanides, and Thiocyanate in Water and Waste Water,* EPA-600/4-83-054. U.S. EPA: Washington, D.C.

Inorganic Chemical Sampling and Analytical Requirements. (1997) *Code of Federal Regulations,* Section 40, Part 141.23.

Izumi, K.; Cassens, R.G.; Greaser, M.L. (1989). Reactions of Nitrite with Ascorbic Acid and Its Significant Role in Nitrite-cured Food. *Meat Sci.* 26, 141.

Jensen, J. N.; Tuan, Y.J.; (1993) Chemical Oxidation of Thiocyanate Ion by Ozone. *Ozone Sci. Eng.* 15, 343.

Kanno, S.; Nojima, K.; Ohya T. (1978). Formation of Cyanide Ion by the Reaction of Organic Compounds with Nitrous Acid or Other Reagents. *Chemosphere* 7 (8) 657.

Kelada (1989) *Kelada Automated Test Method for Total Cyanide*. National Technical Information Service, U.S. Dept. of Commerce, Springfield, VA.

Kooijman, S.A.L.M. (1987a). A Safety Factor for LC50 Values Allowing for Differences in Water Quality Standards. *J. Water Poll. Cont. Fed* 51 2054-2070.

Koshimizu, T. (1975). Formation of Cyanide Ion by the Reaction of Amines with Nitrite Ion. II Formation of Cyanide Ion by the Reaction of Aromatic Amine and Nitrite Ion. *Eisei Kagaku* 21(6), 326-329.

Kovacic, P.; Lowery, M.K.; Field, K.W. (1970). Chemistry of N-Bromamines and N-Chloramines. *Chem. Rev.* 70, 639.

Krasner, S.W.; Hwang, C.J.; Lieu, T.K.; West, M.J. (1991). Development of a Bench-scale Method to Investigate the Factors that Impact Cyanogen Chloride Production in Chloraminated Waters. *Proceedings 1991 American Water Works Association Water Quality Technology Conference,* American Water Works Association, Denver, CO.

Krasner, S.W.; McGuire, M.J.; Jacangelo, J.G.; Patania, N.L.; Reagan, K.M.; Aieta, E.M. (1989) The Occurrence of Disinfection By-products in US Drinking Water. *J. Am. Water Works Assoc.* 80, 41.

Landreth, R.R. (1989). The Use of Cultured Bacteria in a Full Scale Biological System Treating Coke Plant Wasetwater. *Water Sci. Technol.* 21, 1829.

Lewis, R.J., Sr. (1992). *Sax's Dangerous Properties of Industrial Materials*, 8th Ed., Van Nostrand Reinhold, New York.

Luthy, R.G.; Jones, L.D. (1980) Biological Oxidation of Coke Plant Effluent. *J. Environ. Eng. Div. (Am. Soc. Civ. Eng.)* 106(EE4), 847.

Maeda, K.; Ohya, T.; Nojima K.; Kanno, S. (1987). Formation of Cyanide Ion or Cyanogen Chloride through the Cleavage of Aromatic Rings by Nitrous Acid or Chlorine. IX. *Chemosphere* 16(10-12), 2249.

Malcolm Pirnie, Inc. (1997). *Investigation of Cyanide Formation in Sunnyvale Water Pollution Control Plant*. White Plains, NY: Malcolm Pirnie, Inc.

March, J. (1992). *Advanced Organic Chemistry*. Wiley & Sons, New York, pp. 930-931.

Martin, J.L.; McCutcheon, S.C. (1999). *Hydrodynamics and Transport for Water Quality Modeling*. Lewis Publ., Boca Raton, FL.

McFarland, I. (1997). *Evolving Regulatory and Risk Assessment Approaches for Cyanide in Wastewater Discharges, Receiving Waters, and Groundwater.* Presented at WEFTEC National Meeting, Chicago.

McGuire, M.J. (1989). Preparing for the Disinfection By-products Rule: A Water Industry Status Report. *Journal of American Water Works Association* 81, 35.

Meeussen, J.C.L; Temminghoff, E.J.M.; Keizer, M.G.; Novozamsky, I. (1989) Spectrophotometric Determination of Total Cyanide, Iron-Cyanide Complexes, Free Cyanide and Thiocyanate in Water by a Continuous-Flow System. *The Analyst*, 114, 959.

Meeussen, J.C.L; Keizer, M.G.; de Haan, F.A. (1992). Chemical Stability and Decomposition Rate of Iron Cyanide Complexes in Soil Solutions. *Environ. Sci. Technol.* 26(3), 511-516.

Metcalf and Eddy, Inc. (1991). *Wastewater Engineering: Treatment, Disposal and Reuse*, 3rd Ed. McGraw-Hill: New York.

Mills, W.P.; Dean, J.D.; Porcella, D.B.; Gherini, S.A.; Hudson, R.J.M.; Frick, W.E.; Rupp, G.L.; Bowie, G.L (1982) *Water Quality Assessment: A Screening Procedure for Toxic and Conventional Pollutants, Part* I; EPA-600/6-82-004a; Tetra Tech, Inc., Environmental Research lab., U.S. EPA, Athens, GA, p. 570.

Moggi, L.; Bolletta, F.; Balzani, V.; Scandola, F. (1966). Photochemistry of Coordination Compounds. XV. Cyanide Complexes. *J. Inorg. Nucl. Chem.* 28, 2589.

Neufeld, R.D.; Valiknac, T. (1979). Inhibition of Phenol Biodegradation by Thiocyanate. *J. Water Pollut. Contr. Fed.* 51, 2283.

Newman, A.A. (1975). *Chemistry and Biochemistry of Thiocyanic Acid and its Derivatives.* London: Academic Press, Inc., Ltd.

Ohya, T.; Kanno, S. (1984). *Chem. Pharm. Bull.* 32, 280.

Ohya, T.; Kanno S. (1991). Formation of Cyanide Ion or Cyanogen Chloride through the Cleavage of Aromatic Rings by Nitrous Acid or Chlorine XII. *Chemosphere*. 22 (3-4), 267.

Pablo, F., Buckney, R.T.; Lim, R.P. (1997). Toxicity of Cyanide, Iron-Cyanide Complexes, and a Blast-Furnace Effluent to the Banana Prawn, *Penaeus monodon. Bull. Environ. Contam. Toxicol* 58(5), 822-829.

Paruchuri, Y.L.; Shivaraman, N.; Kumaran, P. (1990). Microbial Transformation of Thiocyanate. *Environ. Pollut*. 68, 15.

Pedersen, E.J. III, Urbansky, E.T., Marinas B.J.; Margerum, D.W. (1995). *Formation of Cyanogen Chloride from the Reaction of Monochloramine and Formaldehyde*, Extended Abstract. Division of Environmental Chemistry, American Chemical Society 210th National Meeting, Chicago, Aug. 20-24, 1995.

Pederson, E.J. III; Urbansky, E.T.; Marinas, B.J.; Margerum, D.W. (1999) Formation of Cyanogen Chloride from the Reaction of Monochloramine with Formaldehyde. *Environ. Sci. Technol*. 33, 4239.

Perkin-Elmer. (1994). Analytical Methods for Atomic Absorption Spectrometry. Perkin-Elmer.

Pettet, A.E.; Ware, G.C. (1955). Disposal of Cyanide Wastes. *Chemistry and Industry* 10:1232.

Rapean, J.C.; Johnson, R.A.; Hanson, T.P. (1980). Biodegradation of Cyanide: Nitrite Interference in the Standard Test for Total Cyanide. In *Conference Proceedings from the Purdue Industrial Waste Conference*. May 13-15, 1980.

Resnick, J.D.; Moore, W.A.; Ettinger, M.B. (1958). Behavior of Cyanides in Polluted Water. *Industr. Eng. Chem.* 50(1), 71.

Rittmann, B.E.; McCarty, P.L. (2000*). Environmental Biotechnology: Principles and Applications*. McGraw-Hill, Inc., New York.

Roberts, M.H., Jr.; Casey, B.B. (1984). In *Water Chlorination: Chemistry, Environmental Impact and Health Effects*, Jolley, R.L.; Bull, R.J.; Davis, W.P.; Katz, S.; Roberts, M.H., Jr.; Jacobs, V.A. Eds., Vol. 5, Chap. 40. Lewis Publishers, Chelsea, MI.

Sawyer, B; Zenz, D.R.; Lue-Hing, C.; Lordi, D.T.; Hill, R. (1996). Challenging a Water Quality Based Effluent Limit, *Proceedings of WEFTEC '96,* October 5-9, 1996, pp. 451-461.

Sawyer, B.; Zenz, D.R.; Lue-Hing, C.; Lordi, D.T.; Hill, R. (1998). Realistic Limits for Water Toxics: Greater Chicago Demonstrates Need for Site-Specific Standards. *Water Environ. Technol.* 10, 57.

Scully, F.E. (1990). *Reaction Chemistry of Inorganic Monochloramine: Products and Implications for Drinking Water Disinfection.* Extended Abstract, Division of Environmental Chemistry, American Chemical Society 200th National Meeting, Washington, D.C, Aug. 25-31, 1990.

Shifrin, N.S.; Beck, B.D.; Gauthier, T.D.; Chapnick, S.D.; Goodman, G. (1996). Chemistry, Toxicology, and Human Health Risk of Cyanide Compounds in Soils at Former Manufactured Gas Plant Sites. *Regul. Toxicol. Pharmacol.* 23:106.

Smith, A.; Mudder, T. (1991). *The Chemistry and Treatment of Cyanidation Wastes.* Mining Journal Books Ltd.: London.

Snider, J.S. (2001). A Study of the Environmental Fate of Common Inorganic Cyanide Species in Surface Waters. M.S. Thesis, Department of Civil and Environmental Engineering, Clarkson University, Potsdam NY.

Soto, H.; Nava, F.; Leal, J.; Jara, J. (1995). Regeneration of Cyanide by Ozone Oxidation of Thiocyanate in Cyanidation Tailings. *Miner. Eng.* 8, 273.

Stanbro, W.D.; Smith, W.D. (1979). Kinetics and Mechanism of the Decomposition of N-Chloroalanine in Aqueous Solution. *Environ. Sci. Technol.* 13, 446.

Stephan, C.E.; Mount, D.I.; Hansen, D.I.; Gentile, J.H.; Chapman, G.A.; Brungs, W.A. (1985). *Guidelines for Deriving Numerical National Water Quality Criteria for the Protection of Aquatic Organisms and their Uses;* EPA/822-R-85/100; NTIS No. PB85-227049; U.S. EPA, Environmental Research Laboratory: Duluth, MN.

Teo, W.K.; Tan, T.C. (1987). Hypochlorite Oxidation of Cyanate under Mildly Alkaline Conditions. *Water Res.* 21(6): 677.

Theis, T.L.; Young, T.C.; Schaefer, R.J.; Tudman, S. (1997). Advanced Oxidation of Iron Cyanides, *Proceedings of the Symposium on Remediation of Soil and Groundwater*, WEFTEC 97, Water Environment Federation, Alexandria, VA.

Thomann, R.V.; Mueller, J.A. (1987). *Principles of Surface Water Quality Modeling and Control*. New York: Harper & Row.

Thorn, K.A.; Mikita, M.A. (2000). Nitrite Fixation by Humic Substances: 15N NMR Evidence for Potential Intermediates in Chemodenitrification. *Soil Sci. Soc. Am. J.* 64 (2), 568-582.

Turk, O.; Mavinic, D.S. (1989). Stability of Nitrite Build-up in an Activated Sludge System. *J. Water Pollut. Control Fed.* 61, 1440.

U.S. Department of Health and Human Services, Agency for Toxic Substances and Disease Registry (ATSDR) (1996). *Toxicological Profile for Cyanide* (Draft) ATSDR, U.S. Department of Health and Human Services, Atlanta, GA.

U.S. Environmental Protection Agency (U.S. EPA) (1979). *Water-Related Environmental Fate of 129 Priority Pollutants,* Vol. I. EPA-440/4-79-029a. U.S. EPA, Washington, D.C. 20460.

U.S. EPA (1980). *Ambient Water Quality Criteria for Cyanide*. U.S. EPA, Office of Research and Development, Washington, D.C.

U.S. EPA (1983). *Methods for Chemical Analysis of Water and Wastes.* EPA-600/4-79-020. Cincinnati, OH.

U.S. EPA (1984). *Ambient Water Quality Criteria for Cyanide*, EPA-440/5-84-028. U.S. EPA, Washington, D.C.

U.S. EPA (1986). *Test Methods for Evaluating Solid Waste. SW-846, Physical/Chemical Methods,* 3rd Ed. U.S. EPA, Washington, D.C.

U.S. EPA (1992a) National Primary Drinking Water Regulations for Synthetic Organic Chemicals and Inorganic Chemicals, Phase 5 Chemical Regulations. *Fed. Regist*. Vol. 57, 31776.

U.S. EPA (1992b) Guidelines Establishing Test Procedures for the Analysis of Pollutants and National Primary Drinking Water Regulations; Flexibility in Existing Test Procedures and Streamlined Proposal of New Test Procedures; Proposed Rule (1997). *Code of Federal Regulations*, Section 40, Parts 136 and 141; *Fed. Regist.* Vol. 62 (60), 14976–15049.

U.S. EPA (1993). *Methods for the Determination of Inorganic Substances in Environmental Samples.* EPA-600/R-93-100. U.S. EPA, Washington, D.C. 20460.

U.S. EPA (1994a). Definition and Procedure for the Determination of the Method Detection Limit, Appendix B, *Code of Federal Regulations*, Section 40, Part 136.

U.S. EPA (1994b). Methods for the Determination of Metals in Environmental Samples – Supplement I. EPA-600/R-94-111. Cincinnati, OH

U.S. EPA (1994c). *Technical Notes on Drinking Water Methods*. EPA-600/R-94-173. Washington, D.C.

U.S. EPA (1995). *Great Lakes Water Quality Guidance*, EPA-820/R-95-002, U.S. EPA, Washington, D.C.

U.S. EPA (1997a). *Manual for the Certification of Laboratories Analyzing Drinking Water - Criteria and Procedures Quality Assurance*, 4th Ed.. U.S. EPA, Cincinnati, OH.

U.S. EPA (1997b). *Water Quality Standards; Establishment of Numeric Criteria for Priority Toxic Pollutants for the State of California*; Proposed Rule, 62 FR 42159, August 5, 1997.

U.S. EPA (1999a). *Protocol for EPA Approval of Alternate Test Procedures for Organic and Inorganic Analytes in Wastewater and Drinking Water*. EPA-821/B-98-002 (Revised March 1999).

U.S. EPA (1999b). *Protocol for EPA Approval of New Methods for Organic and Inorganic Analytes in Wastewater and Drinking Water*. EPA-821-B/98-003 (Revised March 1999).

U.S. EPA (1999c). Method OIA-1677: Available Cyanide by Flow Injection with Ligand Exchange, Final Rule, *Fed. Regist.* December 30, 1999 (Section 64, Number 250).

U.S. EPA Water Quality Standards: Establishment of Numeric Criteria for Priority Pollutants for the State of California; Final Rule. (2000). *Code of Federal Regulations*, Section 40, Part 131,.

Vol. 65 (97), 31681–31719.

U.S. EPA (2001). *Code of Federal Regulations*, Section 40, Part 141.23; July 1, 2001

White, G.C.; Beebe, R.D.; Alford, V.F.; Sanders, H.A. (1983). *Water Chlorination: Chemistry, Environmental Impact and Health Effects,* Vol. 4, Chapter 79, Jolley, R.L.; Bull, R.J.; Davis, W.P.; Katz, S.; Roberts, M.H., Jr.; Jacobs, V.A. Eds. Ann Arbor, MI: Ann Arbor Science,

Williams, D. L. H. (1988). *Nitrosation.* Cambridge University Press, Cambridge, UK.

Yapa, P.D.; Zheng, L. (1997). Simulation of Oil Spills from Underwater Accidents I: Model Development. *J Hydraul. Res.* 35 (5), 673-687.

Yapa, P.D.; Zheng, L.; Kobayashi, T. (1996). Linked Lists for Transport Simulations Using Lagrangian Parcels. *J. Comput Civil Eng.* 10(1), 88-90

Yi, Y.; Theis, T.L.; Young, T.C. (2002). The Effect of Chlorination on Organocyanide Compounds. *Water Environ. Res.* 74(1), 51-56.

Young, T.C., (1995). *Issues Pertaining to Environmental Transport, Fate and Biotic Exposure to Complex Cyanides in Surface Waters*. Alcoa Technical Center, Alcoa Center, PA.

Zepp, R. G.; Cline, D.M. (1977). Rates of Direct Photolysis in Aquatic Environment. *Environ. Sci. Technol.* 3711(4), 359-366.

Zheng, A.; Dzombak, D.A.; Luthy, R.G. (2003b). Effects of Thiocyanate on the Formation of Cyanide During Chlorination and UV Disinfection of Municipal Wastewater. *Water Environ. Res.,* in press.

Zheng, A.; Dzombak, D.A.; Luthy, R.G.; Sawyer, B.; Lazouskas, W.; Tata, P.; Delaney, M.F.; Zilitinkevitch, L.; Sebroski, J.R.; Swartling, R.S.; Drop, S.M.; Flaherty, J.M. (2003a). Evaluation and Testing of Analytical Methods for Cyanide Species in Municipal and Industrial Contaminated Waters. *Environ. Sci. Technol.* 37, 107-115.

Zilligh, J.A. (1972). Toxicity of Combined Chlorine Residuals to Freshwater Fish. *J. Water Pollution Control Fed* 44, 212.

WASTEWATER UTILITY

Alabama
Montgomery Water Works & Sanitary Sewer Board

Alaska
Anchorage Water & Wastewater Utility

Arizona
Gila Resources
Glendale, City of, Utilities Department
Mesa, City of
Phoenix Water Services Department
Pima County Wastewater Management

Arkansas
Little Rock Wastewater Utility

California
Calaveras County Water District
Central Contra Costa Sanitary District
Contra Costa Water District
Crestline Sanitation District
Delta Diablo Sanitation District
Dublin San Ramon Services District
East Bay Municipal Utility District
El Dorado Irrigation District
Escondido, City of
Fairfield-Suisun Sewer District
Irvine Ranch Water District
Las Virgenes Municipal Water District
Lodi, City of
Los Angeles, City of
Los Angeles County, Sanitation Districts of
Napa Sanitation District
Orange County Sanitation District
Palo Alto, City of
Riverside, City of
Sacramento Regional County Sanitation District
San Diego Metropolitan Wastewater Department, City of
San Francisco, City & County of
San Jose, City of
Santa Barbara, City of
Santa Rosa, City of
South Bayside System Authority
South Orange County Wastewater Authority
Union Sanitary District
West Valley Sanitation District

Colorado
Boulder, City of
Colorado Springs, City of
Littleton/Englewood Water Pollution Control Plant
Metro Wastewater Reclamation District, Denver

Connecticut
New Haven, City of, WPCA

District of Columbia
District of Columbia Water & Sewer Authority

Florida
Broward, County of
Fort Lauderdale, City of
Gainesville Regional Utilities
JEA
Kissimmee, City of, Department of Water Resources
Miami-Dade Water & Sewer Authority
Orange County Utilities Department
Orlando, City of
Reedy Creek Improvement District
Seminole County Environmental Services
St. Petersburg, City of
Stuart Public Utilities
Tallahassee, City of
Tampa, City of
West Palm Beach, City of

Georgia
Atlanta Department of Watershed Management
Augusta, City of
Clayton County Water Authority
Cobb County Water System
Columbus Water Works
Fulton County
Gwinnett County Department of Public Utilities
Macon Water Authority
Savannah, City of

Hawaii
Honolulu, City and County of

Idaho
Pocatello, City of

Illinois
American Bottoms Wastewater Treatment Plant
Dupage, County of
Greater Peoria Sanitary District
Kankakee River Metropolitan Agency
Metropolitan Water Reclamation District of Greater Chicago
North Shore Sanitary District
Wheaton Sanitary District

Indiana
Fort Wayne, City of

Iowa
Ames, City of
Des Moines Metro Wastewater Reclamation Authority

Kansas
Johnson County Unified Wastewater Districts
Unified Government of Wyandotte County/Kansas City, City of

Kentucky
Louisville & Jefferson County Metropolitan Sewer District

Louisiana
Sewerage & Water Board of New Orleans

Maine
Bangor, City of
Portland Water District

Maryland
Anne Arundel County Bureau of Utility Operations
Howard County Department of Public Works
Washington Suburban Sanitary Commission

Massachusetts
Boston Water & Sewer Commission
Upper Blackstone Water Pollution Abatement District

Michigan
Ann Arbor, City of
Detroit, City of
Grand Rapids, City of
Holland Board of Public Works
Lansing, City of
Owosso Mid-Shiawassee County WWTP
Saginaw, City of
Wayne County Department of Environment
Wyoming, City of

Minnesota
Rochester, City of

Missouri
Independence, City of
Kansas City Missouri Water Services Department
Little Blue Valley Sewer District
Metropolitan St. Louis Sewer District

Nebraska
Lincoln Wastewater System

Nevada
Henderson, City of

New Jersey
Passaic Valley Sewerage Commissioners

New York
New York City Department of Environmental Protection

North Carolina
Cary, Town of
Charlotte/Mecklenburg Utilities
Durham, City of
Metropolitan Sewerage District of Buncombe County
Orange Water & Sewer Authority

Ohio
Akron, City of
Butler County Department of Environmental Services
Columbus, City of
Metropolitan Sewer District of Greater Cincinnati
Northeast Ohio Regional Sewer District
Summit, County of

Oklahoma
Tulsa, City of

Oregon
Clean Water Services
Eugene/Springfield Water Pollution Control
Water Environment Services

Pennsylvania
Philadelphia, City of
University Area Joint Authority, State College

South Carolina
Charleston Commissioners of Public Works
Mount Pleasant Waterworks & Sewer Commission
Spartanburg Sanitary Sewer District

WERF SUBSCRIBERS

Tennessee
Cleveland, City of
Knoxville Utilities Board
Murfreesboro Water & Sewer Department
Nashville Metro Water Services

Texas
Amarillo, City of
Austin, City of
Dallas Water Utilities
Denton, City of
El Paso Water Utilities
Fort Worth, City of
Gulf Coast Waste Disposal Authority
Houston, City of
San Antonio Water System
Trinity River Authority

Utah
Salt Lake City Corporation

Virginia
Alexandria Sanitation Authority
Arlington, County of
Fairfax County Virginia
Hampton Roads Sanitation District
Henrico, County of
Hopewell Regional Wastewater Treatment Facility
Loudoun County Sanitation Authority
Lynchburg Regional WWTP
Prince William County Service Authority
Richmond, City of

Washington
Edmonds, City of
Everett, City of
King County Department of Natural Resources
Seattle Public Utilities
Yakima, City of

Wisconsin
Green Bay Metro Sewerage District
Madison Metropolitan Sewerage District
Milwaukee Metropolitan Sewerage District
Racine, City of
Sheboygan Regional Wastewater Treatment
Wausau Water Works

Australia
South Australian Water Corp.
Sydney Water Corp.
Water Corp. of Western Australia

Canada
Greater Vancouver Regional District
Toronto, City of, Ontario

Mexico
Servicios de Agua y Drenaje de Monterrey, I.P.D.

United Kingdom
Yorkshire Water Services Limited

STORMWATER UTLITY

California
Los Angeles, City of, Department of Public Works
Monterey, City of
Santa Rosa, City of

Colorado
Boulder, City of

Georgia
Griffin, City of

Kansas
Overland Park, City of

Kentucky
Louisville & Jefferson County Metropolitan Sewer District

Maine
Portland Water District

North Carolina
Cary, Town of
Charlotte, City of, Stormwater Services

Pennsylvania
Philadelphia, City of

STATE

Fresno Metropolitan Flood Control District, Calif.
Urban Drainage & Flood Control District, Colo.

CORPORATE

ADS Environmental Services
The ADVENT Group Inc.
Alan Plummer & Associates
Alden Research Laboratory
Alpine Technology Inc.
Aquateam–Norwegian Water Technology Centre A/S
BaySaver Inc.
BioVir Inc.
Black & Veatch
Boyle Engineering Corporation
BPR CSO
Brown & Caldwell
Burns & McDonnell
CABE Associates Inc.
The Cadmus Group
Camp Dresser & McKee Inc.
Carollo Engineers Inc.
Carpenter Environmental Associates Inc.
CDS Technologies Inc.
Chemtrac Systems Inc.
CH2M HILL
Clancy Environmental Consultants Inc.
Damon S. Williams Associates, LLC
David L. Sheridan, P.C.
Earth Tech Inc.
Ecolab Water Care Services
EMA Inc.
The Eshelman Company Inc.
Finkbeiner, Pettis, & Strout Inc.
Frontier Geosciences Inc.
ftn Associates Inc.
Gannett Fleming Inc.
GE Betz
Golder Associates Inc.
Greeley and Hansen LLC
The HACH Company
Hazen & Sawyer, P.C.
HDR Engineering Inc.
HNTB Corporation
HydroQual Inc.
Insituform Technologies Inc.
Institute for Environmental Technology & Industry, Korea
Jacobson Helgoth Consultants Inc.
Jason Consultants Inc.
Jordan, Jones, & Goulding Inc.
KCI Technologies Inc.
Kelly & Weaver, P.C.
Kennedy/Jenks Consultants
Komline Sanderson Engineering Corporation
Lawler, Matusky & Skelly Engineers, LLP
Limno-Tech Inc.
Lombardo Associates Inc.
Malcolm Pirnie Inc.
McKim & Creed
MEC Analytical Systems Inc.
MWH
New England Organics
Odor & Corrosion Technology Consultants Inc. (OCTC)
ONDEO Degremont Inc.
Oswald Green, LLC
PA Government Services Inc.
Parametrix Inc.
Parsons
Post, Buckley, Schuh & Jernigan
R&D Engineering/Conestoga Rover & Associates
The RETEC Group
R.M. Towill Corporation
Ross & Associates Ltd.
Royce Technologies
SAIC Maritime Technical Group
Sear-Brown
Sierra Environmental Services
SYNAGRO
Tetra Tech Inc.
Trojan Technologies Inc.
URS Corporation
USFilter NATC
Vortechnics Inc.
Wade-Trim Inc.
Weston Solutions Inc.
Woodard & Curran
WRc/D&B, LLC
WWETCO, LLC

INDUSTRY

American Electric Power
ChevronTexaco Energy Research & Technology Company
The Coca-Cola Company
Dow Chemical Company
DuPont Company
Eastman Kodak Company
Merck & Company Inc.
ONDEO Services
Procter & Gamble Company
PSEG Services Corp.
Severn Trent Services Inc.
Shell Global Solutions
Thames Water Plc
United Water Services LLC

Note: List as of 5/12/03

WERF Product Order Form

As a benefit of joining the Water Environment Research Foundation, subscribers are entitled to receive one complimentary copy of all final reports and other products.Additional copies are available at cost (usually $10).To order your complimentary copy of a report, please write "free" in the unit price column.WERF keeps track of all orders.If the charge differs from what is shown here, we will call to confirm the total before processing.

Name ____________________ Title ____________________

Organization ____________________

Address ____________________

City ________ State ________ Zip Code ________ Country ________

Phone ________ Fax ________ Email ________

Stock #	Product	Quantity	Unit Price	Total
			Postage & Handling	
			VA Residents Add 4.5% Sales Tax	
			Canadian Residents Add 7% GST	
			TOTAL	

Method of Payment:

(All orders must be prepaid.)

- ❑ Check or Money Order Enclosed
- ❑ Visa ❑ Mastercard ❑ American Express

Account No. ____________________ Exp. Date ________

Signature ____________________

Shipping & Handling:

Amount of Order	United States	Canada & Mexico	All Others
Up to but not more than:	Add:	Add:	Add:
$20.00	$5.00*	$8.00	50% of amount
30.00	5.50	8.00	40% of amount
40.00	6.00	8.00	
50.00	6.50	14.00	
60.00	7.00	14.00	
80.00	8.00	14.00	
100.00	10.00	21.00	
150.00	12.50	28.00	
200.00	15.00	35.00	
More than $200.00	Add 20% of order	Add 20% of order	

*minimum amount for all orders

Note: Please make checks payable to the Water Environment Research Foundation.

To Order (Subscribers Only):

Log on to www.werf.org and click on the "Product Catalog."

Phone: (703) 684-2470
Fax: (703) 299-0742.

WERF
Attn: Subscriber Services
635 Slaters Lane
Alexandria, VA 22314-1177

To Order (Non-Subscribers):

Non-subscribers may be able to order WERF publications either through WEF (www.wef.org) or IWAP (www.iwapublishing.com).Visit WERF's website at www.werf.org for details.

IWA Publishing's authorised EU representative for General Product Safety Regulations is Diane D'Arras, 15 rue Duret, 75116 Paris, France, e-mail: safety@iwap.co.uk.

Printed and bound by CPI Group (UK) Ltd, Croydon, CR0 4YY
12/05/2026
02108893-0006